CONCEPTS OF DATABASE MANAGEMENT

SIXTH EDITION

Philip J. Pratt

Grand Valley State University

Joseph J. Adamski

Grand Valley State University

CENGAGE
Learning™

Australia • Brazil • Japan • Korea • Mexico • Singapore • Spain • United Kingdom • United States

COURSE TECHNOLOGY
CENGAGE Learning™

Concepts of Database Management, Sixth Edition
Philip J. Pratt, Joseph J. Adamski

Product Manager: Kate Hennessy

Developmental Editor: Jessica Evans

Production Editors: Catherine DiMassa, GEX
   Publishing Services

Marketing Manager: Bryant Chrzan

Marketing Specialist: Vicki Ortiz

Editorial Assistant: Patrick Frank

Cover Designer: Marissa Falco

Manufacturing Coordinator: Justin Palmeiro

For product information and technology assistance, contact us at
Cengage Learning Customer & Sales Support 1-800-354-9706

For permission to use material from this text or product, submit all requests online at **www.cengage.com/permissions**
Further permission questions can be emailed to
permissionrequest@cengage.com

ISBN-13: 978-1-4239-0147-1
ISBN-10: 1-4239-0147-9

Course Technology
25 Thomson Place
Boston, Massachusetts, 02210
USA

Cengage Learning is a leading provider of customized learning solutions with office locations around the globe, including Singapore, the United Kingdom, Australia, Mexico, Brazil, and Japan. Locate your office at:
international.cengage.com/region

Cengage Learning products are represented in Canada by Nelson Education, Ltd.

For your lifelong learning solutions, visit **course.cengage.com**

Purchase any of our products at your local college store or at our preferred online store **www.ichapters.com**

Microsoft is a registered trademark of Microsoft Corporation in the United States and/or other countries. Course Technology is an independent entity from the Microsoft Corporation, and not affiliated with Microsoft in any manner.

Oracle is a registered trademark, and Oracle10*g* is a trademark of Oracle Corporation.

Printed in the United States of America
3 4 5 6 7 8 9 11 10 09

TABLE OF CONTENTS

Preface   ix

Chapter 1   *Introduction to Database Management*   1
Premiere Products Background   1
Database Background   4
Database Management Systems   9
Advantages of Database Processing   12
Disadvantages of Database Processing   14
Introduction to the Henry Books Database Case   14
Introduction to the Alexamara Marina Group Database Case   20
Summary   26
Key Terms   26
Review Questions   26
Premiere Products Exercises   27
Henry Books Case   27
Alexamara Marina Group Case   28

Chapter 2   *The Relational Model 1: Introduction, QBE, and Relational Algebra*   29
Relational Databases   29
Query-By-Example (QBE)   32
Simple Queries   33
Simple Criteria   35
Compound Criteria   37
Computed Fields   40
Functions   42
Grouping   44
Sorting   45
  Sorting on Multiple Keys   46
Joining Tables   49
  Joining Multiple Tables   51
Using an Update Query   53
Using a Delete Query   53
Using a Make-Table Query   54
Relational Algebra   56
Select   57
Project   57
Join   58
Normal Set Operations   60
  Union   60
  Intersection   61
  Difference   61
Product   62
Division   63
Summary   64
Key Terms   65
Review Questions   65
Premiere Products Exercises: QBE   66
Premiere Products Exercises: Relational Algebra   67
Henry Books Case   67
Alexamara Marina Group Case   68

Chapter 3 *The Relational Model 2: SQL* 69

Getting Started with SQL 70
    Getting Started with Microsoft Office Access 2003 and 2007 70
    Getting Started with MySQL 70
Table Creation 71
Simple Retrieval 73
Compound Conditions 80
Computed Fields 84
Using Special Operators (LIKE and IN) 86
Sorting 88
Built-In Functions 90
Subqueries 93
Grouping 94
Joining Tables 97
Union 100
Updating Tables 101
Creating a Table from a Query 103
Summary of SQL Commands 105
Summary 112
Key Terms 112
Review Questions 112
Premiere Products Exercises 113
Henry Books Case 113
Alexamara Marina Group Case 114

Chapter 4 *The Relational Model 3: Advanced Topics* 117

Views 117
Indexes 124
Security 128
Integrity Rules 129
    Entity Integrity 129
    Referential Integrity 130
    Legal-Values Integrity 133
Structure Changes 135
    Making Complex Changes 137
System Catalog 137
Stored Procedures 140
Triggers 141
Summary 143
Key Terms 143
Review Questions 144
Premiere Products Exercises 144
Henry Books Case 145
Alexamara Marina Group Case 146

Chapter 5 *Database Design 1: Normalization* 149

Functional Dependence 151
Keys 153
First Normal Form 155
Second Normal Form 156
Third Normal Form 159
Incorrect Decompositions 163
Multivalued Dependencies and Fourth Normal Form 167
Avoiding the Problem with Multivalued Dependencies 170
Application to Database Design 171
Summary 173
Key Terms 173
Review Questions 173
Premiere Products Exercises 174
Henry Books Case 175
Alexamara Marina Group Case 175

Chapter 6   *Database Design 2: Design Method*   177
   User Views   178
   Information-Level Design Method   178
   Represent the User View as a Collection of Tables   179
   Normalize the Tables   180
   Identify All Keys   180
      Types of Primary Keys   181
   Database Design Language (DBDL)   181
   Entity-Relationship (E-R) Diagrams   182
   Merge the Result into the Design   183
   Database Design Examples   185
   Physical-Level Design   195
   Top-Down Versus Bottom-Up   196
   Survey Form   197
   Obtaining Information from Existing Documents   198
   One-to-One Relationship Considerations   203
   Many-to-Many Relationship Considerations   206
   Nulls and Entity Subtypes   208
   Avoiding Problems with Third Normal Form When Merging Tables   213
   The Entity-Relationship Model   213
   Summary   219
   Key Terms   219
   Review Questions   220
   Premiere Products Exercises   221
   Henry Books Case   222
   Alexamara Marina Group Case   223

Chapter 7   *DBMS Functions*   225
   Update and Retrieve Data   226
   Provide Catalog Services   227
   Support Concurrent Update   228
      The Concurrent Update Problem   228
      Avoiding the Lost Update Problem   232
      Two-Phase Locking   233
      Deadlock   236
      Locking on PC-Based DBMSs   237
      Timestamping   238
   Recover Data   238
      Journaling   238
      Forward Recovery   240
      Backward Recovery   241
      Recovery on PC-Based DBMSs   241
   Provide Security Services   242
      Encryption   242
      Authentication   242
      Authorizations   243
      Views   243
      Privacy   243
   Provide Data Integrity Features   244
   Support Data Independence   246
      Adding a Field   246
      Changing the Length of a Field   246
      Creating an Index   246
      Adding or Changing a Relationship   246
   Support Data Replication   247
   Provide Utility Services   248
   Summary   249
   Key Terms   249
   Review Questions   250
   Premiere Products Exercises   251
   Henry Books Case   251
   Alexamara Marina Group Case   252

Chapter 8 *Database Administration* 253
  Database Policy Formulation and Enforcement 254
    Access Privileges 254
    Security 257
    Disaster Planning 258
    Archiving 259
  Other Database Administrative Functions 260
    DBMS Evaluation and Selection 261
    DBMS Maintenance 265
    Data Dictionary Management 265
    Training 265
  Technical Functions 266
    Database Design 266
    Testing 266
    Performance Tuning 267
  Summary 271
  Key Terms 271
  Review Questions 271
  Premiere Products Exercises 272
  Henry Books Case 272
  Alexamara Marina Group Case 273

Chapter 9 *Database Management Approaches* 275
  Distributed Databases 276
  Characteristics of Distributed DBMSs 277
    Location Transparency 277
    Replication Transparency 277
    Fragmentation Transparency 278
  Advantages of Distributed Databases 279
  Disadvantages of Distributed Databases 280
  Rules for Distributed Databases 283
  Client/Server Systems 283
    Advantages of Client/Server Systems 286
  Web Access to Databases 287
  XML 290
  Data Warehouses 293
    Data Warehouse Structure and Access 295
    Rules for OLAP Systems 298
  Object-Oriented DBMSs 299
    What Is an Object-Oriented DBMS? 299
    Objects and Classes 299
    Methods and Messages 301
    Inheritance 302
    Unified Modeling Language (UML) 302
    Rules for OODBMSs 305
  Summary 307
  Key Terms 308
  Review Questions 309
  Premiere Products Exercises 311
  Henry Books Case 311
  Alexamara Marina Group Case 311

Appendix A *Comprehensive Design Example: Marvel College* 313
  Marvel College Requirements 313
    General Description 313
    Report Requirements 314
    Update (Transaction) Requirements 317
  Marvel College Information-Level Design 318
  Final Information-Level Design 336
  Exercises 337

Appendix B *SQL Reference*   345
   ALTER TABLE   345
   Column or Expression List (SELECT Clause)   345
      Computed Fields   345
      Functions   346
   Conditions   346
      Simple Conditions   346
      Compound Conditions   346
      BETWEEN Conditions   346
      LIKE Conditions   347
      IN Conditions   347
   CREATE INDEX   347
   CREATE TABLE   347
   CREATE VIEW   348
   Data Types   348
   DELETE Rows   349
   DROP INDEX   349
   DROP TABLE   350
   GRANT   350
   INSERT   350
   Integrity   350
   JOIN   351
   REVOKE   351
   SELECT   352
   SELECT INTO   352
   Subqueries   353
   UNION   353
   UPDATE   353

Appendix C *"How Do I" Reference*   355

Appendix D *Answers to Odd-Numbered Review Questions*   357
   Chapter 1—Introduction to Database Management   357
   Chapter 2—The Relational Model 1: Introduction, QBE, and Relational Algebra   357
   Chapter 3—The Relational Model 2: SQL   358
   Chapter 4—The Relational Model 3: Advanced Topics 359
   Chapter 5—Database Design 1: Normalization   360
   Chapter 6—Database Design 2: Design Method   360
   Chapter 7—DBMS Functions   361
   Chapter 8—Database Administration   362
   Chapter 9—Database Management Approaches   363

Glossary   365

Index   377

The advent of database management systems for personal computers in the 1980s moved database management beyond the realm of database professionals and into the hands of everyday users from all segments of the population. A field once limited to highly trained users of large, mainframe, database-oriented application systems became an essential productivity tool for such diverse groups as home computer owners, owners of small businesses, and end-users in large organizations.

The major PC-based database software systems have continually added features to increase their ease of use, allowing users to enjoy the benefits of database tools relatively quickly. Truly effective use of such a product, however, requires more than just knowledge of the product itself, although that knowledge is obviously important. It requires a general knowledge of the database environment, including topics such as database design, database administration, and application development using these systems. While the depth of understanding required is certainly not as great for the majority of users as it is for the data processing professional, a lack of any understanding in these areas precludes effective use of the product in all but the most limited applications.

ABOUT THIS BOOK

This book is intended for anyone who is interested in gaining some familiarity with database management. It is appropriate for students in introductory database classes in computer science or information systems programs. It is appropriate for students in database courses in related disciplines, such as business, at either the undergraduate or graduate level. Such students require a general understanding of the database environment. In addition, courses introducing students of any discipline to database management have become increasingly popular over the past few years, and this book is ideal for such courses. It is also appropriate for individuals considering purchasing a PC-based database package and who want to make effective use of such a package.

This book assumes that you have some familiarity with computers; a single introductory course is all the background that is required. While you need not have any background in programming to use this text effectively, there are certain areas where some programming experience will allow you to explore topics in more depth.

Changes to the Sixth Edition

The Sixth Edition includes the following new features and content:

- As in the Fifth Edition, the SQL material is covered using Microsoft Access. Also included are generic forms of all examples that students can use on a variety of platforms, including Oracle and, new to the Sixth Edition, MySQL. The Sixth Edition continues the two appendices that provide a useful reference for anyone wanting to use SQL effectively. Appendix B includes a command reference of all the SQL commands and operators that are taught in the chapters. Students can use this appendix as a quick resource when constructing commands. Each command includes a short description, a table that shows the required and optional clauses and operators, and an example and its results. Appendix C provides students with an opportunity to ask a question, such as "How do I delete rows?," and to identify the appropriate section in Appendix B to use to find the answer. Appendix C is extremely valuable when students know what they want to accomplish, but can't remember the exact SQL command they need.

- In addition to the section of Review Questions, the end of each chapter includes three sets of exercises—one featuring the Premiere Products database and the others featuring the Henry Books database and the Alexamara Marina Group database—that give students "hands-on" experiences with the concepts found in the chapter.

- As in the previous edition, the Sixth Edition covers entity-relationship diagrams. The database design material in the Sixth Edition includes a discussion of the entity-relationship model as a database model. It also includes a discussion of a characterization of various types of primary keys.

- The Sixth Edition contains expanded coverage of the important topics of stored procedures and views.

- The Premiere Products, Henry Books, and Alexamara Marina Group databases are available in Access 2000 format (the default database format for Access 2000, Access 2002, and Access 2003) and in Access 2007 format on the Instructor's Resource CD-ROM and at *course.com*. For those students using database management systems that run scripts (such as Oracle and MySQL), the data files also include the script files that create the tables and add the data to the tables in the databases used in the book.

- For instructors who want to use an Access or SQL text as a companion to the Sixth Edition, the Instructor's Manual for this book includes detailed tips on integrating the Sixth Edition with other books from Course Technology that cover Microsoft Access 2007 and SQL (for more information, see the "Teaching Tools" section in this preface).

SPECIAL FEATURES

Detailed Coverage of the Relational Model, including Query-By-Example (QBE) and SQL

The book features detailed coverage of the important aspects of the relational model, including comprehensive coverage of SQL. It also covers QBE and relational algebra as well as advanced aspects of the model, such as views, the use of indexes, the catalog, and relational integrity rules.

Normalization Coverage

The Sixth Edition covers first normal form, second normal form, third normal form (Boyce-Codd normal form), and fourth normal form. The text describes in detail the update anomalies associated with lower normal forms as part of the motivation for the need for higher normal forms. Finally, the book examines correct and incorrect ways to normalize tables. This book specifically addresses this by showing students some of the mistakes people can make in the normalization process, explaining why the approach is incorrect, demonstrating the problems that would result from incorrect normalizations, and, most importantly, identifying how to avoid these mistakes.

Views Coverage

This text covers the important topic of views. It discusses the creation and use of views as well as the advantages of using views.

Database Design

The important process of database design is given detailed treatment. A highly useful method for designing databases is presented and illustrated through a variety of examples. In addition to the method, this text includes important design topics such as the use of survey forms, obtaining information by reviewing existing documents, special relationship considerations, and entity subtypes. Appendix A contains a comprehensive design example that illustrates how to apply the complete design process to a large and complex set of requirements. After mastering the design method presented in this text, students should be able to produce correct database designs for future database requirements they encounter.

Functions Provided by a Database Management System

With such a wide range of features included in current database management systems, it is important for students to know the functions that such systems should provide. These functions are presented and discussed in detail.

Database Administration

While database administration (DBA) is absolutely essential in the mainframe environment, it is also important in a personal computer environment, especially when the database is shared among several users. Thus, this text includes a detailed discussion of the database administration function.

Database Management System Selection

The process of selecting a database management system is important, considering the number of available systems from which to choose. Unfortunately, selecting the correct database management system is not an easy task. To prepare students to be able to do an effective job in this area, the text includes a detailed discussion of the selection process together with a comprehensive checklist that greatly assists in making such a selection.

Advanced Topics

The text also covers distributed database management systems, client/server systems, data warehouses, object-oriented database management systems, Web access to databases, and XML. Each of these topics encompasses an enormous amount of complex information, but the goal is to introduce students to these important topics. The text also includes detailed coverage of stored procedures and triggers.

Glossary

The glossary included at the end of this text contains definitions to the key terms in the chapters.

Numerous Realistic Examples

The book contains numerous examples illustrating each of the concepts. A running "case" example—Premiere Products—is used throughout the book to illustrate concepts. The examples are realistic and represent the kinds of problems students will encounter in the design, manipulation, and administration of databases. Exercises that use the Premiere Products case are included at the end of each chapter. In addition, there is another complete set of exercises at the end of each chapter that feature a second and third case—Henry Books and Alexamara Marina Group—giving students a chance to apply what they have learned to a database that they have not seen in the chapter material.

Review Material

This text contains a wide variety of questions. At key points within the chapters, students are asked questions to reinforce their understanding of the material before proceeding. The answers to these questions follow the questions. A summary and a list of key terms appear at the end of each chapter, followed by review questions that test the students' knowledge of the important points in the chapter and that occasionally test their ability to apply what they have learned. The answers to the odd-numbered review questions are provided in Appendix D. Each chapter also contains hands-on exercises related to the Premiere Products, Henry Books, and Alexamara Marina Group case examples.

Teaching Tools

When this book is used in an academic setting, instructors may obtain the following teaching tools from Course Technology through their sales representative or by visiting *course.com*:

- **Instructor's Manual** The Instructor's Manual has been carefully prepared and tested to ensure its accuracy and dependability. The Instructor's Manual includes suggestions and strategies for using this text, including the incorporation of companion texts on Access or SQL for those instructors who desire to do so. For instructors who want to use an Access or SQL text as a companion to the Sixth Edition, the Instructor's Manual for this book includes detailed tips on integrating the Sixth Edition with the following books, also published by Course Technology: *New Perspectives on Microsoft Office Access 2007—Comprehensive, New Perspectives on Microsoft Office Access 2007—Introductory,* and *New Perspectives on Microsoft Office Access 2007—Brief,* by Adamski and Finnegan; *Microsoft Office Access 2007: Introductory Concepts and Techniques, Microsoft Office Access 2007: Complete Concepts and Techniques,* and

Microsoft Office Access 2007: Comprehensive Concepts and Techniques, by Shelly, Cashman, Pratt, and Last; *A Guide to SQL, Sixth Edition,* by Pratt; and *A Guide to MySQL,* by Pratt and Last.

- **Data and Solution Files** Data and solution files are available at *www.course.com* and on the Instructor's Resource CD-ROM. Data files consist of copies of the Premiere Products, Henry Books, and Alexamara Marina Group databases in Access 2000 and Access 2007 format and script files to create the tables and data in these databases in other systems, such as Oracle and MySQL.

- **ExamView®** This text is accompanied by ExamView, a powerful testing software package that allows instructors to create and administer printed, computer (LAN-based), and Internet exams. ExamView includes hundreds of questions that correspond to the topics covered in this text, enabling students to generate detailed study guides that include page references for further review. The computer-based and Internet testing components allow students to take exams at their computers, and also save the instructor time by grading each exam automatically.

- **PowerPoint Presentations** Microsoft PowerPoint slides are included for each chapter as a teaching aid for classroom presentations, to make available to students on a network for chapter review, or to be printed for classroom distribution. Instructors can add their own slides for additional topics they introduce to the class. The presentations are included on the Instructor's CD.

- **Figure Files** Figure files are included so that instructors can create their own presentations using figures appearing in the text.

ORGANIZATION OF THE TEXTBOOK

This text includes nine chapters covering general database topics that are relevant to any database management system. A brief description of the organization of topics in the chapters and an overview each chapter's contents follows.

Introduction

Chapter 1 provides a general introduction to the field of database management.

The Relational Model

The relational model is covered in detail in Chapters 2, 3, and 4. Chapter 2 covers the data definition and manipulation aspects of the model using QBE and relational algebra. The text uses Microsoft Office Access 2007 to illustrate the QBE material. The relational algebra section includes the entire relational algebra. (*Note:* The extra material on relational algebra is optional and can be omitted if desired.)

Chapter 3 is devoted exclusively to SQL. The SQL material is illustrated using Access, but the chapter also includes generic versions of all examples that can be used with a variety of platforms, including Oracle and MySQL.

Chapter 4 covers some advanced aspects of the relational model such as views, the use of indexes, the catalog, relational integrity rules, stored procedures, and triggers.

Database Design

Chapters 5 and 6 are devoted to database design. Chapter 5 covers the normalization process, which enables students to identify and correct bad designs. This chapter discusses and illustrates the use of first, second, third, and fourth normal forms. (*Note:* The material on fourth normal form is optional and can be omitted if desired.)

Chapter 6 presents a method for database design using many examples. The material includes entity-relationship diagrams and their role in database design. It also includes discussions of several special design issues as well as the use of survey forms, obtaining information by reviewing existing documents, special relationship considerations, and entity subtypes. After completing Chapter 6, students can further challenge themselves by completing Appendix A, which includes a comprehensive design example that illustrates the application of the complete design process to a large and complex set of requirements. (*Note:* Chapters 5 and 6 can be covered immediately after Chapter 2 if desired.)

Database Management System Functions

Chapter 7 discusses the features that should be provided by a full-functioned PC-based database management system. This chapter includes coverage of journaling, forward recovery, backward recovery, authentication, and authorizations.

Database Administration

Chapter 8 is devoted to the role of database administration. Also included in this chapter is a discussion of the process of selecting a database management system.

Database Management Approaches

Chapter 9 provides an overview of several advanced topics: distributed databases, client/server systems, Web access to databases, XML and related document specification standards, data warehouses, and object-oriented databases. This chapter has been expanded from the previous edition to include more details on Web access to databases and XML.

GENERAL NOTES TO THE STUDENT

Embedded Questions

There are many places in the text where special questions have been embedded. Sometimes the purpose of these questions is to ensure that you understand some crucial material before you proceed. In other cases, the questions are designed to give you the chance to consider some special concept in advance of its actual presentation. In all cases, the answers to these questions follow each question. You could simply read the question and its answer. You will receive maximum benefit from the text, however, if you take the time to work out the answers to the questions and then check your answer against the one provided before continuing.

End-of-Chapter Material

The end-of-chapter material consists of a summary, a list of key terms, review questions, and exercises for the Premiere Products, Henry Books, and Alexamara Marina Group databases. The summary briefly describes the material covered in the chapter. The review questions require you to recall and apply the important material in the chapter. The answers to the odd-numbered review questions appear in Appendix D. The Premiere Products, Henry Books, and Alexamara Marina Group exercises test your knowledge of the chapter material; your instructor will assign one or more of these exercises for you to complete.

ACKNOWLEDGMENTS

We would like to acknowledge the following individuals who all made contributions during the preparation of this book. We appreciate the following individuals who reviewed the text and made many helpful suggestions: Linda Lau, Longwood University; Janine Loveless, Eastern Iowa Community College; Dave Braunschweig, Harper College; and William Wagner, Villanova University. We also appreciate the efforts of the following individuals, who have been invaluable during this book's development: Kate Hennessy, Product Manager; Nicole Ashton, Quality Assurance tester; and Jessica Evans, Developmental Editor.

We have again been privileged to work with Jessica Evans as our Developmental Editor. Thank you, Jess, for your knowledge, skill, guidance, and energy, and for your hard work and positive influence on our work. We appreciate your dedication and many contributions to this book and value you as a friend.

CHAPTER 1

INTRODUCTION TO DATABASE MANAGEMENT

LEARNING OBJECTIVES

Objectives

- Introduce Premiere Products, the company that is used as the basis for many of the examples throughout the text
- Introduce basic database terminology
- Describe database management systems (DBMSs)
- Explain the advantages and disadvantages of database processing
- Introduce Henry Books, the company that is used in a case that appears throughout the text
- Introduce Alexamara Marina Group, the company that is used in another case that appears throughout the text

INTRODUCTION

In this chapter, you will examine the requirements of Premiere Products, a company that will be used in many examples in this chapter and in the rest of the text. You will learn how Premiere Products initially stored its data, what problems employees encountered with the storage method, and why management decided to use a database management system. Then you will study the basic terminology and concepts of databases and database management systems, and learn the advantages and disadvantages of database processing. Finally, you will examine the database requirements for Henry Books and Alexamara Marina Group, the companies featured in the cases that appear at the end of each chapter.

PREMIERE PRODUCTS BACKGROUND

Premiere Products is a distributor of appliances, housewares, and sporting goods. Since its inception, the company has used spreadsheet software to maintain customer, order, inventory, and sales representative (sales rep) data. Management has determined that the company's recent growth means it is no longer feasible to use spreadsheets to maintain the firm's data.

What has led the managers at Premiere Products to this decision? One of the company's spreadsheets, shown in Figure 1-1, displays sample order data and illustrates the company's problems with the spreadsheet approach. For each order, the spreadsheet displays the number and name of the customer placing the order; the number and date of the order; the number, description, number ordered, quoted price, and warehouse number of the item ordered; and the number of the sales rep assigned to the customer. Note that the Ferguson's order (order number 21610) and the first Johnson's Department Store order (order number 21617) appear in two rows because these customers purchased two different items in their orders.

Customer Number	Customer Name	Order Number	Order Date	Part Number	Part Description	Number Ordered	Quoted Price	Warehouse	Rep Number
148	Al's Appliance and Sport	21608	10/20/2010	AT94	Iron	11	$21.95	3	20
148	Al's Appliance and Sport	21619	10/23/2010	DR93	Gas Range	1	$495.00	2	20
282	Brookings Direct	21614	10/21/2010	KT03	Dishwasher	2	$595.00	3	35
356	Ferguson's	21610	10/20/2010	DR93	Gas Range	1	$495.00	2	65
356	Ferguson's	21610	10/20/2010	DW11	Washer	1	$399.99	3	65
408	The Everything Shop	21613	10/21/2010	KL62	Dryer	4	$329.95	1	35
608	Johnson's Department Store	21617	10/23/2010	BV06	Home Gym	2	$794.95	2	65
608	Johnson's Department Store	21617	10/23/2010	CD52	Microwave Oven	4	$150.00	1	65
608	Johnson's Department Store	21623	10/23/2010	KV29	Treadmill	2	$1,290.00	2	65

Orders requiring more than one spreadsheet row

FIGURE 1-1   Sample orders spreadsheet

Redundancy is one problem that employees have with the orders spreadsheet. **Redundancy** is the duplication of data or the storing of the same data in more than one place. In the orders spreadsheet, redundancy occurs in the Customer Name column because the name of a customer is stored in more than one place. All three rows for customer number 608, for example, store "Johnson's Department Store" as the customer name. In the orders spreadsheet, redundancy also occurs in other columns, such as the Order Date and Part Description columns.

Q & A

Question: What problems does redundancy cause?

Answer: Redundancy wastes space because you're storing the same data in multiple places. The extra space results in larger spreadsheets that require more space in memory and on disk and that take longer to save and open.

When you need to change data, redundancy also makes your changes more cumbersome and time-consuming. For example, if you incorrectly enter "Johnson's Department Store" in the Customer Name column, you would need to correct it in three places. Even if you use the global find-and-replace feature, multiple changes require more computer time than does a single change.

Finally, redundancy can lead to inconsistencies. For example, you might enter "Johnson's Department Store," "Johnsons Department Store," and "Johnsons' Department Store" in the Customer Name column, and then not be sure which is the correct spelling. Further, if that customer's name is spelled three different ways and you use the search feature with one of the three values, you'd find a single match instead of three matches.

Difficulty accessing related data is another problem that employees at Premiere Products encounter with their spreadsheets. For example, if you want to see a customer's address and a part's standard price, you must open and search other spreadsheets that contain this data.

Spreadsheets also have limited security features to protect data from being accessed by unauthorized users. A spreadsheet's data sharing features also prevent multiple employees from updating data in one spreadsheet at the same time. Finally, if the increase in sales and growth at Premiere Products continues at its planned rate, spreadsheets have inherent size limitations that will eventually force the company to split its order data into multiple spreadsheets. Splitting the spreadsheets would create further redundancy.

Having decided to replace its spreadsheet software, management has determined that Premiere Products must maintain the following information about its sales reps, customers, and parts inventory:

- The sales rep number, last name, first name, address, total commission, and commission rate for each sales rep.
- The customer number, name, address, current balance, and credit limit for each customer, as well as the number of the sales rep who represents the customer.
- The part number, description, number of units on hand, item class, number of the warehouse where the item is stored, and unit price for each part in inventory.

Premiere Products must also store information about orders. Figure 1-2 shows a sample order.

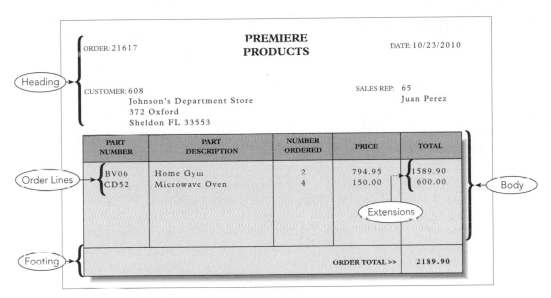

FIGURE 1-2   Sample order

The sample order has three components:

- The heading (top) of the order contains the order number and date; the customer's number, name, and address; and the sales rep's number and name.
- The body of the order contains one or more order lines, sometimes called line items. Each order line contains a part number, a part description, the number of units of the part ordered, and the quoted price for the part. Each order line also contains a total, usually called an extension, which is the result of multiplying the number ordered by the quoted price.
- The footing (bottom) of the order contains the order total.

Premiere Products must also store the following items for each customer's order:

- For each order, the company must store the order number, the date the order was placed, and the number of the customer that placed the order. The customer's name and address and the number of the sales rep who represents the customer are stored with customer information. The name of the sales rep is stored with the sales rep information.
- For each order line, the company must store the order number, the part number, the number of units ordered, and the quoted price. Remember that the part description is stored with the information about parts. The result of multiplying the number of units ordered by the quoted price is not stored because the computer can calculate it when necessary.
- The overall order total is not stored. Instead, the computer calculates the total whenever an order is printed or displayed on the screen.

The problem facing Premiere Products is common to many businesses and individuals that need to store and retrieve data in an efficient and organized way. Furthermore, most organizations are interested in more than one category of information. For example, Premiere Products is interested in categories such as sales reps, customers, orders, and parts. A school is interested in students, faculty, and classes; a real estate agency is interested in clients, houses, and agents; and a used car dealership is interested in customers, vehicles, and manufacturers.

Besides wanting to store data that pertains to more than one category, Premiere Products is also interested in the relationships between the categories. For example, company employees want to be able to associate orders with the customers that ordered them, the sales reps who coordinated the orders, and the parts that the customers requested. Likewise, a real estate agency wants to know not only about clients, houses, and agents, but also about the relationships between clients and houses (which clients have listed which houses and which clients have expressed interest in which houses). A real estate agency also wants to know about the relationships between agents and houses (which agent sold which house, which agent is listing which house, and which agents are receiving commissions for which houses).

DATABASE BACKGROUND

After studying the alternatives to using spreadsheet software, Premiere Products decided to switch to a database system. A database is a structure that contains information about many different categories of information and about the relationships between those categories. The Premiere Products database, for example, will contain information about sales reps, customers, orders, and parts. It will also provide facts that relate sales reps to the customers they represent and customers to the orders they currently have placed.

With the use of a database, employees can enter the number of a particular order and find out which customer placed the order, as well as which parts the customer ordered. Alternately, employees can start with a customer and find all orders the customer placed, together with which parts the customer ordered and the amount of the commission earned by the customer's sales rep. Using a database, Premiere Products can not only maintain its data better, but it can use the data in the database to produce a variety of reports and to answer a variety of questions.

There are some terms and concepts in the database environment that are important for you to know. For instance, the terms *entity*, *attribute*, and *relationship* are fundamental when discussing databases. An **entity** is a person, place, object, event, or idea for which you want to store and process data. The entities of interest to Premiere Products, for example, are sales reps, customers, orders, and parts.

An **attribute** is a characteristic or property of an entity. The term is used in this text exactly as it is used in everyday English. For the entity *person*, for example, the list of attributes might include such things as eye color and height. For Premiere Products, the attributes of interest for the entity *customer* are such things as customer name, street, city, and so on. An attribute is also called a **field** or **column** in many database systems.

Figure 1-3 shows two entities, Rep (short for Sales Rep) and Customer, along with the attributes for each entity. The Rep entity has nine attributes: RepNum, LastName, FirstName, Street, City, State, Zip, Commission, and Rate. The attributes are the same as the columns in a spreadsheet. The Customer entity has nine attributes: CustomerNum, CustomerName, Street, City, State, Zip, Balance, CreditLimit, and RepNum.

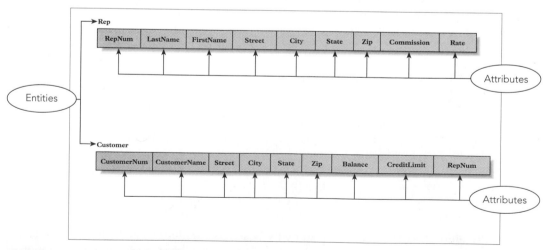

FIGURE 1-3 Entities and attributes

The final key term is relationship. A **relationship** is an association between entities. There is an association between reps and customers, for example, at Premiere Products. A rep is associated with all of his or her customers, and a customer is associated with its rep. Technically, you say that a rep is *related to* all of his or her customers, and a customer is *related to* its rep.

This particular relationship is called a **one-to-many relationship** because each rep is associated with *many* customers, but each customer is associated with only *one* rep. In this type of relationship, the word *many* is used differently than in everyday English; it might not always indicate a large number. In this context, for example, the term *many* means that a rep can be associated with *any* number of customers. That is, a given rep can be associated with zero, one, or more customers.

A one-to-many relationship often is represented visually in the manner shown in Figure 1-4. In such a diagram, entities and attributes are represented in precisely the same way as they are shown in Figure 1-3. A line connecting the entities represents the relationship. The *one* part of the relationship (in this case, Rep) does not have an arrow on its end of the line, and the *many* part of the relationship (in this case, Customer) is indicated by a single-headed arrow.

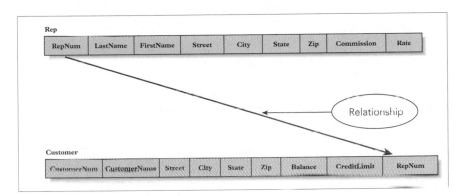

FIGURE 1-4 One-to-many relationship

Spreadsheets, word-processed documents, Web pages, and other computer information sources are stored in files. Basically, a file that is used to store data, which is often called a **data file**, is the computer counterpart to an ordinary paper file you might keep in a file cabinet, an accounting ledger, or other place. A database, however, is more than a file. Unlike a typical data file, a database can store information about multiple entities. There is also another difference. A database holds information about the relationships among the various entities. Not only will the Premiere Products database have information about both reps and customers, it will also hold information relating reps to the customers they service, customers to orders, parts to orders, and so on. Formally, the definition of a database is as follows:

Definition: A **database** is a structure that can store information about multiple types of entities, the attributes of those entities, and the relationships between the entities.

How does a database handle entities, attributes of entities, and relationships between entities? Entities and attributes are fairly simple. Each entity has its own table. In the Premiere Products database, for example, there will be one table for reps, one table for customers, and so on. The attributes of an entity become the columns in the table. In the table for reps, for example, there will be a column for the rep number, a column for the rep last name, and so on.

What about relationships between entities? At Premiere Products, there is a one-to-many relationship between reps and customers. (Each rep is related to the many customers that he or she represents, and each customer is related to the one rep who represents the customer.) How is this relationship handled in a database system? It is handled by using common columns in the two tables. Consider Figure 1-4 again. The RepNum column in the Rep table and the RepNum column in the Customer table are used to implement the relationship between reps and customers. Given a rep, you can use these columns to determine all the customers that he or she represents; given a customer, you can use these columns to find the rep who represents the customer.

How will Premiere Products store its data in a database? Figure 1-5 shows sample data for Premiere Products.

Rep

RepNum	LastName	FirstName	Street	City	State	Zip	Commission	Rate
20	Kaiser	Valerie	624 Randall	Grove	FL	33321	$20,542.50	0.05
35	Hull	Richard	532 Jackson	Sheldon	FL	33553	$39,216.00	0.07
65	Perez	Juan	1626 Taylor	Fillmore	FL	33336	$23,487.00	0.05

Customer

CustomerNum	CustomerName	Street	City	State	Zip	Balance	CreditLimit	RepNum
148	Al's Appliance and Sport	2837 Greenway	Fillmore	FL	33336	$6,550.00	$7,500.00	20
282	Brookings Direct	3827 Devon	Grove	FL	33321	$431.50	$10,000.00	35
356	Ferguson's	382 Wildwood	Northfield	FL	33146	$5,785.00	$7,500.00	65
408	The Everything Shop	1828 Raven	Crystal	FL	33503	$5,285.25	$5,000.00	35
462	Bargains Galore	3829 Central	Grove	FL	33321	$3,412.00	$10,000.00	65
524	Kline's	838 Ridgeland	Fillmore	FL	33336	$12,762.00	$15,000.00	20
608	Johnson's Department Store	372 Oxford	Sheldon	FL	33553	$2,106.00	$10,000.00	65
687	Lee's Sport and Appliance	282 Evergreen	Altonville	FL	32543	$2,851.00	$5,000.00	35
725	Deerfield's Four Seasons	282 Columbia	Sheldon	FL	33553	$248.00	$7,500.00	35
842	All Season	28 Lakeview	Grove	FL	33321	$8,221.00	$7,500.00	20

Orders

OrderNum	OrderDate	CustomerNum
21608	10/20/2010	148
21610	10/20/2010	356
21613	10/21/2010	408
21614	10/21/2010	282
21617	10/23/2010	608
21619	10/23/2010	148
21623	10/23/2010	608

OrderLine

OrderNum	PartNum	NumOrdered	QuotedPrice
21608	AT94	11	$21.95
21610	DR93	1	$495.00
21610	DW11	1	$399.99
21613	KL62	4	$329.95
21614	KT03	2	$595.00
21617	BV06	2	$794.95
21617	CD52	4	$150.00
21619	DR93	1	$495.00
21623	KV29	2	$1,290.00

Part

PartNum	Description	OnHand	Class	Warehouse	Price
AT94	Iron	50	HW	3	$24.95
BV06	Home Gym	45	SG	2	$794.95
CD52	Microwave Oven	32	AP	1	$165.00
DL71	Cordless Drill	21	HW	3	$129.95
DR93	Gas Range	8	AP	2	$495.00
DW11	Washer	12	AP	3	$399.99
FD21	Stand Mixer	22	HW	3	$159.95
KL62	Dryer	12	AP	1	$349.95
KT03	Dishwasher	8	AP	3	$595.00
KV29	Treadmill	9	SG	2	$1,390.00

FIGURE 1-5 Sample data for Premiere Products

In the Rep table, you see that there are three reps whose numbers are 20, 35, and 65. The name of sales rep 20 is Valerie Kaiser. Her street address is 624 Randall. She lives in Grove, FL, and her zip code is 33321. Her total commission is $20,542.50, and her commission rate is 5% (0.05).

Premiere Products has 10 customers that are identified with the numbers 148, 282, 356, 408, 462, 524, 608, 687, 725, and 842. The name of customer number 148 is Al's Appliance and Sport. This customer's address is 2837 Greenway in Fillmore, FL, with a zip code of 33336. The customer's current balance is $6,550.00, and its credit limit is $7,500.00. The number 20 in the RepNum column indicates that Al's Appliance and Sport is represented by sales rep 20 (Valerie Kaiser).

Skipping to the table named Part, you see that there are 10 parts, whose part numbers are AT94, BV06, CD52, DL71, DR93, DW11, FD21, KL62, KT03, and KV29. Part AT94 is an iron, and the company has 50 units of this part on hand. Irons are in item class HW (housewares) and are stored in warehouse 3. The price of an iron is $24.95. Other item classes are AP (appliances) and SG (sporting goods).

Moving back to the table named Orders, you see that there are seven orders, which are identified with the numbers 21608, 21610, 21613, 21614, 21617, 21619, and 21623. Order number 21608 was placed on October 20, 2010, by customer 148 (Al's Appliance and Sport).

> ## NOTE
>
> In some database systems, the word "Order" has a special purpose. Having a table named Order could cause problems in such systems. For this reason, Premiere Products uses the table name Orders instead of Order.

The table named OrderLine might seem strange at first glance. Why do you need a separate table for the order lines? Couldn't the order lines be included in the Orders table? The answer is yes. The Orders table could be structured as shown in Figure 1-6. Notice that this table contains the same orders as shown in Figure 1-5, with the same dates and customers. In addition, each table row in Figure 1-6 contains all the order lines for a given order. Examining the fifth row, for example, you see that order 21617 has two order lines. One of the order lines is for two BV06 parts at $794.95 each, and the other order line is for four CD52 parts at $150.00 each.

Orders

OrderNum	OrderDate	CustomerNum	PartNum	NumOrdered	QuotedPrice
21608	10/20/2010	148	AT94	11	$21.95
21610	10/20/2010	356	DR93	1	$495.00
			DW11	1	$399.99
21613	10/21/2010	408	KL62	4	$329.95
21614	10/21/2010	282	KT03	2	$595.00
21617	10/23/2010	608	BV06	2	$794.95
			CD52	4	$150.00
21619	10/23/2010	148	DR93	1	$495.00
21623	10/23/2010	608	KV29	2	$1,290.00

FIGURE 1-6   Alternative Orders table structure

> ## Q & A
>
> **Question:** How is the information from Figure 1-5 represented in Figure 1-6?
> **Answer:** Examine the OrderLine table shown in Figure 1-5 and note the sixth and seventh rows. The sixth row indicates that there is an order line in order 21617 for two BV06 parts at $794.95 each. The seventh row indicates that there is an order line in order 21617 for four CD52 parts at $150.00 each. Thus, the information in Figure 1-6 is represented in Figure 1-5 in two separate rows rather than in one row.

It might seem inefficient to use two rows to store information that can be represented in one row. There is a problem, however, with the arrangement shown in Figure 1-6—the table is more complicated. In Figure 1-5, there is a single entry at each position in the table. In Figure 1-6, some of the individual positions within the table contain multiple entries, thus making it difficult to track the information between columns. In the row for order number 21617, for example, it is crucial to know that BV06 corresponds to the 2 in the NumOrdered column (not to the 4) and that it corresponds to $794.95 in the QuotedPrice column (not to $150.00). In addition, having a more complex table means that there are practical issues to worry about, such as:

- How much room do you allow for these multiple entries?
- What if an order has more order lines than you have allowed room for?
- Given a part, how do you determine which orders contain order lines for that part?

Certainly, none of these problems is unsolvable. These problems do add a level of complexity, however, that is not present in the arrangement shown in Figure 1-5. In Figure 1-5, there are no multiple entries to worry about, it doesn't matter how many order lines exist for any order, and it is easy to find every order that contains an order line for a given part (just look for all order lines with the given part number in the PartNum column). In general, this simpler structure is preferable, which is why order lines appear in a separate table.

To test your understanding of the Premiere Products data, use the data shown in Figure 1-5 to answer the following questions.

Q & A

Question: What are the numbers of the customers represented by Valerie Kaiser?

Answer: 148, 524, and 842. (Look up the RepNum value for Valerie Kaiser in the Rep table and obtain the number 20. Then find all customers in the Customer table that have the number 20 in the RepNum column.)

Q & A

Question: What is the name of the customer that placed order 21610, and what is the name of the rep who represents this customer?

Answer: Ferguson's is the customer, and Juan Perez is the rep. (Look up the CustomerNum value in the Orders table for order number 21610 and obtain the number 356. Then find the customer in the Customer table with a CustomerNum value of 356. Using this customer's RepNum value, which is 65, find the name of the rep in the Rep table.)

Q & A

Question: List all the parts that appear in order 21610. For each part, give the description, number ordered, and quoted price.

Answer: Part number: DR93; part description: Gas Range; number ordered: 1; and quoted price: $495.00. Also, part number: DW11; part description: Washer; number ordered: 1; and quoted price: $399.99. (Look up each OrderLine table row in which the order number is 21610. Each row contains a part number, the number ordered, and the quoted price. Use the part number to look up the corresponding description in the Part table.)

A visual way to represent a database is with an **entity-relationship (E-R) diagram**. In an E-R diagram, rectangles represent entities, and lines represent relationships between connected entities. The E-R diagram for the Premiere Products database appears in Figure 1-7.

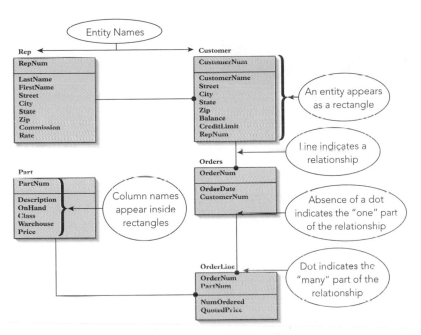

FIGURE 1-7   E-R diagram for the Premiere Products database

Each of the five entities in the Premiere Products database appears as a rectangle in the E-R diagram shown in Figure 1-7. The name of each entity appears above the rectangle. The columns for each entity appear within the rectangle. Because the Rep and Customer entities have a one-to-many relationship, a line connects these two entities; similarly, a line connects the Customer and Orders entities, the Orders and OrderLine entities, and the Part and OrderLine entities. The dot at the end of a line, such as the dot at the Customer end of the line that connects the Rep and Customer entities, indicates the "many" part of the one-to-many relationship between two entities. You will learn more about E-R diagrams in Chapter 6.

DATABASE MANAGEMENT SYSTEMS

Managing a database is inherently a complicated task. Fortunately, software packages, called database management systems, can do the job of manipulating databases for you. A **database management system (DBMS)** is a program, or a collection of programs, through which users interact with a database. The actual manipulation of the underlying database is handled by the DBMS. In some cases, users may interact with the DBMS directly, as shown in Figure 1-8.

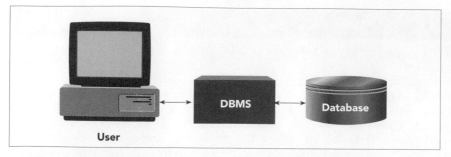

FIGURE 1-8 Using a DBMS directly

In other cases, users may interact with programs such as those created with Visual Basic, Java, Perl, PHP, or C++; these programs, in turn, interact with the DBMS, as shown in Figure 1-9. In either case, only the DBMS actually accesses the database.

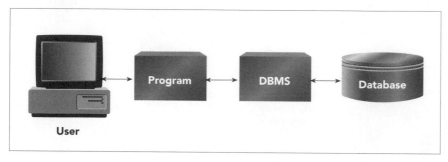

FIGURE 1-9 Using a DBMS through another program

With a DBMS, for example, users at Premiere Products can ask the system to find data about part KV29; the system will either locate the part and provide the data or display a message that no such part exists in the database. All the work involved in this task is performed by the DBMS. If part KV29 is in the database, users then can ask for the order lines that contain the part; and again the system will perform all the work involved in locating the order lines. Likewise, when users add data about a new customer to the database, the DBMS performs all the tasks necessary to ensure that the customer data is added and that the customer is related to the appropriate rep.

Popular DBMSs include Access, Oracle, DB2, MySQL, and SQL Server. Because Premiere Products uses the Microsoft Office suite of programs, which includes Access, management elects to use Access as its DBMS initially. Using the tables shown in Figure 1-5 as the starting point, a database expert at Premiere Products determines the structure of the required database—this process is called **database design**. Then this person enters the design in the DBMS and creates several **forms**, which are screen objects used to maintain, view, and print data from a database. Employees then begin to enter data.

The form that employees use to process part data is shown in Figure 1-10. Using this form, employees can enter a new part; view, change, or delete an existing part; and print the information for a part. No one at Premiere Products needs to write a program to create this form; instead, the DBMS creates the form based on answers provided in response to the DBMS's questions about the form's content and appearance.

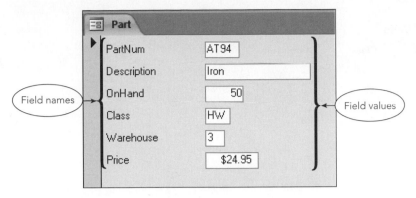

FIGURE 1-10 Part form

In this same way, the DBMS creates the other forms that Premiere Products needs. A more complicated form for processing order data is shown in Figure 1-11. This form displays data about an order and its order lines, using data from the Orders table and related data from the OrderLine table.

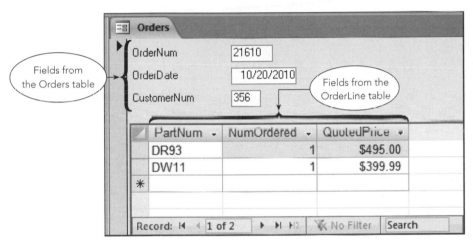

FIGURE 1-11 Orders form

Premiere Products can create the reports it needs in a similar way—the DBMS asks questions about the desired content and appearance of each report and then creates the reports automatically based on the answers. The Parts report, listing each part in stock, is shown in Figure 1-12.

Parts

PartNum	Description	OnHand	Class	Warehouse	Price
AT94	Iron	50	HW	3	$24.95
BV06	Home Gym	45	SG	2	$794.95
CD52	Microwave Oven	32	AP	1	$165.00
DL71	Cordless Drill	21	HW	3	$129.95
DR93	Gas Range	8	AP	2	$495.00
DW11	Washer	12	AP	3	$399.99
FD21	Stand Mixer	22	HW	3	$159.95
KL62	Dryer	12	AP	1	$349.95
KT03	Dishwasher	8	AP	3	$595.00
KV29	Treadmill	9	SG	2	$1,390.00

FIGURE 1-12   Parts report

ADVANTAGES OF DATABASE PROCESSING

The database approach to processing offers nine clear advantages over alternative data management methods. These advantages are listed in Figure 1-13 and are discussed on the following pages.

1. Getting more information from the same amount of data
2. Sharing data
3. Balancing conflicting requirements
4. Controlling redundancy
5. Facilitating consistency
6. Improving integrity
7. Expanding security
8. Increasing productivity
9. Providing data independence

FIGURE 1-13   Advantages of database processing

1. *Getting more information from the same amount of data.* The primary goal of a computer system is to turn data (recorded facts) into information (the knowledge gained by processing those facts). In a nondatabase, file-oriented environment, data often is partitioned into several disjointed systems, with each system having its own collection of files. Any request for information that necessitates accessing data from more than one of these collections can be extremely difficult. In some cases, for all practical purposes, it is impossible. Thus, the desired information is unavailable—it has been stored in the computer, but it is scattered across multiple files. When all the data for the various systems is stored in a single database, however, the information becomes available. Given the power of a DBMS, the information is available, and the process of getting it is quick and easy.

2. *Sharing data.* The data of various users can be combined and shared among authorized users, allowing all users access to a greater pool of data. Several users can have access to the same piece of data—for example, a customer's address—and still use it in a variety of ways. When one user

changes a customer's address, the new address immediately becomes available to all users. In addition, the existing data can be used in new ways, such as generating new types of reports, without having to create additional data files, as is the case in the nondatabase approach.

3. ***Balancing conflicting requirements.*** For the database approach to function adequately within an organization, a person or group should be in charge of the database, especially if the database will serve many users. This person or group is often called the **database administrator** or **database administration (DBA)**, respectively. By keeping the overall needs of the organization in mind, a DBA can structure the database in such a way that it benefits the entire organization, not just a single group. Although this approach might mean that an individual user group is served less well than it would have been if it had its own isolated system, the organization as a whole is better off. Ultimately, when the organization benefits, so do the individual groups of users.

4. ***Controlling redundancy.*** With database processing, data that formerly was kept separate in a nondatabase, file-oriented system is integrated into a single database; so multiple copies of the same data no longer exist. With the nondatabase approach, each user group at Premiere Products has its own copy of each customer's address. With the database approach, each customer's address would occur only once, thus eliminating redundancy.

    Eliminating redundancy not only saves space but also makes the process of updating data much simpler. With the database approach, changing a customer's address means making one change. With the nondatabase approach, in which data for each customer might be stored in three different places, the same address change means that three changes have to be made.

    Although eliminating redundancy is the ideal, it is not always possible. Sometimes, for reasons having to do with performance, you might choose to introduce a limited amount of redundancy into a database. However, even in these cases, you would be able to keep the redundancy under tight control, thus obtaining the same advantages. This is why it is better to say that you *control* redundancy rather than *eliminate* it.

5. ***Facilitating consistency.*** Suppose an individual customer's address appears in more than one place. Customer 148, for example, might be listed at 2837 Greenway in one place and at 2856 Wisner in another place. In this case, the data for the customer is inconsistent. Because the potential for this sort of problem is a direct result of redundancy and because the database approach reduces redundancy, there is less potential for this sort of inconsistency occurring with the database approach.

6. ***Improving integrity.*** An **integrity constraint** is a rule that data must follow in the database. For example, the rep number given for any customer must be one that is already in the database. In other words, users cannot enter an incorrect or nonexistent rep number for a customer. A database has **integrity** when the data in it satisfies all established integrity constraints. A good DBMS should provide an opportunity for users to incorporate these integrity constraints when they design the database. The DBMS then should ensure that the constraints are not violated. According to the integrity constraint about customers, the DBMS should *not allow* you to store data about a given customer when the rep number you enter is not the number of a rep that already is in the database.

7. ***Expanding security.*** **Security** is the prevention of unauthorized access to the database. A DBMS has many features that help ensure the enforcement of security measures. For example, a DBA can assign passwords to authorized users; then only those users who enter an acceptable password can gain access to the data in the database. Further, a DBMS lets you assign users to groups, with some groups permitted to view and update data in the database and other groups permitted only to view certain data in the database. With the nondatabase approach, you have limited security features and are more vulnerable to intentional and accidental access and changes to data.

8. ***Increasing productivity.*** A DBMS frees the programmers who are writing database access programs from having to engage in mundane data manipulation activities, such as adding new data and deleting existing data, thus making the programmers more productive. A good DBMS has many features that allow users to gain access to data in a database without having to do any programming. These features increase the productivity of programmers, who may not need to write complex programs in order to perform certain tasks, and nonprogrammers, who may be able to get the results they seek from the data in a database without waiting for a program to be written for them.

9. ***Providing data independence.*** The structure of a database often needs to be changed. For example, changing user requirements might necessitate the addition of an entity, an attribute, or a relationship, or a change might be required to improve performance. A good DBMS provides

data independence, which is a property that lets you change the structure of a database without requiring you to change the programs that access the database; examples of these programs are the forms you use to interact with the database and the reports that provide information from the database. Without data independence, a great deal of effort can be expended in changing programs to match the new database structure. The presence of many programs in the system may make this effort so prohibitive that management might decide to avoid changing the database, even though the change might improve the database's performance or add valuable data. With data independence, management is more likely to make the decision to change the database.

DISADVANTAGES OF DATABASE PROCESSING

As you would expect, when there are advantages to doing something in a certain way, there are also disadvantages. Database processing is no exception. In terms of numbers alone, the advantages outweigh the disadvantages; the latter are listed in Figure 1-14 and explained next.

1. **Larger file size**
2. **Increased complexity**
3. **Greater impact of failure**
4. **More difficult recovery**

FIGURE 1-14   Disadvantages of database processing

1. *Larger file size.* To support all the complex functions that it provides to users, a DBMS must be a large program that occupies a great deal of disk space as well as a substantial amount of internal memory. In addition, because all of the data that the database manages for you is stored in one file, the database file requires a large amount of disk space and internal memory.
2. *Increased complexity.* The complexity and breadth of the functions provided by a DBMS make it a complex product. Users of the DBMS must learn a great deal to understand the features of the system in order to take full advantage of it. In the design and implementation of a new system that uses a DBMS, many choices have to be made; it is possible to make incorrect choices, especially with an insufficient understanding of the system. Unfortunately, a few incorrect choices can spell disaster for the whole project. A sound database design is critical to the successful use of a DBMS.
3. *Greater impact of failure.* In a nondatabase, file-oriented system, each user has a completely separate system; the failure of any single user's system does not necessarily affect any other user. On the other hand, if several users are sharing the same database, a failure on the part of any one user that damages the database in some way might affect all the other users.
4. *More difficult recovery.* Because a database inherently is more complex than a simple file, the process of recovering it in the event of a catastrophe also is more complicated. This is particularly true when the database is being updated by many users at the same time. The database must first be restored to the condition it was in when it was last known to be correct; any updates made by users since that time must be redone. The greater the number of users involved in updating the database, the more complicated this task becomes.

INTRODUCTION TO THE HENRY BOOKS DATABASE CASE

Similar to the management of Premiere Products, Ray Henry, the owner of a bookstore chain named Henry Books, has decided to store his data in a database. He wants to achieve the same benefits as Premiere Products; that is, he wants to ensure that his data is current and accurate. He also needs to create forms to interact with the data and to produce reports from that data. In addition, he wants to be able to ask questions concerning the data and to obtain answers to those questions easily and quickly.

In running his chain of bookstores, Ray gathers and organizes information about branches, publishers, authors, and books. Figure 1-15 shows sample branch and publisher data for Henry Books. Each branch has a number that uniquely identifies the branch. In addition, Ray tracks the branch's name, location, and number of employees. Each publisher has a code that uniquely identifies the publisher. In addition, Ray tracks the publisher's name and city.

Branch

BranchNum	BranchName	BranchLocation	NumEmployees
1	Henry Downtown	16 Riverview	10
2	Henry On The Hill	1289 Bedford	6
3	Henry Brentwood	Brentwood Mall	15
4	Henry Eastshore	Eastshore Mall	9

Publisher

PublisherCode	PublisherName	City
AH	Arkham House	Sauk City WI
AP	Arcade Publishing	New York
BA	Basic Books	Boulder CO
BP	Berkley Publishing	Boston
BY	Back Bay Books	New York
CT	Course Technology	Boston
FA	Fawcett Books	New York
FS	Farrar Straus & Giroux	New York
HC	HarperCollins Publishers	New York
JP	Jove Publications	New York
JT	Jeremy P. Tarcher	Los Angeles
LB	Lb Books	New York
MP	McPherson and Co.	Kingston
PE	Penguin USA	New York
PL	Plume	New York
PU	Putnam Publishing Group	New York
RH	Random House	New York
SB	Schoken Books	New York
SC	Scribner	New York
SS	Simon & Schuster	New York
ST	Scholastic Trade	New York
TA	Taunton Press	Newtown CT
TB	Tor Books	New York
TH	Thames and Hudson	New York
TO	Touchstone Books	Westport CT
VB	Vintage Books	New York
WN	W.W. Norton	New York
WP	Westview Press	Boulder CO

FIGURE 1-15 Sample branch and publisher data for Henry Books

Figure 1-16 shows sample author data for Henry Books. Each author has a number that uniquely identifies the author. In addition, Ray records each author's last and first names.

Author

AuthorNum	AuthorLast	AuthorFirst
1	Morrison	Toni
2	Solotaroff	Paul
3	Vintage	Vernor
4	Francis	Dick
5	Straub	Peter
6	King	Stephen
7	Pratt	Philip
8	Chase	Truddi
9	Collins	Bradley
10	Heller	Joseph
11	Wills	Gary
12	Hofstadter	Douglas R.
13	Lee	Harper
14	Ambrose	Stephen E.
15	Rowling	J.K.
16	Salinger	J.D.
17	Heaney	Seamus
18	Camus	Albert
19	Collins, Jr.	Bradley
20	Steinbeck	John
21	Castelman	Riva
22	Owen	Barbara
23	O'Rourke	Randy
24	Kidder	Tracy
25	Schleining	Lon

FIGURE 1-16 Sample author data for Henry Books

Figure 1-17 shows sample book data for Henry Books. Each book has a code that uniquely identifies the book. For each book, Ray also tracks the title, publisher, type of book, and price, and whether the book is a paperback.

Book

BookCode	Title	PublisherCode	Type	Price	Paperback
0180	A Deepness in the Sky	TB	SFI	7.19	Yes
0189	Magic Terror	FA	HOR	7.99	Yes
0200	The Stranger	VB	FIC	8.00	Yes
0378	Venice	SS	ART	24.50	No
079X	Second Wind	PU	MYS	24.95	No
0808	The Edge	JP	MYS	6.99	Yes
1351	Dreamcatcher: A Novel	SC	HOR	19.60	No
1382	Treasure Chests	TA	ART	24.46	No
138X	Beloved	PL	FIC	12.95	Yes
2226	Harry Potter and the Prisoner of Azkaban	ST	SFI	13.96	No
2281	Van Gogh and Gauguin	WP	ART	21.00	No
2766	Of Mice and Men	PE	FIC	6.95	Yes
2908	Electric Light	FS	POE	14.00	No
3350	Group: Six People in Search of a Life	BP	PSY	10.40	Yes
3743	Nine Stories	LB	FIC	5.99	Yes
3906	The Soul of a New Machine	BY	SCI	11.16	Yes
5163	Travels with Charley	PE	TRA	7.95	Yes
5790	Catch-22	SC	FIC	12.00	Yes
6128	Jazz	PL	FIC	12.95	Yes
6328	Band of Brothers	TO	HIS	9.60	Yes
669X	A Guide to SQL	CT	CMP	37.95	Yes
6908	Franny and Zooey	LB	FIC	5.99	Yes
7405	East of Eden	PE	FIC	12.95	Yes
7443	Harry Potter and the Goblet of Fire	ST	SFI	18.16	No
7559	The Fall	VB	FIC	8.00	Yes
8092	Godel, Escher, Bach	BA	PHI	14.00	Yes
8720	When Rabbit Howls	JP	PSY	6.29	Yes
9611	Black House	RH	HOR	18.81	No
9627	Song of Solomon	PL	FIC	14.00	Yes
9701	The Grapes of Wrath	PE	FIC	13.00	Yes
9882	Slay Ride	JP	MYS	6.99	Yes
9883	The Catcher in the Rye	LB	FIC	5.99	Yes
9931	To Kill a Mockingbird	HC	FIC	18.00	No

FIGURE 1-17   Sample book data for Henry Books

Q & A

Question: To check your understanding of the relationship between publishers and books, answer the following questions: Who published *Jazz*? Which books did Jove Publications publish?
Answer: Plume published *Jazz*. In the row in the Book table for *Jazz* (see Figure 1-17), find the publisher code PL. Examining the Publisher table (see Figure 1-15), you see that PL is the code assigned to Plume.

Jove Publications published *The Edge*, *When Rabbit Howls*, and *Slay Ride*. To find the books published by Jove Publications, find its code (JP) in the Publisher table. Next, find all rows in the Book table for which the publisher code is JP.

The table named Wrote, shown in Figure 1-18, is used to relate books and authors. The Sequence field indicates the order in which the authors of a particular book are listed on the cover. The table named Inventory in the same figure is used to indicate the number of copies of a particular book that are currently on hand at a particular branch of Henry Books. The first row, for example, indicates that there are two copies of the book with the number 0180; they are on hand at branch 1.

Wrote

BookCode	AuthorNum	Sequence
0180	3	1
0189	5	1
0200	18	1
0378	11	1
079X	4	1
0808	4	1
1351	6	1
1382	23	2
1382	25	1
138X	1	1
2226	15	1
2281	9	2
2281	19	1
2766	20	1
2908	17	1
3350	2	1
3743	16	1
3906	24	1
5163	20	1
5790	10	1
6128	1	1
6328	14	1
669x	7	1
6908	16	1

Inventory

BookCode	BranchNum	OnHand
0180	1	2
0189	2	2
0200	1	1
0200	2	3
0378	3	2
079X	2	1
079X	3	2
079X	4	3
0808	2	1
1351	2	4
1351	3	2
1382	2	1
138X	2	3
2226	1	3
2226	3	2
2226	4	1
2281	4	3
2766	3	2
2908	1	3
2908	4	1
3350	1	2
3743	2	1
3906	2	1
3906	3	2

(continued)

FIGURE 1-18 Sample data that relates books to authors and books to branches for Henry Books

Wrote

BookCode	AuthorNum	Sequence
7405	20	1
7443	15	1
7559	18	1
8092	12	1
8720	8	1
9611	5	2
9611	6	1
9627	1	1
9701	20	1
9882	4	1
9883	16	1
9931	13	1

Inventory

BookCode	BranchNum	OnHand
5163	1	1
5790	4	2
6128	2	4
6128	3	3
6328	2	2
669X	1	1
6908	2	2
7405	3	2
7443	4	1
7559	2	2
8092	3	1
8720	1	3
9611	1	2
9627	3	5
9627	4	2
9701	1	2
9701	2	1
9701	3	3
9701	4	2
9882	3	3
9883	2	3
9883	4	2
9931	1	2

FIGURE 1-18   Sample data that relates books to authors and books to branches for Henry Books (continued)

Q & A

Question: To check your understanding of the relationship between authors and books, answer the following questions: Who wrote *Black House*? (Make sure you list the authors in the correct order.) Which books did Toni Morrison write?

Answer: Stephen King and Peter Straub wrote *Black House*. To determine who wrote *Black House*, first examine the Book table (see Figure 1-17) to find its book code (9611). Next, look for all rows in the Wrote table in which the book code is 9611. There are two such rows. In one row, the author number is 5; and in the other row, the author number is 6. Then look in the Author table to find the authors who have been assigned the numbers 5 and 6. The answer is Peter Straub (5) and Stephen King (6). The sequence number for author number 5 is 2, and the sequence number for author number 6 is 1. Thus, listing the authors in the proper order results in Stephen King and Peter Straub.

Toni Morrison wrote *Beloved*, *Jazz*, and *Song of Solomon*. To find the books written by Toni Morrison, look up her author number in the Author table and find that it is 1. Then look for all rows in the Wrote table for which the author number is 1. There are three such rows. The corresponding book codes are 138X, 6128, and 9627. Looking up these codes in the Book table, you find that Toni Morrison wrote *Beloved*, *Jazz*, and *Song of Solomon*.

Q & A

Question: A customer in branch 1 wants to purchase *The Soul of a New Machine*. Is it currently in stock at branch 1?

Answer: No. Looking up the code for *The Soul of a New Machine* in the Book table, you find it is 3906. To find out how many copies are in stock at branch 1, look for a row in the Inventory table with 3906 in the Book-Code column and 1 in the BranchNum column. Because there is no such row, branch 1 doesn't have any copies of *The Soul of a New Machine*.

Q & A

Question: You would like to obtain a copy of *The Soul of a New Machine* for this customer. Which other branches currently have it in stock, and how many copies does each branch have?

Answer: Branch 2 has one copy, and branch 3 has two copies. You already know that the code for *The Soul of a New Machine* is 3906. (If you didn't know the book code, you would look it up in the Book table.) To find out which branches currently have copies, look for rows in the Inventory table with 3906 in the BookCode column. There are two such rows. The first row indicates that branch number 2 currently has one copy. The second row indicates that branch number 3 currently has two copies.

The E-R diagram for the Henry Books database appears in Figure 1-19.

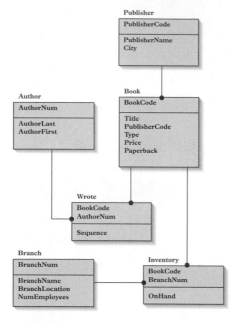

FIGURE 1-19   E-R diagram for the Henry Books database

INTRODUCTION TO THE ALEXAMARA MARINA GROUP DATABASE CASE

Alexamara Marina Group offers in-water boat storage to owners by providing boat slips that boat owners can rent on an annual basis. Alexamara owns two marinas: Alexamara East and Alexamara Central. In addition to boat slips, Alexamara also provides a variety of boat repair and maintenance services to the boat owners who rent the slips. Alexamara stores in a database the data it needs to manage its operations.

In the Marina table shown in Figure 1-20, Alexamara stores information about its two marinas. A marina number uniquely identifies each marina. The table also includes the marina name, address, city, state, and zip code.

Marina

MarinaNum	Name	Address	City	State	Zip
1	Alexamara East	108 2nd Ave	Brinman	FL	32273
2	Alexamara Central	283 Branston	W Brinman	FL	32274

FIGURE 1-20   Sample marina data for Alexamara Marina Group

Alexamara stores information about the boat owners to whom it rents slips in the Owner table shown in Figure 1-21. An owner number that consists of two uppercase letters followed by a two-digit number uniquely identifies each owner. For each owner, the table also includes the last name, first name, address, city, state, and zip code.

Owner

OwnerNum	LastName	FirstName	Address	City	State	Zip
AD57	Adney	Bruce and Jean	208 Citrus	Bowton	FL	31313
AN75	Anderson	Bill	18 Wilcox	Glander Bay	FL	31044
BL72	Blake	Mary	2672 Commodore	Bowton	FL	31313
EL25	Elend	Sandy and Bill	462 Riverside	Rivard	FL	31062
FE82	Feenstra	Daniel	7822 Coventry	Kaleva	FL	32521
JU92	Juarez	Maria	8922 Oak	Rivard	FL	31062
KE22	Kelly	Alyssa	5271 Waters	Bowton	FL	31313
NO27	Norton	Peter	2811 Lakewood	Lewiston	FL	32765
SM72	Smeltz	Becky and Dave	922 Garland	Glander Bay	FL	31044
TR72	Trent	Ashton	922 Crest	Bay Shores	FL	30992

FIGURE 1-21   Sample owner data for Alexamara Marina Group

Each marina contains slips that are identified by slip numbers. Marina 1 (Alexamara East) has two sections named A and B. Slips are numbered within each section. Thus, slip numbers at marina 1 consist of the letter A or B followed by a number (for example, A3 or B2). At marina 2 (Alexamara Central), a number (1, 2, 3) identifies each slip.

Information about the slips in the marinas is contained in the MarinaSlip table shown in Figure 1-22. The table contains the slip ID, the marina number and slip number, the length of the slip (in feet), the annual rental fee, the name of the boat currently occupying the slip, the type of boat, and the boat owner's number.

MarinaSlip

SlipID	MarinaNum	SlipNum	Length	RentalFee	BoatName	BoatType	OwnerNum
1	1	A1	40	$3,800.00	Anderson II	Sprite 4000	AN75
2	1	A2	40	$3,800.00	Our Toy	Ray 4025	EL25
3	1	A3	40	$3,600.00	Escape	Sprite 4000	KE22
4	1	B1	30	$2,400.00	Gypsy	Dolphin 28	JU92
5	1	B2	30	$2,600.00	Anderson III	Sprite 3000	AN75
6	2	1	25	$1,800.00	Bravo	Dolphin 25	AD57
7	2	2	25	$1,800.00	Chinook	Dolphin 22	FE82
8	2	3	25	$2,000.00	Listy	Dolphin 25	SM72
9	2	4	30	$2,500.00	Mermaid	Dolphin 28	BL72
10	2	5	40	$4,200.00	Axxon II	Dolphin 40	NO27
11	2	6	40	$4,200.00	Karvel	Ray 4025	TR72

FIGURE 1-22   Sample data about marina slips for Alexamara Marina Group

Q & A

Question: To check your understanding of the relationship between owners and marina slips, answer the following questions: Who owns Axxon II? What are the names of the boats owned by Bill Anderson?
Answer: Peter Norton owns Axxon II. In the row in the MarinaSlip table for Axxon II (see Figure 1-22), find owner number NO27. Examining the Owner table (see Figure 1-21), you see that NO27 is the code assigned to Peter Norton.

Bill Anderson owns Anderson II and Anderson III. To find the boats owned by Bill Anderson, find his owner code (AN75) in the Owner table. Next, find all rows in the MarinaSlip table for which the owner code is AN75.

Alexamara provides boat maintenance service at its two marinas. The types of service provided are stored in the ServiceCategory table shown in Figure 1-23. A category number uniquely identifies each service that Alexamara performs. The table also contains a description of the category.

ServiceCategory

CategoryNum	CategoryDescription
1	Routine engine maintenance
2	Engine repair
3	Air conditioning
4	Electrical systems
5	Fiberglass repair
6	Canvas installation
7	Canvas repair
8	Electronic systems (radar, GPS, autopilots, etc.)

FIGURE 1-23   Sample data about service categories for Alexamara Marina Group

Information about the services requested by owners is stored in the ServiceRequest table shown in Figure 1-24. Each row in the table contains a service ID that identifies each service request. The slip ID identifies the location (marina number and slip number) of the boat to be serviced. For example, the slip ID on the second row is 5. As indicated in the MarinaSlip table, the slip ID 5 identifies the boat in marina 1 and slip number B2. The ServiceRequest table also contains the category number of the service to be performed, a description of the specific service to be performed, and a description of the current status of the service. The table also contains the estimated number of hours required to complete the service. For completed jobs, the table contains the actual number of hours it took to complete the service. If another appointment is required to complete additional service, the appointment date appears in the NextServiceDate column.

ServiceRequest

Service ID	Slip ID	Category Num	Description	Status	Est Hours	Spent Hours	Next ServiceDate
1	1	3	Air conditioner periodically stops with code indicating low coolant level. Diagnose and repair.	Technician has verified the problem. Air conditioning specialist has been called.	4	2	7/12/2010
2	5	4	Fuse on port motor blown on two occasions. Diagnose and repair.	Open	2	0	7/12/2010
3	4	1	Oil change and general routine maintenance (check fluid levels, clean sea strainers, etc.).	Service call has been scheduled.	1	0	7/16/2010
4	1	2	Engine oil level has been dropping drastically. Diagnose and repair.	Open	2	0	7/13/2010
5	3	5	Open pockets at base of two stantions.	Technician has completed the initial filling of the open pockets. Will complete the job after the initial fill has had sufficient time to dry.	4	2	7/13/2010
6	11	4	Electric-flush system periodically stops functioning. Diagnose and repair.	Open	3	0	
7	6	2	Engine overheating. Loss of coolant. Diagnose and repair.	Open	2	0	7/13/2010
8	6	2	Heat exchanger not operating correctly	Technician has determined that the exchanger is faulty. New exchanger has been ordered.	4	1	7/17/2010
9	7	6	Canvas severely damaged in windstorm. Order and install new canvas.	Open	8	0	7/16/2010
10	2	8	Install new GPS and chart plotter.	Scheduled	7	0	7/17/2010
11	2	3	Air conditioning unit shuts down with "HHH" showing on the control panel.	Technician not able to replicate the problem. Air conditioning unit ran fine through multiple tests. Owner to notify technician if the problem recurs.	1	1	

FIGURE 1-24   Sample data about service requests for Alexamara Marina Group

ServiceRequest *(continued)*

Service ID	Slip ID	Category Num	Description	Status	Est Hours	Spent Hours	Next ServiceDate
12	4	8	Both speed and depth readings on data unit are significantly less than the owner thinks they should be.	Technician has scheduled appointment with owner to attempt to verify the problem.	2	0	7/16/2010
13	8	2	Customer describes engine as making a "clattering" sound.	Technician suspects problem with either propeller or shaft and has scheduled the boat to be pulled from the water for further investigation.	5	2	7/12/2010
14	7	5	Owner accident caused damage to forward portion of port side.	Technician has scheduled repair.	6	0	7/13/2010
15	11	7	Canvas leaks around zippers in heavy rain. Install overlap around zippers to prevent leaks.	Overlap has been created. Installation has been scheduled.	8	3	7/17/2010

FIGURE 1-24　Sample data about service requests for Alexamara Marina Group (continued)

Q & A

Question: To check your understanding of the relationship between service requests and boats, answer the following questions: What is the name of the boat that had a fuse blown on its port motor, and who owns this boat? What service has been performed on the boat named Gypsy?

Answer: Anderson III, owned by Bill Anderson, had a fuse blown on its port motor. To determine which boat had a fuse blown on its port motor, first examine the service descriptions in the ServiceRequest table (see Figure 1-24) to find the slip ID of the boat (5). Next, look for the row in the MarinaSlip table (see Figure 1-22) in which the slip ID is 5. Anderson III is the boat name in that row, and AN75 is the owner number for that boat. Finally, to identify the owner, Bill Anderson, look for the row in the Owner table (see Figure 1-21) in which the owner number is AN75.

The Gypsy had services for an oil change and routine maintenance and for an inspection of the speed and depth readings on its data unit. To find the service request for the Gypsy, look up the slip ID for the Gypsy in the MarinaSlip table and find that it is 4. Then look for all rows in the ServiceRequest table for which the slip ID is 4. There are two such rows. The corresponding service IDs are 3 for an oil change and routine maintenance and 12 for an inspection of the data unit.

The E-R diagram for the Alexamara Marina Group database appears in Figure 1-25.

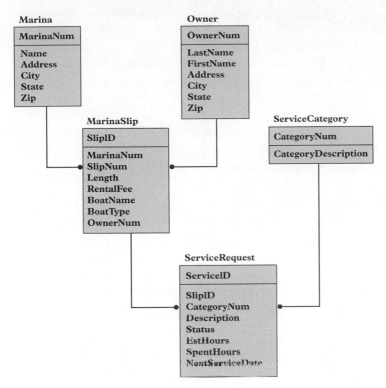

FIGURE 1-25   E-R diagram for the Alexamara Marina Group database

Summary

- Problems with nondatabase approaches to data management include redundancy, difficulties accessing related data, limited security features, limited data sharing features, and potential size limitations.

- An entity is a person, place, object, event, or idea for which you want to store and process data. An attribute, field, or column is a characteristic or property of an entity. A relationship is an association between entities.

- A one-to-many relationship between two entities exists when each occurrence of the first entity is related to many occurrences of the second entity and each occurrence of the second entity is related to only one occurrence of the first entity.

- A database is a structure that can store information about multiple types of entities, the attributes of the entities, and the relationships among the entities.

- Premiere Products is an organization whose requirements include information about the following entities: reps, customers, parts, orders, and order lines.

- An entity-relationship (E-R) diagram represents a database visually by using a rectangle for each entity that includes the entity's name above the rectangle and the entity's columns inside the rectangle, using a line to connect two entities that have a relationship, and placing a dot at the end of a line to indicate the "many" part of a one-to-many relationship.

- A database management system (DBMS) is a program, or a collection of programs, through which users interact with a database. DBMSs let you create forms and reports quickly and easily, as well as obtain answers to questions about the data stored in a database.

- Database processing offers the following advantages: getting more information from the same amount of data, sharing data, balancing conflicting requirements, controlling redundancy, facilitating consistency, improving integrity, expanding security, increasing productivity, and providing data independence. The disadvantages of database processing include the following: larger file size, increased complexity, greater impact of failure, and more difficult recovery.

- Henry Books is a company whose requirements include information about the following entities: branches, publishers, authors, books, inventory, and author sequence.

- Alexamara Marina Group is a company whose requirements include information about the following entities: marinas, owners, marina slips, service categories, and service requests.

Key Terms

attribute	entity-relationship (E-R) diagram
column	field
data file	form
data independence	integrity
database	integrity constraint
database administration (DBA)	one-to-many relationship
database administrator	redundancy
database design	relationship
database management system (DBMS)	security
entity	

Review Questions

1. What is redundancy? What problems are associated with redundancy?
2. Besides redundancy, what other problems are associated with the nondatabase approach to processing data?
3. What is an entity? An attribute?
4. What is a relationship? A one-to-many relationship?
5. What is a database?
6. How do you create a one-to-many relationship in a database system?

7. What is an E-R diagram?

8. What is a DBMS?

9. What is database design?

10. What is a form?

11. How is it possible to get more information from the same amount of data by using a database approach as opposed to a nondatabase approach?

12. What is meant by the sharing of data?

13. What is a DBA? What kinds of responsibilities does a DBA have in a database environment?

14. How does consistency result from controlling redundancy?

15. What is an integrity constraint? When does a database have integrity?

16. What is security? How does a DBMS provide security?

17. What is data independence? Why is it desirable?

18. How is file size a disadvantage in a database environment?

19. How can the complexity of a DBMS be a disadvantage?

20. Why can a failure in a database environment be more serious than an error in a nondatabase environment?

21. Why might recovery of data be more difficult in a database environment?

Premiere Products Exercises

Answer each of the following questions using the Premiere Products data shown in Figure 1-5. No computer work is required.

1. List the names of all customers that have a credit limit less than $10,000.

2. List the descriptions of all parts in item class AP and located in warehouse number 3.

3. List the order numbers for orders placed by customer number 608 on October 23, 2010.

4. List the part number, part description, and on-hand value (OnHand * Price) for each part in item class SG.

5. List the name of each customer that placed an order for two different parts in the same order.

6. List the name of each customer that has a credit limit of $5,000 and is represented by Richard Hull.

7. Find the sum of the balances for all customers represented by Juan Perez.

8. For each order, list the order number, order date, customer number, and customer name.

9. For each order placed on October 21, 2010, list the order number, order date, customer number, and customer name.

10. For each order placed on October 20, 2010, list the order number and customer name, along with the name of the rep who represents the customer.

Henry Books Case

In later chapters, you will be asked to perform many tasks to help Ray Henry manage and manipulate his database. To familiarize yourself with the database in preparation for those tasks, answer each of the following questions using the Henry Books data shown in Figures 1-15 through 1-18. No computer work is required.

1. How many total employees work at all branches of Henry Books?

2. List the name of each publisher that's not located in New York.

3. List the title of each book published by Jove Publications.

4. List the title of each book that has the type PSY.

5. List the title of each book that has the type MYS and that is in paperback.

6. List the title of each book that has the type HOR or whose publisher code is SC.

7. List the titles of paperback books that have the type FIC and a price of less than $8.

8. List the title of each book that has the type HIS or TRA.

9. List the title and publisher name for all books with a price greater than $20.

Introduction to Database Management

10. For each book with coauthors, list the title, price, and author names (in the order listed on the cover).

11. For each book with more than six total copies on hand across all branches, list the book's title and price.

12. List the title of each book that has the type FIC and that was written by John Steinbeck.

Alexamara Marina Group Case

Answer each of the following questions using the Alexamara Marina Group data shown in Figures 1-20 through 1-24. No computer work is required.

1. List the owner number, last name, and first name of every boat owner.

2. List the names of all boats that are stored in a slip with a length of less than 30 feet.

3. List the last name, first name, and street address of every owner located in Glander Bay.

4. List the last name, first name, and city of every owner who has more than one boat stored at the marina.

5. List the last name, first name, and city of every owner with a boat located in a slip whose rental fee is greater than $2,500.

6. List the boat name and boat type of all boats that are stored at the Alexamara East marina.

7. List the last name and first name of the owner and the boat name of all boats that have a completed or open service request.

8. List the boat name and boat type of all boats that have two or more completed or open service requests.

9. List the boat name and boat type of all boats that have a completed or open service request for a fiberglass or canvas-related service.

10. List the boat name and the last and first names of the owners for all boats that have engine repair service requests.

CHAPTER 2

THE RELATIONAL MODEL 1: INTRODUCTION, QBE, AND RELATIONAL ALGEBRA

LEARNING OBJECTIVES

OBJECTIVES

- Describe the relational model
- Understand Query-By-Example (QBE)
- Use criteria in QBE
- Create calculated columns in QBE
- Use functions in QBE
- Sort data in QBE
- Join tables in QBE
- Update data using QBE
- Understand relational algebra

INTRODUCTION

The database management approach implemented by most personal computer DBMSs (and by many mainframe systems as well) is the relational model. In this chapter, you will study the relational model and examine a method of retrieving data from relational databases called Query-By-Example (QBE). Finally, you will learn about relational algebra, which is one of the original ways of manipulating a relational database.

RELATIONAL DATABASES

A relational database is a collection of tables like the ones you viewed for Premiere Products in Chapter 1. These tables also appear in Figure 2-1. You might wonder why this type of database is not called a "table" database, or something similar, if a database is nothing more than a collection of tables. Formally, these tables are called relations, and this is where this type of database gets its name.

Rep

RepNum	LastName	FirstName	Street	City	State	Zip	Commission	Rate
20	Kaiser	Valerie	624 Randall	Grove	FL	33321	$20,542.50	0.05
35	Hull	Richard	532 Jackson	Sheldon	FL	33553	$39,216.00	0.07
65	Perez	Juan	1626 Taylor	Fillmore	FL	33336	$23,487.00	0.05

Customer

CustomerNum	CustomerName	Street	City	State	Zip	Balance	CreditLimit	RepNum
148	Al's Appliance and Sport	2837 Greenway	Fillmore	FL	33336	$6,550.00	$7,500.00	20
282	Brookings Direct	3827 Devon	Grove	FL	33321	$431.50	$10,000.00	35
356	Ferguson's	382 Wildwood	Northfield	FL	33146	$5,785.00	$7,500.00	65
408	The Everything Shop	1828 Raven	Crystal	FL	33503	$5,285.25	$5,000.00	35
462	Bargains Galore	3829 Central	Grove	FL	33321	$3,412.00	$10,000.00	65
524	Kline's	838 Ridgeland	Fillmore	FL	33336	$12,762.00	$15,000.00	20
608	Johnson's Department Store	372 Oxford	Sheldon	FL	33553	$2,106.00	$10,000.00	65
687	Lee's Sport and Appliance	282 Evergreen	Altonville	FL	32543	$2,851.00	$5,000.00	35
725	Deerfield's Four Seasons	282 Columbia	Sheldon	FL	33553	$248.00	$7,500.00	35
842	All Season	28 Lakeview	Grove	FL	33321	$8,221.00	$7,500.00	20

Orders

OrderNum	OrderDate	CustomerNum
21608	10/20/2010	148
21610	10/20/2010	356
21613	10/21/2010	408
21614	10/21/2010	282
21617	10/23/2010	608
21619	10/23/2010	148
21623	10/23/2010	608

OrderLine

OrderNum	PartNum	NumOrdered	QuotedPrice
21608	AT94	11	$21.95
21610	DR93	1	$495.00
21610	DW11	1	$399.99
21613	KL62	4	$329.95
21614	KT03	2	$595.00
21617	BV06	2	$794.95
21617	CD52	4	$150.00
21619	DR93	1	$495.00
21623	KV29	2	$1,290.00

Part

PartNum	Description	OnHand	Class	Warehouse	Price
AT94	Iron	50	HW	3	$24.95
BV06	Home Gym	45	SG	2	$794.95
CD52	Microwave Oven	32	AP	1	$165.00
DL71	Cordless Drill	21	HW	3	$129.95
DR93	Gas Range	8	AP	2	$495.00
DW11	Washer	12	AP	3	$399.99
FD21	Stand Mixer	22	HW	3	$159.95
KL62	Dryer	12	AP	1	$349.95
KT03	Dishwasher	8	AP	3	$595.00
KV29	Treadmill	9	SG	2	$1,390.00

FIGURE 2-1 Sample data for Premiere Products

How does a relational database handle entities, attributes of entities, and relationships between entities? Each entity is stored in its own table. For example, the Premiere Products database has a table for sales reps, a table for customers, and so on. The attributes of an entity become the fields or columns in the table. In the table for sales reps, for example, there is a column for the rep number, a column for the rep's last name, and so on.

What about relationships? At Premiere Products, there is a one-to-many relationship between sales reps and customers. (Each sales rep is related to the *many* customers he or she represents, and each customer is related to the *one* sales rep who represents it.) How is this relationship implemented in a relational database? The answer is through common columns in two or more tables. Consider Figure 2-1 again. The RepNum columns in the Rep and Customer tables implement the relationship between sales reps and customers. For any sales rep, you can use these columns to determine all the customers the sales rep represents; and for any customer, you can use these columns to find the sales rep who represents the customer. If the Customer table did not include the sales rep number, you would not be able to identify the sales rep for a given customer and the customers for a given sales rep.

More formally, a relation is essentially just a two-dimensional table. If you consider the tables shown in Figure 2-1, however, you might see certain restrictions that you can place on relations. Each column in a table should have a unique name, and all entries in each column should be consistent with this column name. For example, in the CreditLimit column, all entries should, in fact, *be* credit limits. In addition, each row should be unique. After all, when two rows in a table contain identical data, the second row doesn't provide any information that you don't already have. In addition, for maximum flexibility, the order in which columns and rows appear in a table should be immaterial. Finally, a table's design is less complex when you restrict each location in the table to a single value; that is, you should not permit multiple entries (often called **repeating groups**) in the table. These ideas lead to the following definitions:

Definition: A **relation** is a two-dimensional table in which:

1. The entries in the table are single-valued; that is, each location in the table contains a single value.
2. Each column has a distinct name (technically called the attribute name).
3. All values in a column are values of the same attribute (that is, all entries must match the column name).
4. The order of columns is immaterial.
5. Each row is distinct.
6. The order of rows is immaterial.

Definition: A **relational database** is a collection of relations.

Later in this text, you will encounter situations in which a structure satisfies all the properties of a relation *except for the first item*; that is, some of the entries contain repeating groups and thus are not single-valued. Such a structure is called an **unnormalized relation**. This jargon is a little strange in that an unnormalized relation is really not a relation at all. This term is used for such a structure, however. The table shown in Figure 2-2 is an example of an unnormalized relation.

Orders

OrderNum	OrderDate	CustomerNum	PartNum	NumOrdered	QuotedPrice
21608	10/20/2010	148	AT94	11	$21.95
21610	10/20/2010	356	DR93	1	$495.00
			DW11	1	$399.99
21613	10/21/2010	408	KL62	4	$329.95
21614	10/21/2010	282	KT03	2	$595.00
21617	10/23/2010	608	BV06	2	$794.95
			CD52	4	$150.00
21619	10/23/2010	148	DR93	1	$495.00
21623	10/23/2010	608	KV29	2	$1,290.00

FIGURE 2-2   Sample structure of an unnormalized relation

There is a commonly accepted shorthand representation that shows the structure of a relational database: You write the name of the table and then, within parentheses, list all the columns in the table. In addition, each table should appear on its own line. Using this method, you would write the Premiere Products database as follows:

```
Rep (RepNum, LastName, FirstName, Street, City, State, Zip, Commission, Rate)
Customer (CustomerNum, CustomerName, Street, City, State, Zip, Balance, CreditLimit, RepNum)
Orders (OrderNum, OrderDate, CustomerNum)
OrderLine (OrderNum, PartNum, NumOrdered, QuotedPrice)
Part (PartNum, Description, OnHand, Class, Warehouse, Price)
```

The Premiere Products database contains some duplicate column names. For example, the RepNum column appears in *both* the Rep table *and* the Customer table. Suppose a situation exists wherein the two columns might be confused. If you write RepNum, how would the computer or another programmer know which RepNum column in which table you intend to use? That could be a problem.

When duplicate column names exist in a database, you need a way to indicate the column to which you are referring. One common approach to this problem is to write both the table name and the column name, separated by a period. Thus, you would write the RepNum column in the Customer table as Customer.RepNum and the RepNum column in the Rep table as Rep.RepNum. Technically, when you combine a column name with a table name, you say that you **qualify** the column names. It is *always* acceptable to qualify column names, even when there is no possibility of confusion. If confusion may arise, however, it is *essential* to qualify column names.

The **primary key** of a table (relation) is the column or collection of columns that uniquely identifies a given row in that table. In the Rep table, the sales rep's number uniquely identifies a given row. For example, rep 35 occurs in only one row of the table. Thus, RepNum is the primary key.

The primary key provides an important way of distinguishing one row in a table from another. Primary keys are usually represented by underlining the column or collection of columns that comprises the primary key for each table in the database. Thus, the complete representation for the Premiere Products database is as follows:

```
Rep (RepNum, LastName, FirstName, Street, City, State, Zip, Commission, Rate)
Customer (CustomerNum, CustomerName, Street, City, State, Zip, Balance, CreditLimit, RepNum)
Orders (OrderNum, OrderDate, CustomerNum)
OrderLine (OrderNum, PartNum, NumOrdered, QuotedPrice)
Part (PartNum, Description, OnHand, Class, Warehouse, Price)
```

Q & A

Question: Why does the primary key of the OrderLine table consist of two columns, not just one?
Answer: No single column uniquely identifies a given row in the OrderLine table. It requires a combination of two columns: OrderNum and PartNum.

QUERY-BY-EXAMPLE (QBE)

When you ask Access or any other DBMS a question about the data in a database, the question is called a query. A **query** is simply a question represented in a way that the DBMS can recognize and process. In this section, you will investigate **Query-By-Example (QBE)**, an approach to writing queries that is very visual. With QBE, users ask their questions by entering column names and other criteria using an on-screen grid, and data appears on the screen in tabular form.

In Access, you create queries using the Query window, which has two panes. The upper portion of the window contains a field list for each table you want to query. The lower pane contains the **design grid**, the area in which you specify the format of your output, the fields to be included in the query results, a sort order for the query results, and any criteria the records you are looking for must satisfy.

The following figures and examples will show you how to retrieve data using the Access version of QBE.

SIMPLE QUERIES

To include a field in the results of a query, you place it in the design grid.

EXAMPLE 1

List the number, name, balance, and credit limit of all customers in the database.

To include a field in an Access query, double-click the field in the field list to place it in the design grid, as shown in Figure 2-3. The check marks in the Show check boxes indicate the fields that will appear in the query results. To omit a field from the query results, remove the check mark from the field's Show check box.

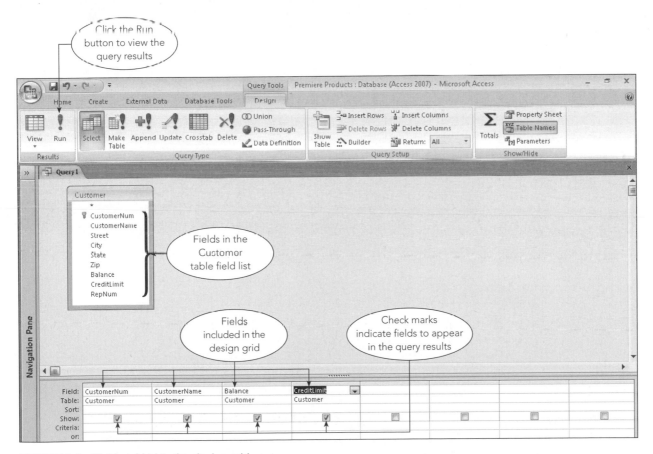

FIGURE 2-3 Fields added to the design grid

Clicking the Run button in the Results group on the Query Tools Design tab runs the query and displays the query results, as shown in Figure 2-4.

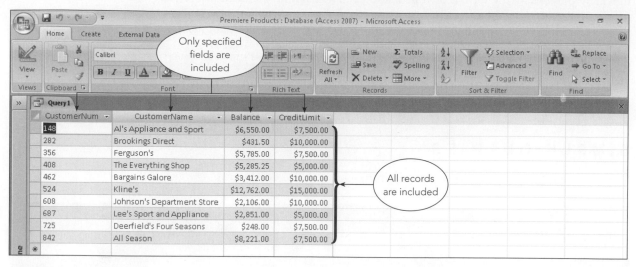

FIGURE 2-4 Query results

EXAMPLE 2

List all fields and all rows in the Orders table.

To display all fields and all rows in the Orders table, you could add each field to the design grid. There is a shortcut, however. In Access, you can add all fields from a table to the design grid by double-clicking the asterisk in the table's field list. As shown in Figure 2-5, the asterisk appears in the design grid, indicating that all fields will be included in the query results.

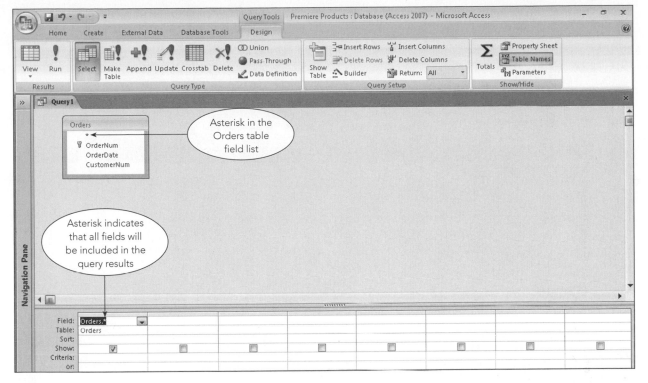

FIGURE 2-5 Query that includes all fields from the Orders table

The query results appear in Figure 2-6.

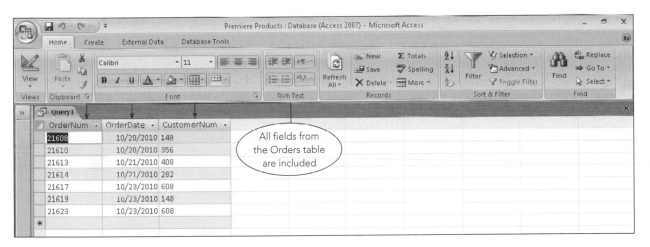

FIGURE 2-6 Query results

SIMPLE CRITERIA

When the records that you want to display in a query's results must satisfy a condition, you enter the condition in the appropriate column in the design grid. Conditions that data must satisfy are also called **criteria**. (A single condition is called a **criterion**.) The following example illustrates the use of a criterion to select data.

EXAMPLE 3

Find the name of customer 148.

To enter a criterion for a field, include the field in the design grid, and then enter the criterion in the row labeled "Criteria" for that field, as shown in Figure 2-7.

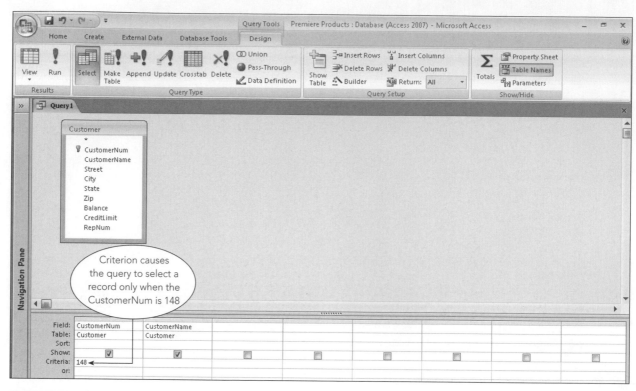

FIGURE 2-7 Query to find the name of customer 148

NOTE

When you enter a criterion for a Text field, such as CustomerNum, Access automatically adds double quotation marks around the value when you run the query or move the insertion point to another box in the design grid. Typing the quotation marks is optional. (Some DBMSs use single quotation marks to enclose Text values.)

Q & A

Question: Why is the CustomerNum field a Text field? Doesn't it contain numbers?
Answer: Fields such as the CustomerNum field that contain numbers, but are not involved in calculations, are usually assigned the Text data type.

The query results shown in Figure 2-8 show an exact match; the query selects a record only when CustomerNum equals 148.

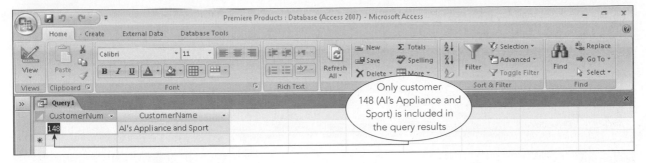

FIGURE 2-8 Query results

If you want something other than an exact match, you must enter the appropriate **comparison operator**, also called a **relational operator**, as you will see in the next example. The comparison operators are = (equal to), > (greater than), < (less than), >= (greater than or equal to), <= (less than or equal to), and NOT (not equal to).

NOTE

It is common in QBE to omit the = symbol in "equal to" comparisons, although you can use it every time.

COMPOUND CRITERIA

You can use the comparison operators by themselves to create conditions. You can also combine criteria to create **compound criteria**, or **compound conditions**. In many query languages, you create compound criteria by including the word *AND* or *OR* between the separate criteria. In an **AND criterion**, both criteria must be true for the compound criterion to be true. In an **OR criterion**, the overall criterion is true if either of the individual criteria is true.

In QBE, to create an AND criterion, place the criteria for multiple fields on the same Criteria row in the design grid; to create an OR criterion, place the criteria for multiple fields on different Criteria rows in the design grid.

EXAMPLE 4

List the description, on hand value, and warehouse number for all parts that have more than 10 units on hand *and* that are located in warehouse 3.

To indicate that two criteria must both be true to select a record, place the conditions for each field in the same Criteria row, as shown in Figure 2-9. In this case, you want the query to select those parts where the value in the OnHand field is greater than 10 (which requires the use of the > comparison operator) and the value in the Warehouse field is 3.

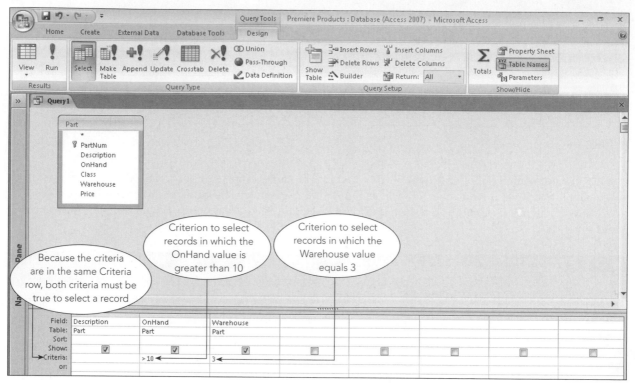

FIGURE 2-9 Query that uses an AND criterion

The query results appear in Figure 2-10.

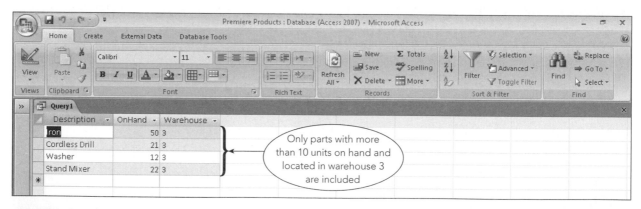

FIGURE 2-10 Query results

EXAMPLE 5

List the description, on hand value, and warehouse number for all parts that have more than 10 units on hand *or* that are located in warehouse 3.

To indicate that either of two conditions must be true to select a record, place the first criterion in the Criteria row for the first column and place the second criterion in the row labeled "or," as shown in Figure 2-11.

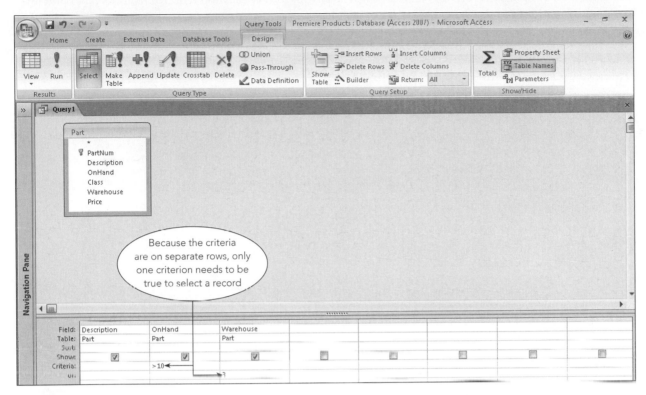

FIGURE 2-11 Query that uses an OR criterion

The query results appear in Figure 2-12.

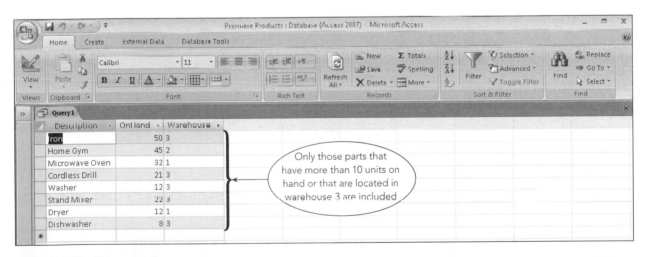

FIGURE 2-12 Query results

EXAMPLE 6

List the number, name, and balance for each customer whose balance is between $1,000 and $5,000.

This example requires you to search for a range of values to find all customers with balances between $1,000 and $5,000. When you ask this kind of question, you are looking for all balances that are greater than $1,000 *and* all balances that are less than $5,000; the answer to this question requires using a compound criterion, or two criteria in the same field.

To place two criteria in the same field, separate the criteria with the AND operator to create an AND condition. Figure 2-13 shows the AND condition to select all records with a value of more than 1000 and less than 5000 in the Balance field.

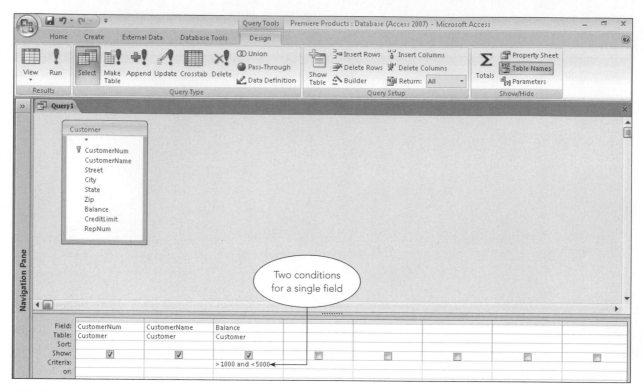

FIGURE 2-13 Query that uses an AND condition for a single field

The query results appear in Figure 2-14.

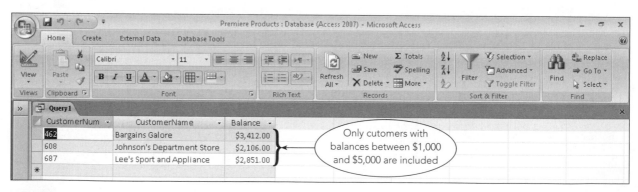

FIGURE 2-14 Query results

COMPUTED FIELDS

Sometimes you'll need to include calculated fields that are not in the database in queries. A **computed field** or **calculated field** is a field that is the result of a calculation using one or more existing fields. Example 7 illustrates the use of a calculated field.

EXAMPLE 7

List the number, name, and available credit for all customers.

Available credit is computed by subtracting the balance from the credit limit. Because there is no available credit field in the Customer table, you must calculate it from the existing Balance and CreditLimit fields. To include a computed field in a query, you enter a name for the computed field, followed by a colon, and then followed by an expression in one of the columns in the Field row.

To calculate available credit, you can enter the expression *AvailableCredit:CreditLimit-Balance* in the desired Field row in the design grid. When entering an expression in the design grid, the default column size prevents you from being able to see the complete expression. An alternative method is to right-click the column in the Field row to display the shortcut menu, and then click Zoom to open the Zoom dialog box. Then you can type the expression in the Zoom dialog box, as shown in Figure 2-15.

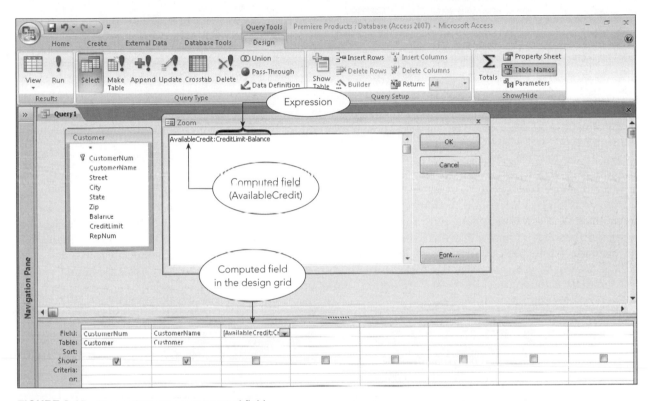

FIGURE 2-15 Query that uses a computed field

NOTE

When a field name contains spaces or SQL reserved words, you must enclose it in square brackets ([ ]). For example, if the field name was Credit Limit instead of CreditLimit, you would enter the expression as *[Credit Limit]-Balance*. You can also enclose a field name that does not contain spaces in square brackets, but you do not need to do so.

The query results appear in Figure 2-16.

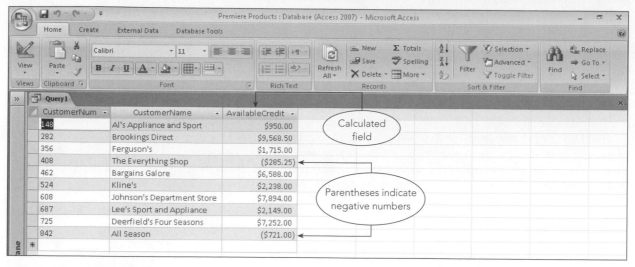

FIGURE 2-16 Query results

You are not restricted to subtraction in computations. You can also use addition (+), multiplication (*), or division (/). You can include parentheses in your expressions to indicate which computations Access should perform first.

FUNCTIONS

All products that support QBE, including Access, support the following built-in **functions** (called **aggregate functions** in Access): Count, Sum, Avg (average), Max (largest value), Min (smallest value), StDev (standard deviation), Var (variance), First, and Last. To use any of these functions in a query, you include them in the Total row for the desired column in the design grid. By default, the Total row does not appear automatically in the design grid. To include it, you must click the Totals button in the Show/Hide group on the Query Tools Design tab.

Example 8 illustrates how to use a function in a query by counting the number of customers represented by sales rep 35.

EXAMPLE 8

How many customers are represented by sales rep 35?

To count the number of rows in the Customer table that have the value 35 in the RepNum column, you select the Count function in the Total row for the CustomerNum column. In the RepNum column, you select the Where operator in the Total row to indicate that there will also be a criterion. In the Criteria row for the RepNum column, the entry 35 selects only those records for sales rep number 35, as shown in Figure 2-17.

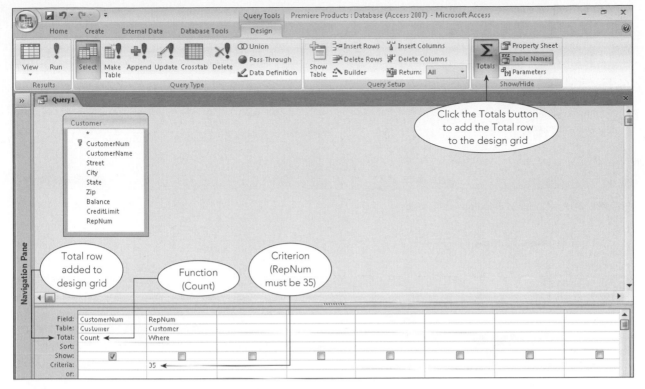

FIGURE 2-17 Query to count records

The query results appear in Figure 2-18. Notice that Access used the default name, CountOfCustomerNum, for the new column. You could create your own column name by preceding the field name with the desired column name and a colon in the query design (for example, NumberOfCustomers: CustomerNum changes the default name to "NumberOfCustomers").

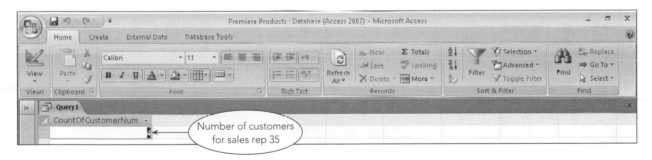

FIGURE 2-18 Query results

EXAMPLE 9

What is the average balance of all customers of sales rep 35?

To calculate the average balance, use the Avg function, as shown in Figure 2-19.

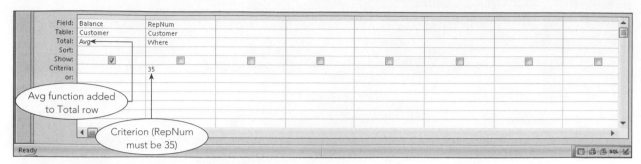

FIGURE 2-19 Query to calculate an average

The query results appear in Figure 2-20.

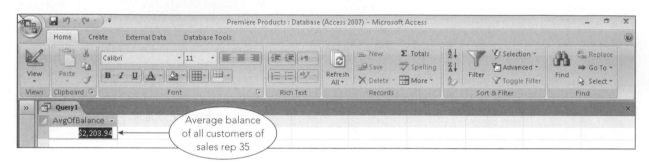

FIGURE 2-20 Query results

GROUPING

You can also use functions in combination with grouping, where calculations affect groups of records. For example, you might need to calculate the average balance for all customers of each sales rep. **Grouping** simply means creating groups of records that share some common characteristic. In grouping by RepNum, for example, the customers of sales rep 20 would form one group, the customers of sales rep 35 would form a second group, and the customers of sales rep 65 would form a third group. The calculations are then made for each group. To group records in Access, select the Group By operator in the Total row for the field on which to group.

EXAMPLE 10

What is the average balance for all customers of each sales rep?

In this example, include the RepNum and Balance fields in the design grid. To group the customer records for each sales rep, select the Group By operator in the Total row for the RepNum column. To calculate the average balance for each group of customers, select the Avg function in the Total row for the Balance column, as shown in Figure 2-21.

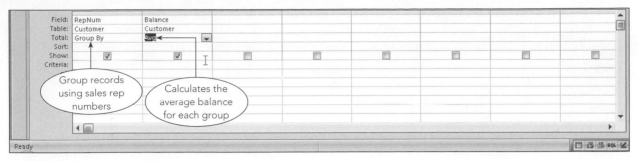

FIGURE 2-21 Query to group records

The query results appear in Figure 2-22.

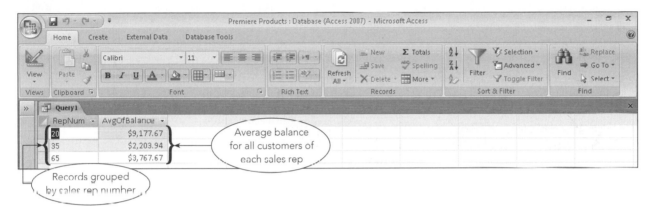

FIGURE 2-22 Query results

SORTING

In most queries, the order in which records appear doesn't matter. In other queries, however, the order in which records appear can be very important. You might want to see customers listed alphabetically by customer name or listed by rep number. Further, you might want to see customer records listed alphabetically by customer name *and* grouped by sales rep number.

To list the records in query results in a particular way, you need to **sort** the records. The field on which records are sorted is called the **sort key**; you can sort records using more than one field when necessary. When you are sorting records by more than one field (such as sorting by rep number and then by customer name), the first sort field (RepNum) is called the **major sort key** (also called the **primary sort key**) and the second sort field (CustomerName) is called the **minor sort key** (also called the **secondary sort key**).

To sort in Access, specify the sort order in the Sort row of the design grid for the sort key field.

EXAMPLE 11

List the customer number, name, balance, and rep number for each customer. Sort the output alphabetically by customer name.

To sort the records alphabetically using the CustomerName field, select the Ascending sort order in the Sort row for the CustomerName column, as shown in Figure 2-23. (To sort the records in reverse alphabetical order, select the Descending sort order.)

FIGURE 2-23 Query to sort records

The query results appear in Figure 2-24. Notice that the customer names appear in alphabetical order.

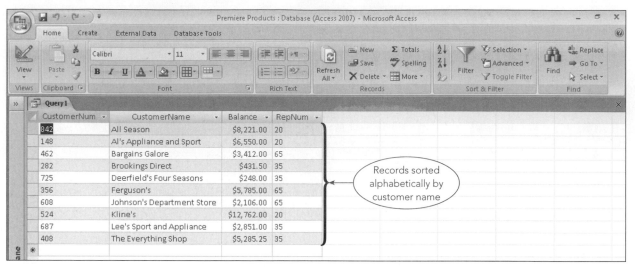

FIGURE 2-24 Query results

Sorting on Multiple Keys

You can specify more than one sort key in a query; in this case, the sort key on the left in the design grid will be the major (primary) sort key and the sort key on the right will be the minor (secondary) sort key. Example 12 illustrates the process.

EXAMPLE 12

List the customer number, name, balance, and rep number for each customer. Sort the output by sales rep number. Within the customers of each sales rep, sort the output by customer name.

To sort records by sales rep number and then by customer name, RepNum is the major sort key and CustomerName is the minor sort key. You might be tempted to select the sort orders for these fields in the design grid, but your results would not be sorted correctly. Figure 2-25 shows an incorrect query design.

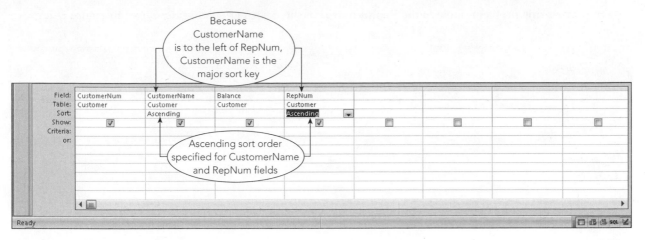

FIGURE 2-25   Incorrect query design to sort by RepNum and then by CustomerName

In Figure 2-25, the CustomerName field is to the left of the RepNum field in the design grid. With this order, CustomerName is the major sort key, so the data is sorted by customer name first, and not by sales rep number, as shown in Figure 2-26. If two customers had the same name, the data for these customers would be further sorted by sales rep number because RepNum is the minor sort key.

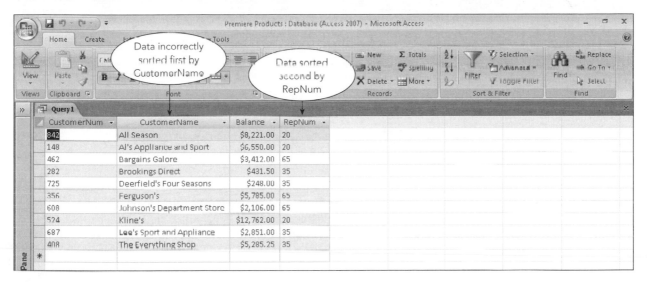

FIGURE 2-26   Query results

To correct this problem, include the RepNum field in the design grid twice, as shown in Figure 2-27.

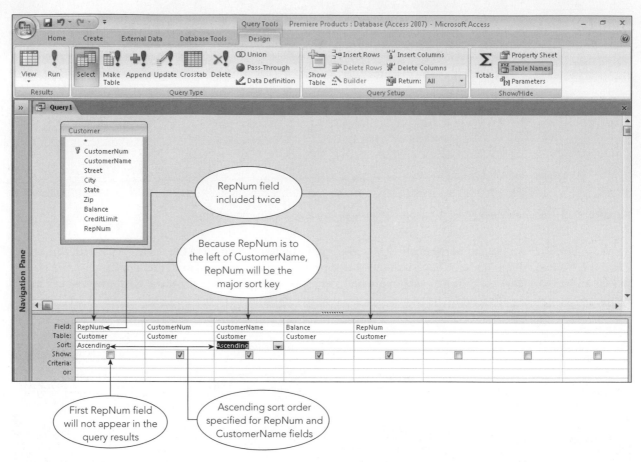

FIGURE 2-27 Correct query design to sort by RepNum and then by CustomerName

The correct query design shows the Ascending sort order selected for the first RepNum column, which will not appear in the query results because the check mark was removed from its Show check box. Because this RepNum column is to the left of the minor sort key (CustomerName), RepNum is the major sort key. The second RepNum column in the design grid will display the rep numbers in the query results in the desired position, as shown in Figure 2-28.

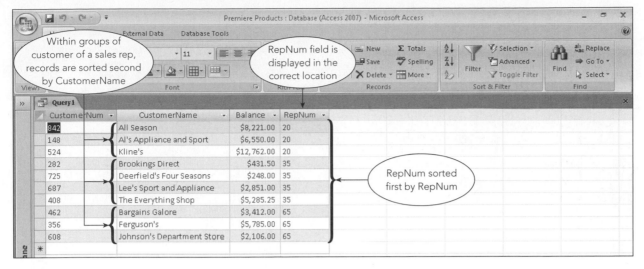

FIGURE 2-28 Query results

JOINING TABLES

So far, the queries used in the examples have displayed records from a single table. In many cases, you'll need to create queries to select data from more than one table. To do so, it is necessary to **join** the tables based on matching fields in corresponding columns. To join tables in Access, first you add the field lists for both tables to the upper pane of the Query window. Access will draw a line, called a **join line**, between matching fields in the two tables, indicating that the tables are related. (If the corresponding fields have the same field name and at least one of the fields is the primary key of the table that contains it, Access will join the tables automatically.) Then you can select fields from either or both tables, as you will see in the next example.

EXAMPLE 13

List each customer's number and name, along with the number, last name, and first name of each customer's sales rep.

You cannot create this query using a single table—the customer name is in the Customer table and the sales rep name is in the Rep table. The sales rep number can come from either table because it is the matching field. To select the correct data, you need to join the tables by adding both the Customer and Rep table field lists to the upper pane and then adding the desired fields from the field lists to the design grid, as shown in Figure 2-29.

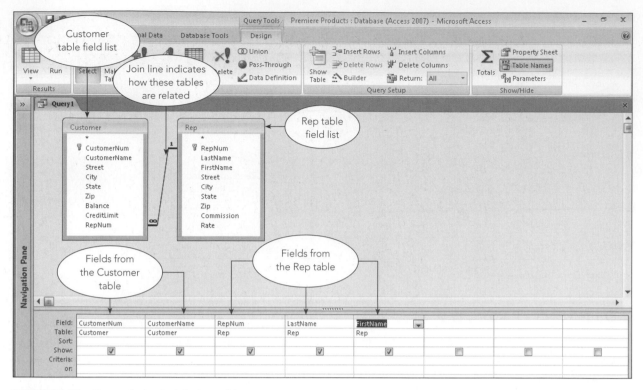

FIGURE 2-29 Query design to join two tables

Notice that the Table row in the design grid indicates the table with which each field is associated. The query results appear in Figure 2-30.

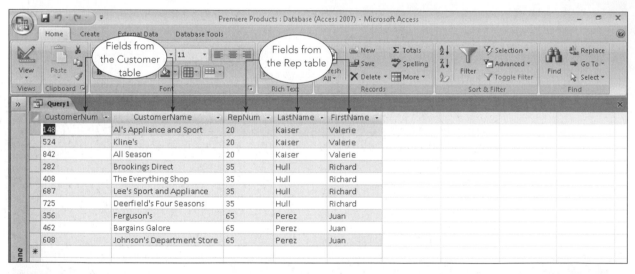

FIGURE 2-30 Query results

EXAMPLE 14

For each customer whose credit limit is $10,000, list the customer's number and name, along with the number, last name, and first name of the corresponding sales rep.

The only difference between this query and the one illustrated in Example 13 is that there is an extra restriction—the credit limit must be $10,000. To include this new condition, add the CreditLimit field to the design grid, enter 10000 as the criterion, and remove the check mark from the CreditLimit field's Show check box (because the CreditLimit column should not appear in the query results). The query design appears in Figure 2-31.

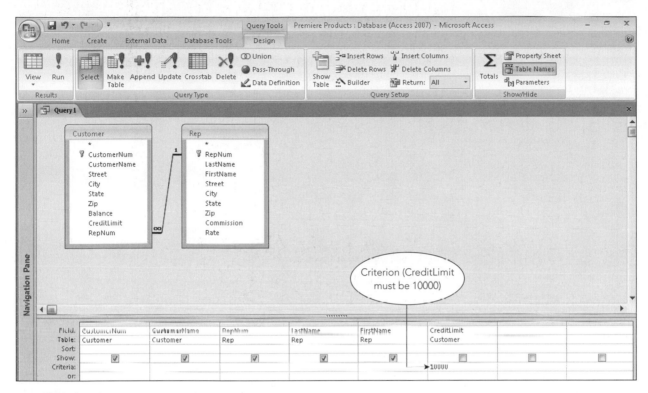

FIGURE 2-31 Query to restrict records in a join

Only customers with credit limits of $10,000 are included in the query results, as shown in Figure 2-32.

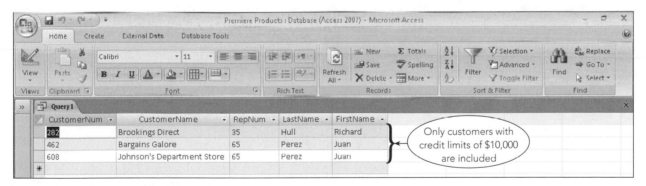

FIGURE 2-32 Query results

Joining Multiple Tables

Joining three or more tables is similar to joining two tables. First you add the field lists for all the tables involved in the join to the upper pane, and then you add the fields to appear in the query results to the design grid in the desired order.

EXAMPLE 15

For each order, list the order number, order date, customer number, and customer name. In addition, for each order line within the order, list the part number, description, number of units ordered, and quoted price.

This query requires data from four tables: Orders (for basic order data), Customer (for the customer number and name), OrderLine (for the part number, number ordered, and quoted price), and Part (for the description). Figure 2-33 shows the query design.

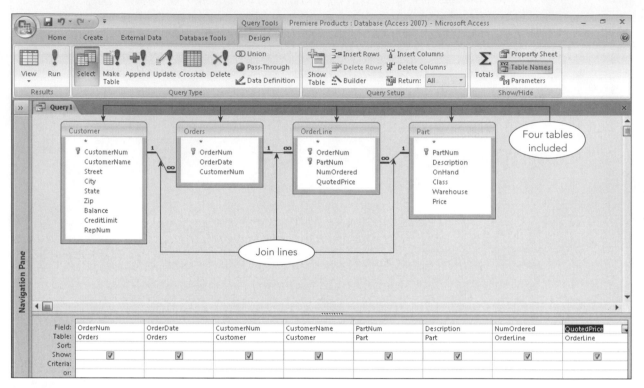

FIGURE 2-33 Query to join multiple tables

The query results appear in Figure 2-34.

OrderNum	OrderDate	CustomerNum	CustomerName	PartNum	Description	NumOrdered	QuotedPrice
21608	10/20/2010	148	Al's Appliance and Sport	AT94	Iron	11	$21.95
21610	10/20/2010	356	Ferguson's	DR93	Gas Range	1	$495.00
21610	10/20/2010	356	Ferguson's	DW11	Washer	1	$399.99
21613	10/21/2010	408	The Everything Shop	KL62	Dryer	4	$329.95
21614	10/21/2010	282	Brookings Direct	KT03	Dishwasher	2	$595.00
21617	10/23/2010	608	Johnson's Department Store	BV06	Home Gym	2	$794.95
21617	10/23/2010	608	Johnson's Department Store	CD52	Microwave Oven	4	$150.00
21619	10/23/2010	148	Al's Appliance and Sport	DR93	Gas Range	1	$495.00
21623	10/23/2010	608	Johnson's Department Store	KV29	Treadmill	2	$1,290.00

FIGURE 2-34 Query results

USING AN UPDATE QUERY

In addition to retrieving data, you can use a query to update data. A query that changes data is called an **update query**. An update query makes a specified change to all records satisfying the criteria in the query. To change a query to an update query, click the Update button in the Query Type group on the Query Tools Design tab. A new row, called the Update To row, is added to the design grid when you create an update query. You use this row to indicate how to update the data selected by the query.

EXAMPLE 16

The zip code for customers located in the city of Fillmore is incorrect; it should be 33363. Change the zip code for these customers to the correct value.

To change the zip code for only those customers located in Fillmore, include the City column in the design grid and enter a criterion of Fillmore in the Criteria row. To indicate the new value for the zip code, include the Zip column in the design grid and enter the new zip code value in the Update To row for the Zip column, as shown in Figure 2-35. When you click the Run button in the Results group on the Query Tools Design tab, Access indicates how many rows the query will change and gives you a chance to cancel the update, if necessary. When you click the Yes button, the query is executed and updates the data specified in the query design. Because the result of an update query is to change data in the records selected by the query, running the query does not produce a query datasheet.

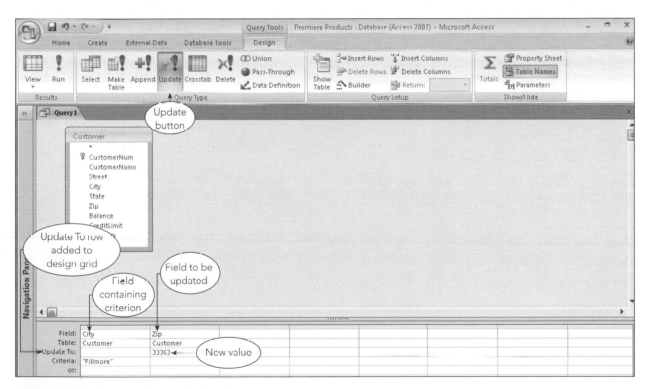

FIGURE 2-35 Query design to update data

USING A DELETE QUERY

You can also use queries to delete one or more records at a time based on criteria that you specify. A **delete query** permanently deletes all the records satisfying the criteria entered in the query. For example, you can delete all the order lines associated with a certain order in the OrderLine table by using a single delete query.

EXAMPLE 17

Delete all order lines in which the order number is 21610.

You enter the criteria that will determine the records to be deleted just as you would enter any other criteria. In this example, include the OrderNum field in the design grid and enter the order number 21610 in the Criteria row, as shown in Figure 2-36. To change the query type to a delete query, click the Delete button in the Query Type group on the Query Tools Design tab. Notice that a new row, called the Delete row, is added to the design grid, indicating that this is a delete query. When you click the Run button, Access indicates how many rows will be deleted and gives you a chance to cancel the deletions, if necessary. If you click the Yes button, the query will delete all rows in the OrderLine table in which the order number is 21610. Because the result of a delete query permanently deletes the records it selects, you should take extra care to make sure that the query design selects the correct records.

Q & A

Question: What happens if you run a delete query that does not include a criterion?
Answer: Because there is no criterion to select records, the query selects all records in the table and then deletes all of them from the table.

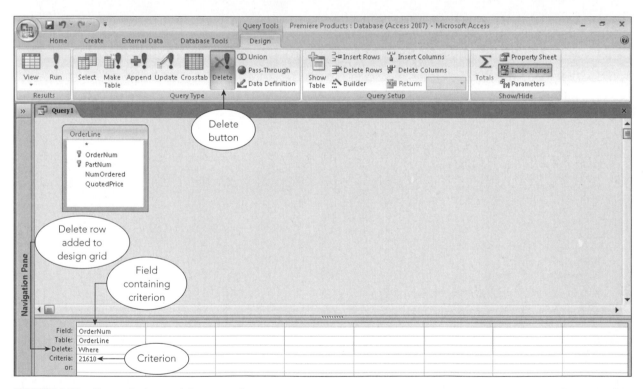

FIGURE 2-36 Query design to delete records

USING A MAKE-TABLE QUERY

You can use a query to create a new table in either the current database or another database. A **make-table query** creates a new table using the results of a query. The records added to the new table are separate from the original table in which they appear; in other words, you don't move the records to a new table; you create a new table using the records selected by the query.

EXAMPLE 18

Create a new table containing the customer number and customer name, and the number, first name, and last name of the customer's sales rep. Name the new table CustomerRep.

Figure 2-37 shows the query design to select records from the Customer and Rep tables.

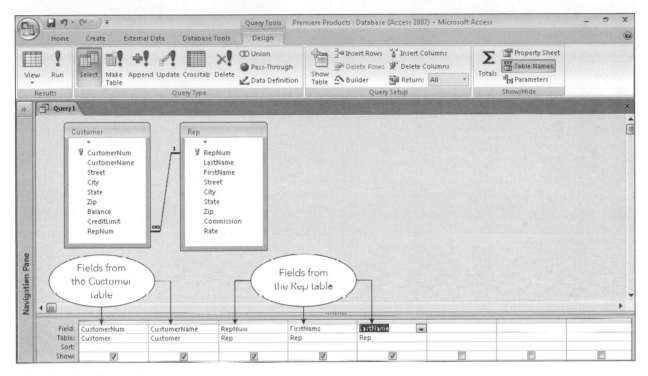

FIGURE 2-37   Make-table query design

After you create and test the query to make sure it selects the correct records, change the query type to a make-table query by clicking the Make Table button in the Query Type group on the Query Tools Design tab. In the Make Table dialog box that opens, enter the new table's name and choose where to create it, as shown in Figure 2-38.

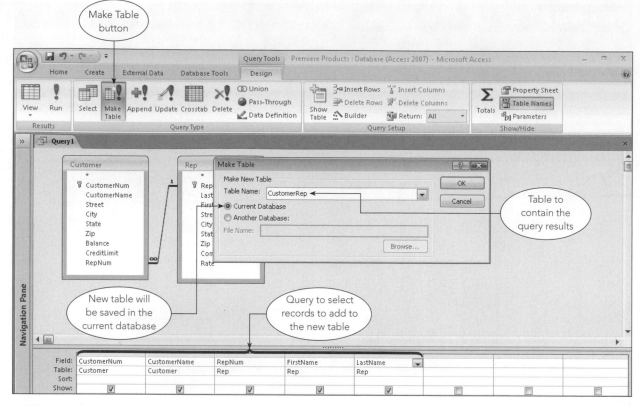

FIGURE 2-38 Make Table dialog box

After closing the Make Table dialog box and running the make-table query, the records it selects are added to a new table named CustomerRep in the current database. Figure 2-39 shows the new CustomerRep table created by the make-table query.

FIGURE 2-39 Table created by the make-table query

RELATIONAL ALGEBRA

Relational algebra is a theoretical way of manipulating a relational database. Relational algebra includes operations that act on existing tables to produce new tables, similar to the way the operations of addition and subtraction act on numbers to produce new numbers in the mathematical algebra with which you are familiar.

Retrieving data from a relational database through the use of relational algebra involves issuing relational algebra commands to operate on existing tables to form a new table containing the desired information. Sometimes you might need to execute a series of commands to obtain the desired result.

> **NOTE**
>
> Unlike QBE, relational algebra is not used in current DBMS systems. Its importance is the theoretical base it furnishes to the relational model and the benchmark it provides. Other approaches to querying relational databases are judged by this benchmark.

> **NOTE**
>
> There is no "standard" method for representing relational algebra commands; this section illustrates one possible approach. What is important is not the particular way the commands are represented, but the results they provide.

As you will notice in the following examples, each command ends with a GIVING clause, followed by a table name. This clause requests the result of the command to be placed in a temporary table with the specified name.

SELECT

In relational algebra, the **SELECT** command takes a horizontal subset of a table; that is, it retrieves certain rows from an existing table (based on some user-specified criteria) and saves them as a new table. The SELECT command includes the word *WHERE* followed by a condition. The rows retrieved are the rows in which the condition is satisfied.

EXAMPLE 19

List all information from the Customer table about customer 282.

```
SELECT Customer WHERE CustomerNum=282
    GIVING Answer
```

This command creates a new table named Answer that contains only the single row in which the customer number is 282, because that is the only row in which the condition is true. All the columns from the Customer table are included in the new Answer table.

EXAMPLE 20

List all information from the Customer table about those customers with credit limits of $7,500.

```
SELECT Customer WHERE CreditLimit=7500
    GIVING Answer
```

This command creates a new table named Answer that contains all the columns from the Customer table, but only those rows in which the credit limit is $7,500.

PROJECT

In relational algebra, the **PROJECT** command takes a vertical subset of a table; that is, it causes only certain columns to be included in the new table. The PROJECT command includes the word *OVER* followed by a list of the columns to be included.

EXAMPLE 21

List the number and name of all customers.

```
PROJECT Customer OVER (CustomerNum, CustomerName)
    GIVING Answer
```

This command creates a new table named Answer that contains the CustomerNum and CustomerName columns for all the rows in the Customer table.

EXAMPLE 22

List the number and name of all customers with credit limits of $7,500.

This example requires a two-step process. You first use a SELECT command to create a new table that contains only those customers with credit limits of $7,500. Then you project the new table to restrict the result to only the indicated columns.

```
SELECT Customer WHERE CreditLimit=7500
    GIVING Temp
PROJECT Temp OVER (CustomerNum, CustomerName)
    GIVING Answer
```

The first command creates a new table named Temp that contains all the columns from the Customer table, but only those rows in which the credit limit is $7,500. The second command creates a new table named Answer that contains all the rows from the Temp table (that is, only customers whose credit limit is $7,500), but only the CustomerNum and CustomerName columns.

JOIN

The join operation is the core operation of relational algebra because it is the command that allows you to extract data from more than one table. In the most common form of the join, two tables are combined based on the values in matching columns, creating a new table containing the columns in both tables. Rows in this new table are the **concatenation** (combination) of a row from the first table and a row from the second table that match on the common column (often called the **join column**). In other words, two tables are joined *on* the join column.

For example, suppose you want to join the two tables shown in Figure 2-40 on RepNum (the join column), creating a new table named Temp.

Customer

CustomerNum	CustomerName	RepNum
148	Al's Appliance and Sport	20
282	Brookings Direct	35
356	Ferguson's	65
408	The Everything Shop	35
462	Bargains Galore	65
524	Kline's	20
608	Johnson's Department Store	65
687	Lee's Sport and Appliance	35
725	Deerfield's Four Seasons	35
842	All Season	20
701	Peters	05

Rep

RepNum	LastName	FirstName
20	Kaiser	Valerie
35	Hull	Richard
65	Perez	Juan
75	Lewis	Joan

FIGURE 2-40 Customer and Rep tables

The result of joining the Customer and Rep tables creates the table shown in Figure 2-41. The column that joins the tables (RepNum) appears only once. Other than that, all columns from both tables appear in the result.

Temp

CustomerNum	CustomerName	RepNum	LastName	FirstName
148	Al's Appliance and Sport	20	Kaiser	Valerie
282	Brookings Direct	35	Hull	Richard
356	Ferguson's	65	Perez	Juan
408	The Everything Shop	35	Hull	Richard
462	Bargains Galore	65	Perez	Juan
524	Kline's	20	Kaiser	Valerie
608	Johnson's Department Store	65	Perez	Juan
687	Lee's Sport and Appliance	35	Hull	Richard
725	Deerfield's Four Seasons	35	Hull	Richard
842	All Season	20	Kaiser	Valerie

FIGURE 2-41 Table produced by joining the Customer and Rep tables

If a row in one table does not match any row in the other table, that row will not appear in the result of the join. Thus, the row for sales rep 75 (Joan Lewis) from the Rep table and the row for customer 701 (Peters) from the Customer table do not appear in the join table because their rows are not common to both tables.

You can restrict the output from the join to include only certain columns by using the PROJECT command, as shown in the following example.

EXAMPLE 23

For each customer, list the customer number, customer name, sales rep number, and sales rep's last name.

```
JOIN Customer Rep
     WHERE Customer.RepNum=Rep.RepNum
     GIVING Temp
PROJECT Temp OVER (CustomerNum, CustomerName, RepNum, LastName)
     GIVING Answer
```

In the WHERE clause of the JOIN command, the matching fields are both named RepNum—the field in the Rep table named RepNum is supposed to match the field in the Customer table named RepNum. Because two fields are named RepNum, you must qualify the field names. Just like in QBE, the RepNum field in the Rep table is written as Rep.RepNum and the RepNum field in the Customer table is written as Customer.RepNum.

In this example, the JOIN command joins the Rep and Customer tables to create a new table named Temp. The PROJECT command creates a new table named Answer that contains all the rows from the Temp table, but only the CustomerNum, CustomerName, RepNum, and LastName columns.

The type of join used in Example 23 is called a **natural join**. Although this type of join is most common, there is another possibility. The other type of join, the **outer join**, is similar to the natural join except that it also includes records from each original table that are not common in both tables. In a natural join, these unmatched records do not appear in the new table. In the outer join, unmatched records are included and the values of the fields are vacant, or **null**, for the records that do not have data common in both tables. Performing an outer join for Example 23 produces the table shown in Figure 2-42.

Temp

CustomerNum	CustomerName	RepNum	LastName	FirstName
148	Al's Appliance and Sport	20	Kaiser	Valerie
282	Brookings Direct	35	Hull	Richard
356	Ferguson's	65	Perez	Juan
408	The Everything Shop	35	Hull	Richard
462	Bargains Galore	65	Perez	Juan
524	Kline's	20	Kaiser	Valerie
608	Johnson's Department Store	65	Perez	Juan
687	Lee's Sport and Appliance	35	Hull	Richard
725	Deerfield's Four Seasons	35	Hull	Richard
842	All Season	20	Kaiser	Valerie
701	Peters	05	-	-
-	-	75	Lewis	Joan

FIGURE 2-42 Table produced by an outer join of the Customer and Rep tables

NORMAL SET OPERATIONS

Relational algebra includes set operations for union, intersection, and difference. The **union** of tables A and B is a table containing all rows that are in either table A or table B or in both table A and table B. The **intersection** of two tables is a table containing all rows that are common in both table A and table B. The **difference** of two tables A and B (referred to as A minus B) is the set of all rows that are in table A but that are not in table B.

Union

There is a restriction on set operations. It does not make sense, for example, to talk about the union of the Rep table and the Customer table because the tables do not contain the same columns. The two tables *must* have the same structure for a union to be appropriate; the formal term is *union compatible*. Two tables are **union compatible** when they have the same number of columns and when their corresponding columns represent the same type of data. If, for example, the first column in table A contains customer numbers, the first column in table B must also contain customer numbers.

EXAMPLE 24

List the number and name of those customers that have orders or are represented by sales rep 65, or both.

You can create a table containing the number and name of all customers that have orders by joining the Orders table and the Customer table (Temp1 in the following example) and then projecting the result over CustomerNum and CustomerName (Temp2). You can also create a table containing the number and name of all customers represented by sales rep 65 by selecting from the Customer table (Temp3) and then projecting the result (Temp4). The two tables ultimately created by this process (Temp2 and Temp4) have the same structure. They each have two fields: CustomerNum and CustomerName. Because these two tables are union compatible, it is appropriate to take the union of these two tables. This process is accomplished in relational algebra using the following code:

```
JOIN Orders, Customer
     WHERE Orders.CustomerNum=Customer.CustomerNum
     GIVING Temp1
PROJECT Temp1 OVER CustomerNum, CustomerName
     GIVING Temp2
SELECT Customer WHERE RepNum='65'
     GIVING Temp3
PROJECT Temp3 OVER CustomerNum, CustomerName
     GIVING Temp4
UNION Temp2 WITH Temp4 GIVING Answer
```

Intersection

As you would expect, using the intersection operation is very similar to using the union operation. The only difference is that you replace the UNION command with the **INTERSECT** command, as illustrated in the following example.

EXAMPLE 25

List the number and name of customers that have orders *and* that are represented by sales rep 65.

In this example, you need to intersect the two tables instead of taking their union. The code to accomplish this is as follows:

```
JOIN Orders, Customer
     WHERE Orders.CustomerNum=Customer.CustomerNum
     GIVING Temp1
PROJECT Temp1 OVER CustomerNum, CustomerName
     GIVING Temp2
SELECT Customer WHERE RepNum='65'
     GIVING Temp3
PROJECT Temp3 OVER CustomerNum, CustomerName
     GIVING Temp4
INTERSECT Temp2 WITH Temp4 GIVING Answer
```

Difference

The difference operation is performed by the **SUBTRACT** command in relational algebra.

EXAMPLE 26

List the number and name of those customers that have orders but that are *not* represented by sales rep 65.

This process is virtually identical to the one you encountered in the union and intersection examples; but in this case, you subtract one of the tables from the other instead of taking their union or intersection. This process is accomplished in relational algebra using the following command:

```
JOIN Orders, Customer
     WHERE Orders.CustomerNum=Customer.CustomerNum
     GIVING Temp1
PROJECT Temp1 OVER CustomerNum, CustomerName
     GIVING Temp2
SELECT Customer WHERE RepNum='65'
     GIVING Temp3
PROJECT Temp3 OVER CustomerNum, CustomerName
     GIVING Temp4
SUBTRACT Temp4 FROM Temp2 GIVING Answer
```

The next two sections present the final two important but infrequently used commands in relational algebra: product and division.

PRODUCT

The **product** of two tables (mathematically called the **Cartesian product**) is the table obtained by concatenating every row in the first table with every row in the second table. Thus, the product of the Orders table and the Part table, which are both shown in Figure 2-43, appears in the figure as the table labeled "Product of Orders and Part."

Orders

OrderNum	OrderDate
21608	10/20/2010
21610	10/20/2010
21613	10/21/2010

Part

PartNum	Description
DR93	Gas Range
DW11	Washer

Product of Orders and Part

OrderNum	OrderDate	PartNum	Description
21608	10/20/2010	DR93	Gas Range
21610	10/20/2010	DR93	Gas Range
21613	10/21/2010	DR93	Gas Range
21608	10/20/2010	DW11	Washer
21610	10/20/2010	DW11	Washer
21613	10/21/2010	DW11	Washer

FIGURE 2-43   Product of two tables

Every row in the Orders table is matched with every row in the Part table. If the Orders table has m rows and the Part table has n rows, there would be m *times* n rows in the product. If, as is typically the case, the tables have many rows, the number of rows in the product can be so great that it is not practical to form the product. Usually you want only those combinations that satisfy certain restrictions; thus, you would almost always use the join operation instead of the product operation.

DIVISION

The **division** process is best illustrated by considering the division of a table with two columns by a table with a single column, which is the most common situation in which to use this operation. Consider the first two tables shown in Figure 2-44. The first table contains two columns: OrderNum and PartNum. The second table contains only a single column, PartNum.

OrderLine

OrderNum	PartNum
21608	AT94
21610	DR93
21610	DW11
21613	KL62
21614	KT03
21617	BV06
21617	CD52
21619	DR93
21623	KV29

Part

PartNum
DR93
DW11

Result of dividing
OrderLine by Part

OrderNum
21610

FIGURE 2-44 Dividing one table by another

The quotient (the result of the division) is a new table with a single column named OrderNum (the column from table A that is *not* in table B). The rows in this new table contain those order numbers from the OrderLine table that "match" *all* the parts appearing in the Part table. For an order number to appear in the quotient, a row in the OrderLine table must have that order number in the OrderNum column and DR93 in the PartNum column. Also, the OrderLine table must have a row with this same order number in the OrderNum column and DW11 in the PartNum column. It doesn't matter if other rows in the OrderLine table contain the same order number as long as the rows with DR93 and DW11 are present. With the sample data, only order number 21610 qualifies. Thus, the result is the final table shown in the figure.

Summary

- A relation is a two-dimensional table in which the entries are single-valued, each field has a distinct name, all the values in a field are values of the same attribute (the one identified by the field name), the order of fields is immaterial, each row is distinct, and the order of rows is immaterial.

- A relational database is a collection of relations.

- An unnormalized relation is a structure in which entries need not be single-valued but that satisfies all the other properties of a relation.

- A field name is qualified by preceding it with the table name and a period (for example, Rep.RepNum).

- A table's primary key is the field or fields that uniquely identify a given row within the table.

- Query-By-Example (QBE) is a visual tool for manipulating relational databases. QBE queries are created by completing on-screen forms.

- To include a field in an Access query, place the field in the design grid and make sure a check mark appears in the field's Show check box.

- To indicate criteria in an Access query, place the criteria in the appropriate columns in the design grid of the Query window.

- To indicate AND criteria in an Access query, place both criteria in the same Criteria row of the design grid; to indicate OR criteria, place the criteria on separate Criteria rows of the design grid.

- To create a computed field in Access, enter an appropriate expression in the desired column of the design grid.

- To use functions to perform calculations in Access, include the appropriate function in the Total row for the appropriate column of the design grid.

- To sort query results in Access, select Ascending or Descending in the Sort row for the field or fields that are sort keys.

- When sorting query results using more than one field, the leftmost sort key in the design grid is the major sort key (also called the primary sort key) and the sort key to its right is the minor sort key (also called the secondary sort key).

- To join tables in Access, place field lists for both tables in the upper pane of the Query window.

- To make the same change to all records that satisfy certain criteria, use an update query.

- To delete all records that satisfy certain criteria, use a delete query.

- To save the results of a query as a table, use a make-table query.

- Relational algebra is a theoretical method of manipulating relational databases.

- The SELECT command in relational algebra selects only certain rows from a table.

- The PROJECT command in relational algebra selects only certain columns from a table.

- The JOIN command in relational algebra combines data from two or more tables based on common columns.

- The UNION command in relational algebra forms the union of two tables. For a union operation to make sense, the tables must be union compatible.

- Two tables are union compatible when they have the same number of columns and their corresponding columns represent the same type of data.

- The INTERSECT command in relational algebra forms the intersection of two tables.

- The SUBTRACT command in relational algebra forms the difference of two tables.

- The product of two tables (mathematically called the Cartesian product) is the table obtained by concatenating every row in the first table with every row in the second table.

- The division process in relational algebra divides one table by another table.

Key Terms

aggregate function	natural join
AND criterion	null
attribute	OR criterion
calculated field	outer join
Cartesian product	primary key
comparison operator	primary sort key
compound condition	product
compound criteria	PROJECT
computed field	qualify
concatenation	query
criteria	Query-By-Example (QBE)
criterion	record
delete query	relation
design grid	relational algebra
difference	relational database
division	relational operator
field	repeating group
function	secondary sort key
grouping	SELECT
INTERSECT	sort
intersection	sort key
join	SUBTRACT
join column	tuple
join line	union
major sort key	union compatible
make-table query	unnormalized relation
minor sort key	update query

Review Questions

1. What is a relation?

2. What is a relational database?

3. What is an unnormalized relation? Is it a relation according to the definition of the word *relation*?

4. How is the term *attribute* used in the relational model? What is a more common name for *atttribute*?

5. Describe the shorthand representation of the structure of a relational database. Illustrate this technique by representing the database for Henry Books as shown in Chapter 1.

6. What does it mean to qualify a field name? How would you qualify the Street field in the Customer table?

7. What is a primary key? What is the primary key for each table in the Henry Books database shown in Chapter 1?

8. How do you include a field in an Access query?

9. How do you indicate criteria in an Access query?

10. How do you use an AND criterion to combine criteria in an Access query? How do you use an OR criterion to combine criteria?

11. How do you create a computed field in an Access query?

12. In which row of the Access design grid do you include functions? What functions can you use in Access queries?

13. How do you sort data in an Access query?

14. When sorting data on more than one field in an Access query, which field is the major sort key? Which field is the minor sort key? What effect do these keys have on the order in which the rows are displayed?

15. How do you join tables in an Access query?

16. When do you use an update query?

17. When do you use a delete query?

18. When do you use a make-table query?

19. What is relational algebra?

20. Describe the purpose of the SELECT command in relational algebra.

21. Describe the purpose of the PROJECT command in relational algebra.

22. Describe the purpose of the JOIN command in relational algebra.

23. Describe the purpose of the UNION command in relational algebra.

24. Are there any restrictions on the tables when using the UNION command? If so, what are these restrictions?

25. Describe the purpose of the INTERSECT command in relational algebra.

26. Describe the purpose of the SUBTRACT command in relational algebra.

27. Describe the purpose of the product process in relational algebra.

28. Describe the results of the division process in relational algebra.

Premiere Products Exercises: QBE

In the following exercises, you will use the data in the Premiere Products database shown in Figure 2-1. (If you use a computer to complete these exercises, use a copy of the Premiere Products database so you will still have the original data when you complete Chapter 3.) In each step, use QBE to obtain the desired results. You can use the query feature in a DBMS to complete the exercises using a computer, or you can simply write a description of how you would complete the task. Check with your instructor if you are uncertain about which approach to take.

1. List the number and name of all customers.

2. List the complete Part table.

3. List the number and name of all customers represented by sales rep 35.

4. List the number and name of all customers that are represented by sales rep 35 and that have a credit limit of $10,000.

5. List the number and name of all customers that are represented by sales rep 35 or that have a credit limit of $10,000.

6. For each order, list the order number, order date, number of the customer that placed the order, and name of the customer that placed the order.

7. List the number and name of all customers represented by Juan Perez.

8. How many customers have a credit limit of $10,000?

9. Find the total of the balances for all customers represented by sales rep 35.

10. Give the part number, description, and on-hand value (OnHand * Price) for each part in item class HW.

11. List all columns and all records in the Part table. Sort the results by part description.

12. List all columns and all records in the Part table. Sort the results by part number within item class.

13. List the item class and the sum of the value of parts on hand. Group the results by item class.

14. Create a new table named SportingGoods to contain the columns PartNum, Description, OnHand, Warehouse, and Price for all rows in which the item class is SG.

15. In the SportingGoods table, change the description of part BV06 to "Fitness Gym."

16. In the SportingGoods table, delete every row in which the price is greater than $1,000.

Premiere Products Exercises: Relational Algebra

In the following exercises, you will use the data in the Premiere Products database shown in Figure 2-1. In each step, indicate how to use relational algebra to obtain the desired results.

1. List the number and name of all sales reps.

2. List all information from the Part table for part FD21.

3. List the order number, order date, customer number, and customer name for each order.

4. List the order number, order date, customer number, and customer name for each order placed by any customer represented by the sales rep whose last name is Kaiser.

5. List the number and date of all orders that were placed on 10/20/2010 or that were placed by a customer whose rep number is 20.

6. List the number and date of all orders that were placed on 10/20/2010 by a customer whose rep number is 20.

7. List the number and date of all orders that were placed on 10/20/2010 but not by a customer whose rep number is 20.

Henry Books Case

Ray Henry is aware that the ability to query a database easily is one of the most important benefits of using a good DBMS. Now that you have helped him computerize his database, he is anxious to have you help him obtain answers to a variety of questions. In the following exercises, you will use the data in the Henry Books database shown in Figures 1-15 through 1-18 in Chapter 1. (If you use a computer to complete these exercises, use a copy of the Henry Books database so you will still have the original data when you complete Chapter 3.) In each step, use QBE to obtain the desired results. You can use the query feature in a DBMS to complete the exercises using a computer, or you can simply write a description of how you would complete the task. Check with your instructor if you are uncertain about which approach to take.

1. List the author number and last name for every author.

2. List the complete Branch table (all rows and all columns).

3. List the name of every publisher located in Boston.

4. List the name of every publisher not located in Boston.

5. List the name of every branch that has at least nine employees.

6. List the book code and title of every book that has the type SFI.

7. List the book code and title of every book that has the type SFI and that is a paperback.

8. List the book code and title of every book that has the type SFI or that has the publisher code PE.

9. List the book code, title, and price for each book with a price that is greater than $5 but less than $10.

10. List the book code and title of every book that has the type FIC and a price of less than $10.

11. Customers who are part of a special program get a 15% discount off regular book prices. To determine the discounted prices, list the book code, title, and discounted price of every book. (Your calculated column should determine 85% of the current price, which is 100% less a 15% discount.)

12. List the book code and title of every book that has the type SFI, HOR, or ART.

13. List the book code, title, and publisher code for all books. Sort the results by title within publisher code.

14. How many books have the type SFI?

15. Calculate the average price for each type of book.

16. For every book, list the book code, book title, publisher code, and publisher name.

17. For every book published by Taunton Press, list the book title and book price.

18. List the book title and book code for every book published by Putnam Publishing Group that has a book price greater than $15.

19. Create a new table named Fiction using the data in the BookCode, Title, PublisherCode, and Price columns in the Book table for those books that have the type FIC.

20. Use an update query to change the price of any book in the Fiction table with a current price of $14.00 to $14.50.

21. Use a delete query to delete all books in the Fiction table that have the publisher code VB.

Alexamara Marina Group Case

In the following exercises, you will use the data in the Alexamara database shown in Figures 1-20 through 1-24 in Chapter 1. (If you use a computer to complete these exercises, use a copy of the Alexamara database so you will still have the original data when you complete Chapter 3.) In each step, use QBE to obtain the desired results. You can use the query feature in a DBMS to complete the exercises using a computer, or you can simply write a description of how you would complete the task. Check with your instructor if you are uncertain about which approach to take.

1. List the owner number, last name, and first name of every boat owner.

2. List the complete Marina table (all rows and all columns).

3. List the last name and first name of every owner located in Bowton.

4. List the last name and first name of every owner not located in Bowton.

5. List the marina number and slip number for every slip whose length is equal to or less than 30 feet.

6. List the marina number and slip number for every boat with the type Dolphin 28.

7. List the slip number for every boat with the type Dolphin 28 that is located in marina 1.

8. List the boat name for each boat located in a slip whose length is between 25 and 30 feet.

9. List the slip number for every slip in marina 1 whose annual rental fee is less than $3,000.

10. Labor is billed at the rate of $60 per hour. List the slip ID, category number, estimated hours, and estimated labor cost for every service request. To obtain the estimated labor cost, multiply the estimated hours by 60. Use the column name "EstimatedCost" for the estimated labor cost.

11. List the marina number and slip number for all slips containing a boat with the type Sprite 4000, Sprite 3000, or Ray 4025.

12. List the marina number, slip number, and boat name for all boats. Sort the results by boat name within the marina number.

13. How many Dolphin 25 boats are stored at both marinas?

14. Calculate the total rental fees Alexamara receives each year based on the length of the slip.

15. For every boat, list the marina number, slip number, boat name, owner number, owner's first name, and owner's last name.

16. For every completed or open service request for routine engine maintenance, list the slip ID, description, and status.

17. For every service request for routine engine maintenance, list the slip ID, marina number, slip number, estimated hours, spent hours, owner number, and owner's last name.

18. Create a new table named LargeSlip using the data in the MarinaNum, SlipNum, RentalFee, BoatName, and OwnerNum columns in the MarinaSlip table for slips with lengths of 40 feet.

19. Use an update query to change the rental fee of any slip in the LargeSlip table whose fee is currently $3,800 to $3,900.

20. Use a delete query to delete slips in the LargeSlip table whose rental fee is $3,600.

THE RELATIONAL MODEL 2: SQL

LEARNING OBJECTIVES

Objectives

- Introduce Structured Query Language (SQL)
- Use simple and compound conditions in SQL
- Use computed fields in SQL
- Use built-in SQL functions
- Use subqueries in SQL
- Group records in SQL
- Join tables using SQL
- Perform union operations in SQL
- Use SQL to update database data
- Use an SQL query to create a table in a database

INTRODUCTION

In this chapter, you will examine the language called **SQL (Structured Query Language)**. Like Access and Query-By-Example (QBE), SQL provides users with the capability of querying a relational database. However, in SQL, you must enter **commands** to obtain the desired results rather than complete an on-screen form as you do in Access and QBE. SQL uses commands to create and update tables and to retrieve data from tables. The commands used to retrieve table data are usually called queries.

SQL was developed under the name SEQUEL at the IBM San Jose research facilities as the data manipulation language for IBM's prototype relational DBMS, System R, in the mid-1970s. In 1980, it was renamed SQL (but still pronounced "sequel," although the equally popular pronunciation of "S-Q-L" ("ess-cue-ell") is used in this book) to avoid confusion with an unrelated hardware product called SEQUEL. Most relational DBMSs use a version of SQL as a data manipulation language. SQL is the standard language for relational database manipulation. The SQL version used in the examples in this chapter is Microsoft Office Access 2007. Although the various versions of SQL are not identical, the differences are relatively minor. After you have mastered one version of SQL, you can apply your skills to learn another version of SQL.

You will begin studying SQL by examining how to use it to create a table. You will examine simple retrieval methods and compound conditions. You will use computed fields in SQL and learn how to sort data. Then you will learn how to use built-in functions, subqueries, and grouping. You will learn how to join tables and use the UNION operator. Finally, you will use SQL to update data in a database. The end of this chapter includes generic versions of all the SQL commands presented in the chapter.

GETTING STARTED WITH SQL

In this chapter, you will be reading the material and examining the figures to understand how to use SQL to manipulate a relational database. You might also be using a DBMS to practice database manipulation at the same time. If you are completing the work in this chapter using Microsoft Office Access 2003, Microsoft Office Access 2007, or MySQL version 4.1 or higher, you should read the following information about your DBMS to learn more about how to start SQL and to learn specific details about differences you might encounter as you complete your work.

Getting Started with Microsoft Office Access 2003 and 2007

If you are using the Access 2003 or 2007 version of the Premiere Products database provided with the Data Files for this text, the tables in the database have already been created. You will not need to execute the CREATE TABLE commands to create the tables or the INSERT commands to add records to the tables.

To execute the SQL commands shown in the figures in Access 2003, open the Premiere Products database, click Queries on the Objects bar in the Database window, and then double-click the "Create query in Design view" option. Click the Close button in the Show Table dialog box, click the View button list arrow on the Query Design toolbar, and then click SQL View. The query opens in SQL view, ready for you to type your SQL commands. To run the SQL command, click the View button on the Query Design toolbar. To return to SQL view for the query, click the View button list arrow, and then click SQL View.

To execute SQL commands shown in the figures in Access 2007, open the Premiere Products database, click the Create tab on the Ribbon, click the Query Design button in the Other group, click the Close button in the Show Table dialog box, click the View button arrow in the Results group on the Query Design Tools tab, and then click SQL View. The Query1 tab displays the query in SQL view, ready for you to type your SQL commands. To run the SQL command, click the Run button in the Results group on the Query Tools Design tab. To return to SQL view, click the View button arrow in the Views group on the Home tab, and then click SQL View.

Unlike other SQL implementations, Access doesn't have a DECIMAL data type. To create numbers with decimals, you must use either the CURRENCY or NUMBER data type. Use the CURRENCY data type for fields that will contain currency values; use the NUMBER data type for all other numeric fields.

In Access, you can correct typing errors in a command just as you would correct errors in a document by using the keyboard arrow keys to move the insertion point and using the Backspace or Delete keys to delete text. After making your corrections, you can run the query again.

Some of the examples in this text change the data in the database. If you plan to work through the examples using Access, you should use a *copy* of the original Premiere Products database because the version of the database that is used in subsequent chapters does not reflect these changes.

Getting Started with MySQL

If you use the MySQL-Premiere script provided with the Data Files for this text to create and activate the Premiere (Products) database, the script will activate the database, create the tables, and insert the records for you. You will not need to execute the CREATE TABLE commands to create the tables or the INSERT commands to add records to the tables. (*Note:* This script file assumes you have not previously created the database or any of the tables in the database. If you have created any of the tables, you should run the MySQL-DropPremiere script provided with the Data Files for this book prior to running the MySQL-Premiere script.)

To run a script in MySQL, type the SOURCE command followed by the name of the file and then press the Enter key. For example, to run a script named MySQL-Premiere, you would type the following command:

```
SOURCE MySQL-Premiere
```

Before typing commands in MySQL, you must activate the database by typing the USE command followed by the name of the database; for example, to activiate the Premiere Products database, the command is USE PREMIERE. After the database is activated, all commands are assumed to pertain to the activated database. To activate a different database during the current session, you execute the USE command again with the new database name. After typing any MySQL command, press the Enter key. MySQL moves the cursor to the next line and displays the continuation indicator (->). After typing the last line of a command, type a semicolon and then press the Enter key to execute the command and display the results.

As you are working in MySQL, the most recent command you entered is stored in a special area of memory called the **statement history**. You can edit the command in the statement history by using the editing keys shown in Table 3-1.

Activity	Editing Key
Move up a line in the statement history	Up arrow
Move down a line in the statement history	Down arrow
Move left one character within a line	Left arrow
Move right one character within a line	Right arrow
Move to beginning of a line	Ctrl + A
Move to end of a line	Ctrl + E
Delete previous character	Backspace
Delete character below cursor	Delete

TABLE 3-1   MySQL editing commands

For example, to make a correction in the first line of a command, you can use the Up arrow key to bring the first line of the incorrect command to the screen, make any necessary changes, and then press the Enter key. You can then bring the second line to the screen, make any necessary changes, and then press the Enter key. You can repeat this process for all the lines in the command. If you need to add a new line, just type it at the appropriate position.

Some of the examples used in this text change the data in the database. If you plan to work through the examples using MySQL, you should follow the instructions in this section for re-creating the Premiere database prior to starting the next chapter. The version of the database used in subsequent chapters does not reflect these changes.

TABLE CREATION

You use the SQL **CREATE TABLE** command to create a table by describing its layout. The word *TABLE* is followed by the name of the table to be created and then by the names and data types of the columns (fields) that make up the table. The rules for naming tables and columns vary slightly from one version of SQL to another. If you have any doubts about the validity of any of the names you have chosen, you should consult the manual for your SQL version.

Some common restrictions placed on table and column names by DBMSs are as follows:

- The names cannot exceed 18 characters.
- The names must start with a letter.
- The names can contain only letters, numbers, and underscores (_).
- The names cannot contain spaces.

The Relational Model 2: SQL

NOTE

Unlike some other versions of SQL, Access SQL permits the use of spaces within table and column names. There is a restriction, however, on the way such names are used in SQL commands. When you use a name containing a space in Access SQL, you must enclose it in square brackets. For example, if the name of the CreditLimit column were changed to Credit Limit (with a space between *Credit* and *Limit*), you would write the column as [Credit Limit] because the name includes a space.

NOTE

In systems that permit the use of uppercase and lowercase letters in table and column names, you can avoid using spaces by capitalizing the first letter of each word in the name and using lowercase letters for the remaining letters in the words. For example, the name of the credit limit column would be CreditLimit. In systems that do not permit the use of spaces or mixed-case letters, some programmers use an underscore to separate words. For example, the name of the credit limit column would be CREDIT_LIMIT.

For each column in a table, you must specify the type of data that the column can store. Although the actual data types will vary slightly from one implementation of SQL to another, the following list indicates the data types you will often encounter:

- **INTEGER:** Stores integers, which are numbers without a decimal part. The valid data range is -2147483648 to 2147483647. You can use the contents of INTEGER fields for calculations.
- **SMALLINT:** Stores integers, but uses less space than the INTEGER data type. The valid data range is -32768 to 32767. SMALLINT is a better choice than INTEGER when you are certain that the field will store numbers within the indicated range. You can use the contents of SMALLINT fields for calculations.
- **DECIMAL(p,q):** Stores a decimal number *p* digits long with *q* of these digits being decimal places. For example, DECIMAL(5,2) represents a number with three places to the left and two places to the right of the decimal. You can use the contents of DECIMAL fields for calculations. (Unlike other SQL implementations, Access doesn't have a DECIMAL data type. To create numbers with decimals, you must use either the CURRENCY or NUMBER data type. Use the CURRENCY data type for fields that will contain currency values; use the NUMBER data type for all other numeric fields.)
- **CHAR(n):** Stores a character string *n* characters long. You use the CHAR type for fields that contain letters and other special characters and for fields that contain numbers that will not be used in calcualtions. Because neither sales rep numbers nor customer numbers will be used in any calculations, for example, both of them are assigned CHAR as the data type. (Some DBMSs, such as Access, use TEXT rather than CHAR, but the two data types mean the same thing.)
- **DATE:** Stores dates in the form DD-MON-YYYY or MM/DD/YYYY. For example, May 12, 2010, could be stored as 12-MAY-2010 or 5/12/2010.

EXAMPLE 1

Use SQL to create the Rep table by describing its layout.

The general CREATE TABLE command for the Rep table is as follows:

```
CREATE TABLE Rep
(RepNum CHAR(2),
LastName CHAR(15),
FirstName CHAR(15),
Street CHAR(15),
City CHAR(15),
State CHAR(2),
Zip CHAR(5),
Commission DECIMAL(7,2),
Rate DECIMAL(3,2) )
;
```

In this SQL command, which uses the data definition features of SQL, you're describing a table that will be named Rep. It contains nine fields: RepNum, LastName, FirstName, Street, City, State, Zip, Commission, and Rate. RepNum is a character field that is two positions in length. LastName is a character field with 15 characters. Commission is a numeric field that stores seven digits, including two decimal places. Similarly, Rate is a numeric field that stores three digits, including two decimal places. Because many versions of SQL require you to end a command with a semicolon, commands in this text will end with semicolons.

NOTE

In SQL, commands are free-format. No rule says that a particular word must begin in a particular position on the line. The previous SQL command could have been written as follows:

```
CREATE TABLE Rep (RepNum CHAR(2),LastName CHAR(15),
FirstName CHAR(15),Street CHAR(15),City CHAR(15),
State CHAR(2),Zip CHAR(5),Commission DECIMAL(7,2),
Rate DECIMAL(3,2) )
;
```

NOTE

The manner in which the first CREATE TABLE command was written simply makes the command more readable. In general, you should strive for such readability when you write SQL commands.

SIMPLE RETRIEVAL

The basic form of an SQL retrieval command is simply SELECT-FROM-WHERE. After the word *SELECT*, you list those fields you want to display in the query results. This portion of the command is called the **SELECT clause**. The fields will appear in the query results in the order in which they are listed in the SELECT clause. After the word *FROM*, you list the table or tables that contain the data to display in the query results. This portion of the command is called the **FROM clause**. Finally, after the word *WHERE*, you list any conditions that you want to apply to the data you want to retrieve, such as indicating that the credit limit must be $10,000. This portion of the command, which is optional, is called the **WHERE clause**.

There are no special formatting rules in SQL—the examples in this text include the SELECT, FROM, and WHERE clauses on separate lines to make the commands more readable. In addition, this text uses a common style in which words that are part of the SQL language, called **reserved words**, appear in all uppercase letters. All other words in commands appear in a combination of uppercase and lowercase letters.

EXAMPLE 2

List the number, name, and balance of all customers.

Because you want to list all customers, you won't need to use the WHERE clause—you don't need to put any restrictions on the data to retrieve. Figure 3-1a shows the query to select the number, name, and balance of all customers, using the SQL implementation in Access 2007.

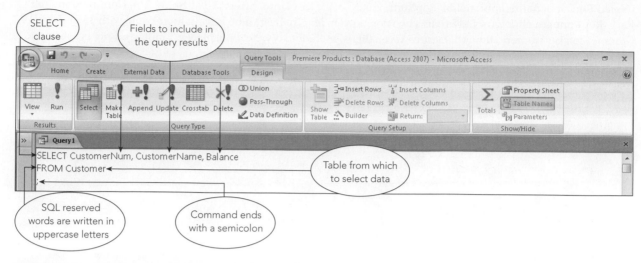

FIGURE 3-1a SQL query to select customer data (Access)

Figure 3-1b shows the MySQL query to select the number, name, and balance of all customers.

```
mysql> SELECT CustomerNum, CustomerName, Balance
    -> FROM Customer
    -> ;
```

FIGURE 3-1b SQL query to select customer data (My SQL)

The results of executing the query shown in Figure 3-1a in Access 2007 appear in Figure 3-2a.

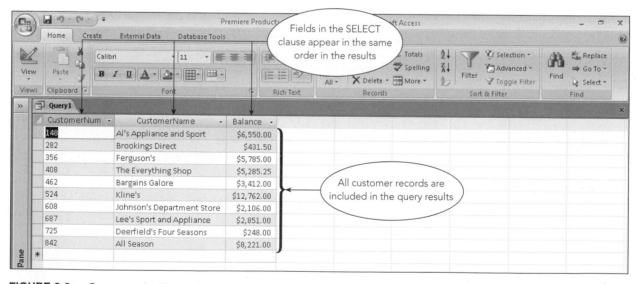

FIGURE 3-2a Query results (Access)

The results of executing the query shown in Figure 3-2a in My SQL appear in Figure 3-2b.

```
+--------------+----------------------------+----------+
| CustomerNum  | CustomerName               | Balance  |
+--------------+----------------------------+----------+
| 148          | Al's Appliance and Sport   | 6550.00  |
| 282          | Brookings Direct           |  431.50  |
| 356          | Ferguson's                 | 5785.00  |
| 408          | The Everything Shop        | 5285.25  |
| 462          | Bargains Galore            | 3412.00  |
| 524          | Kline's                    | 12762.00 |
| 608          | Johnson's Department Store | 2106.00  |
| 687          | Lee's Sport and Appliance  | 2851.00  |
| 725          | Deerfield's Four Seasons   |  248.00  |
| 842          | All Season                 | 8221.00  |
+--------------+----------------------------+----------+
10 rows in set (0.02 sec)
```

FIGURE 3-2b   Query results (MySQL)

EXAMPLE 3

List the complete Part table.

You could use the same approach shown in Example 2 by listing each field in the Part table in the SELECT clause. However, there is a shortcut. Instead of listing all the field names in the SELECT clause, you can use the * symbol. When used after the word *SELECT*, the * symbol indicates that you want to include all fields in the query results in the order in which you described them to the DBMS when you created the table. If you wanted to include all the fields in the query results but in a different order, you could type the names of the fields in the order in which you want them to appear. In this case, assuming the default order is appropriate, the query design appears in Figure 3-3.

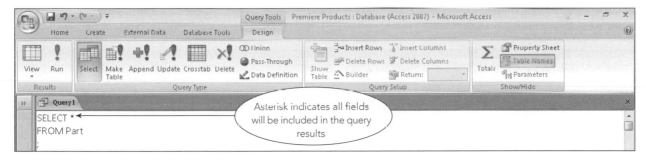

FIGURE 3-3   SQL query to list the complete Part table

The query results appear in Figure 3-4.

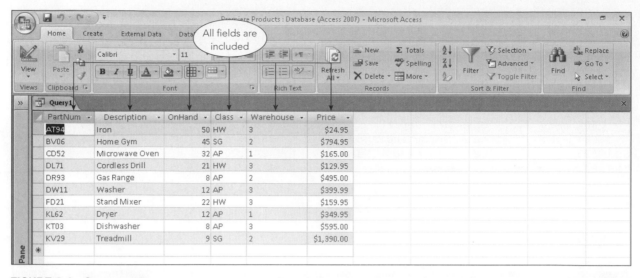

FIGURE 3-4 Query results

EXAMPLE 4

List the name of every customer with a $10,000 credit limit.

You include the following condition in the WHERE clause to restrict the query results to only those customers with a credit limit of $10,000.

```
WHERE CreditLimit=10000
```

Notice that you do not type commas or dollar signs in numbers. The query design appears in Figure 3-5.

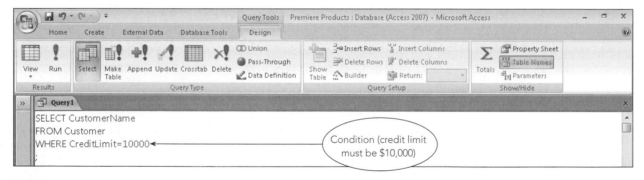

FIGURE 3-5 SQL query with WHERE condition

The query results appear in Figure 3-6.

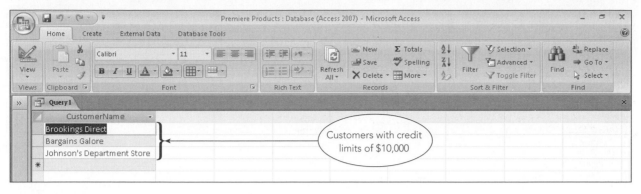

FIGURE 3-6 Query results

Figure 3-5 is called a simple condition. A **simple condition** includes the field name, a comparison operator, and either another field name or a value, such as CreditLimit = 10000 or CreditLimit > Balance. Figure 3-7 lists the comparison operators that you can use in SQL commands. Notice that there are two versions of the "not equal to" operator: < > and !=. You must use the correct one for your version of SQL. If you use the wrong one, your system will generate an error, in which case, you'll know to use the other version.

Comparison Operator	Meaning
=	Equal to
<	Less than
>	Greater than
<=	Less than or equal to
>=	Greater than or equal to
< >	Not equal to (used by most implementations of SQL)
!=	Not equal to (used by some implementations of SQL)

FIGURE 3-7 Comparison operators used in SQL commands

In Example 4, the WHERE clause compared a numeric field (CreditLimit) to a number (10000). When a query involves a character field, such as CustomerNum or CustomerName, you must enclose the value to which the field is being compared in single quotation marks, as illustrated in Examples 5 and 6.

EXAMPLE 5

Find the name of customer 148.

The query design appears in Figure 3-8. Because CustomerNum is a character field, the value 148 is enclosed in single quotation marks.

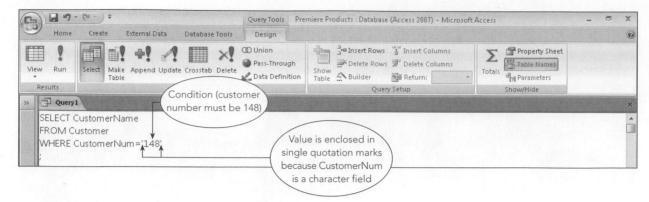

FIGURE 3-8 SQL query to find the name of customer 148

The query results appear in Figure 3-9. Only a single record appears in the query results because the CustomerNum field is the primary key for the Customer table and there can be only one customer with the number 148.

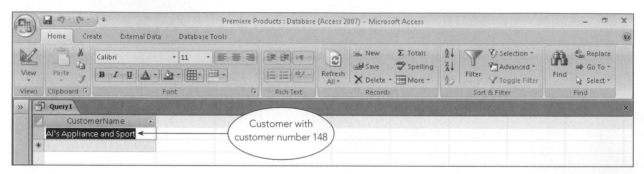

FIGURE 3-9 Query results

EXAMPLE 6

Find the customer name for every customer located in the city of Grove.

The query design appears in Figure 3-10.

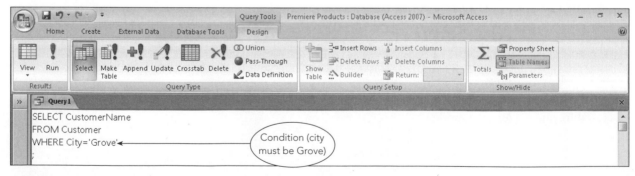

FIGURE 3-10 SQL query to find all customers located in Grove

The query results appear in Figure 3-11. Because more than one customer is located in Grove, there are multiple records in the query results.

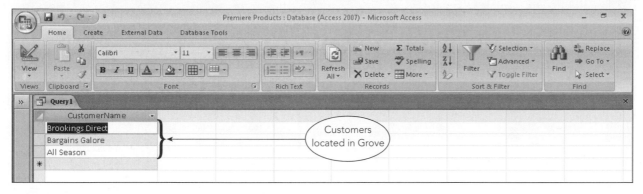

FIGURE 3-11   Query results

You can also use dates in conditions. The way you do so varies slightly from one implementation of SQL to another. In Access, you place number signs around the date (for example, #11/15/2010#). In other programs, you enter the day of the month, a hyphen, the three-character abbreviation for the month, a hyphen, and the year, all enclosed in single quotation marks (for example, '15-NOV-2010').

EXAMPLE 7

List the number, name, credit limit, and balance for all customers with credit limits that exceed their balances.

The query design appears in Figure 3-12. Notice that the condition in the WHERE clause compares the contents of two fields.

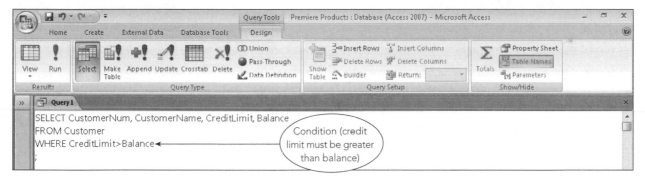

FIGURE 3-12   SQL query to find all customers with credit limits that exceed their balances

The query results appear in Figure 3-13.

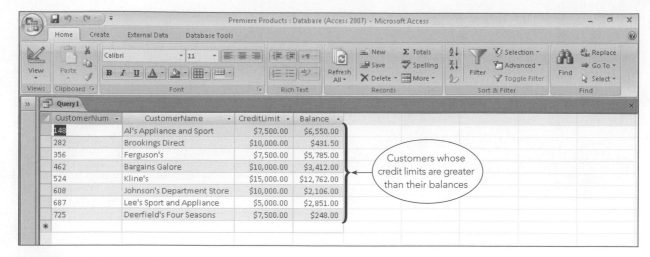

FIGURE 3-13 Query results

COMPOUND CONDITIONS

The conditions you've seen so far are called simple conditions. The following examples require compound conditions. A **compound condition** is formed by connecting two or more simple conditions using one or both of the following operators: AND and OR. You can also precede a single condition with the NOT operator to negate a condition.

When you connect simple conditions using the AND operator, all the simple conditions must be true for the compound condition to be true. When you connect simple conditions using the OR operator, the compound condition will be true whenever any of the simple conditions are true. Preceding a condition by the NOT operator reverses the truth or falsity of the original condition. That is, if the original condition is true, the new condition will be false; if the original condition is false, the new one will be true.

EXAMPLE 8

List the descriptions of all parts that are located in warehouse 3 and for which there are more than 20 units on hand.

In this example, you want to list those parts for which *both* the warehouse number is equal to 3 *and* the number of units on hand is greater than 20. Thus, you form a compound condition using the AND operator, as shown in Figure 3-14.

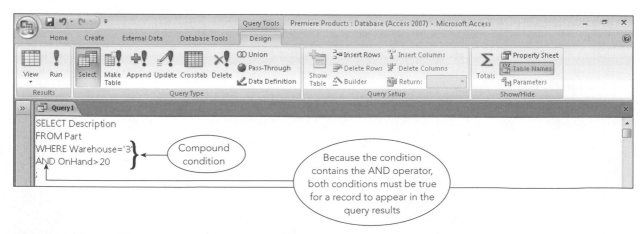

FIGURE 3-14 Compound condition that uses the AND operator

The query results appear in Figure 3-15.

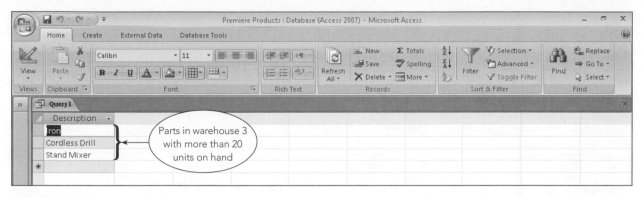

FIGURE 3-15 Query results

EXAMPLE 9

List the descriptions of all parts that are located in warehouse 3 or for which there are more than 20 units on hand or both.

As you would expect, you form compound conditions with the OR operator similar to the way you use the AND operator. The compound condition shown in Figure 3-16 uses the OR operator instead of the AND operator.

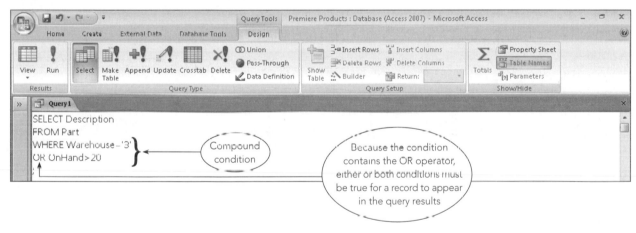

FIGURE 3-16 Compound condition that uses the OR operator

The query results appear in Figure 3-17.

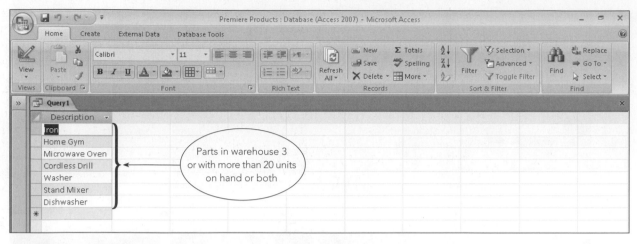

FIGURE 3-17   Query results

EXAMPLE 10

List the descriptions of all parts that are not in warehouse 3.

For this example, you could use a simple condition and the "not equal to" operator (< >). As an alternative, you could use the "equals" operator (=) in the condition, but precede the entire condition with the NOT operator, as shown in Figure 3-18.

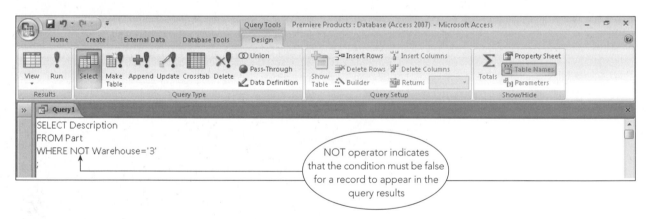

FIGURE 3-18   SQL query with the NOT operator

The query results appear in Figure 3-19.

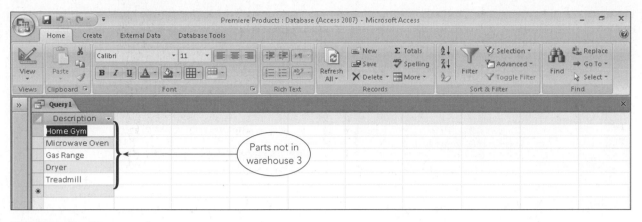

FIGURE 3-19 Query results

EXAMPLE 11

List the number, name, and balance of all customers with balances greater than or equal to $1,000 and less than or equal to $5,000.

You could use a WHERE clause and the AND operator (Balance>=1000 AND Balance<=5000). An alternative to this approach uses the BETWEEN operator, as shown in Figure 3-20.

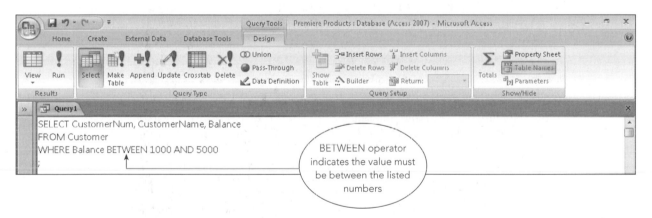

FIGURE 3-20 SQL query with the BETWEEN operator

The query results appear in Figure 3-21.

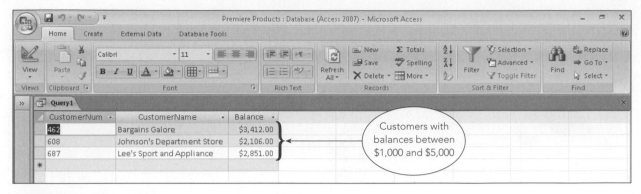

FIGURE 3-21 Query results

The BETWEEN operator is not an essential feature of SQL; you can use the AND operator to obtain the same results. Using the BETWEEN operator, however, does make certain SELECT commands easier to construct.

COMPUTED FIELDS

Similar to QBE, you can include fields in queries that are not in the database but whose values you can compute from existing database fields. A field whose values you derive from existing fields is called a computed field or calculated field. Computed fields can involve addition (+), subtraction (-), multiplication (*), or division (/). The query in Example 12, for example, uses subtraction.

EXAMPLE 12

List the number, name, and available credit for all customers.

There is no field in the database that stores available credit, but you can compute it using two fields that *are* present in the database: CreditLimit and Balance. The query design shown in Figure 3-22 creates a new field named AvailableCredit, which is computed by subtracting the value in the Balance field from the value in the CreditLimit field (AvailableCredit = CreditLimit - Balance). By using the word *AS* after the computation, followed by AvailableCredit, you can assign a name to the computed field.

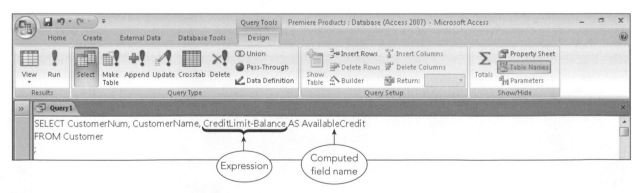

FIGURE 3-22 SQL query with a computed field

The query results appear in Figure 3-23. The column heading for the computed field is the name that you specified in the SELECT clause.

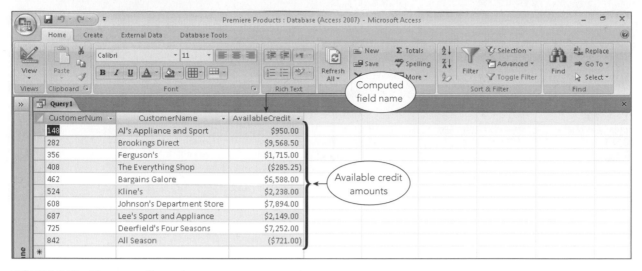

FIGURE 3-23 Query results

Computations are not limited to values in number fields. You can combine values in character fields as well. For example, in Access you can combine the values in the FirstName and LastName fields into a single computed field by using the & operator. The expression would be FirstName&' '&LastName, which places a space between the first name and the last name. The formal term is that you are concatenating the FirstName and LastName fields.

In MySQL, you use the CONCAT function to concatenate fields. To concatenate FirstName and LastName, for example, the expression would be CONCAT(FirstName, LastName).

EXAMPLE 13

List the number, name, and available credit for all customers with credit limits that exceed their balances.

The only difference between Examples 12 and 13 is that Example 13 includes a condition, as shown in Figure 3-24.

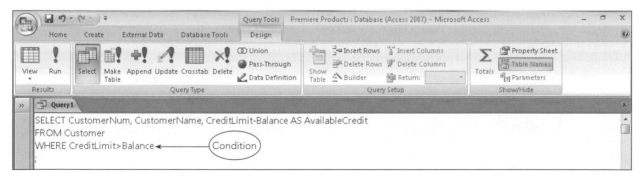

FIGURE 3-24 SQL query with a computed field and condition

The query results appear in Figure 3-25.

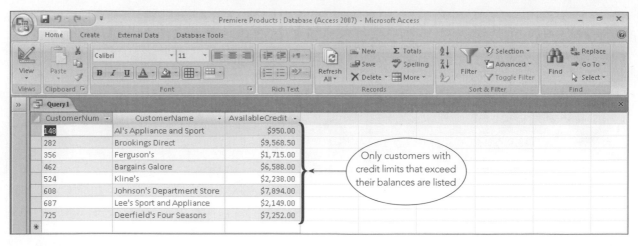

FIGURE 3-25 Query results

USING SPECIAL OPERATORS (LIKE AND IN)

In most cases, your conditions will involve exact matches, such as finding all customers located in the city of Sheldon. In some cases, however, exact matches will not work. For example, you might know only that the desired value contains a certain collection of characters. In such cases, you use the LIKE operator with a wildcard, as shown in Example 14.

EXAMPLE 14

List the number, name, and complete address of every customer located on a street that contains the letters *Oxford*.

All you know is that the addresses that you want contain a certain collection of characters (Oxford) somewhere in the Street field, but you don't know where. In Access SQL, the asterisk (*) is used as a wildcard to represent any collection of characters. (In MySQL, the percent sign (%) is used as a wildcard to represent any collection of characters.) To use a wildcard, include the LIKE operator in the WHERE clause. The query design shown in Figure 3-26 will retrieve information for every customer whose street contains some collection of characters followed by the letters *Oxford*, followed potentially by some additional characters.

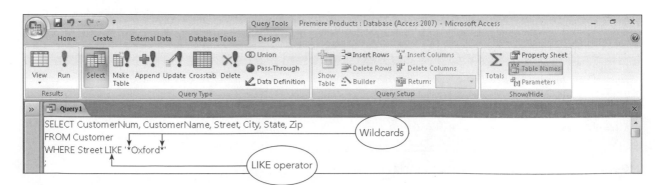

FIGURE 3-26 SQL query with a LIKE operator

The query results appear in Figure 3-27.

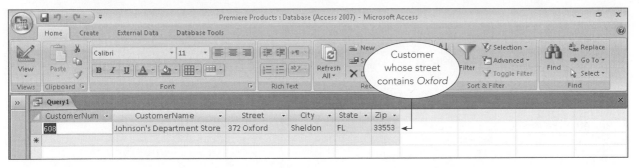

FIGURE 3-27 Query results

Another wildcard in Access SQL is the question mark (?), which represents any individual character. For example, "T?m" represents the letter *T* followed by any single character, followed by the letter *m* and would retrieve records that include the words *Tim*, *Tom*, or *T3m*, for example. Many versions of SQL, including MySQL, use the underscore (_) instead of the question mark to represent any individual character.

N O T E

In a large database, you should use wildcards only when absolutely necessary. Searches involving wildcards can be extremely slow to process.

Another operator, IN, provides a concise way of phrasing certain conditions, as Example 15 illustrates.

E X A M P L E 1 5

List the number, name, street, and credit limit for every customer with a credit limit of $7,500, $10,000, or $15,000.

In this query, you can use the SQL IN operator to determine whether a credit limit is $7,500, $10,000, or $15,000. You can obtain the same result by using the condition WHERE CreditLimit = 7500 OR Credit Limit = 10000 OR CreditLimit = 15000. The approach shown in Figure 3-28 is simpler, however—the IN clause contains the collection of values 7500, 10000, and 15000. The condition is true for those rows in which the value in the CreditLimit column is in this collection.

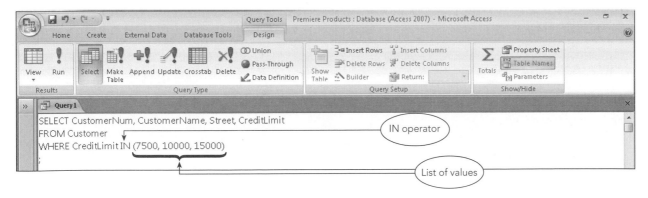

FIGURE 3-28 SQL query with an IN operator

The query results appear in Figure 3-29.

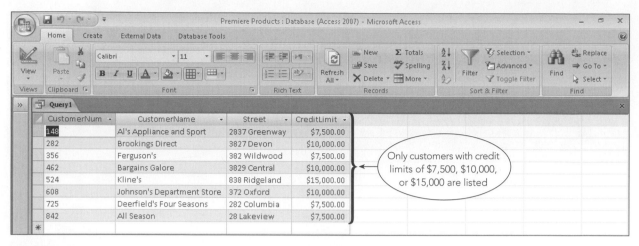

FIGURE 3-29   Query results

SORTING

Recall that the order of rows in a table is considered to be immaterial. From a practical standpoint, this means that when you query a relational database, there are no guarantees concerning the order in which the results will be displayed. The results might appear in the order in which the data was originally entered, but even this is not certain. Thus, if the order in which the data is displayed is important, you should *specifically* request that the results be displayed in a desired order. In SQL, you sort data using the **ORDER BY clause**.

EXAMPLE 16

List the number, name, street, and credit limit of all customers. Order (sort) the customers by name.

The field on which to sort data is called a sort key. To sort the output, you include the words *ORDER BY* in the SQL query, followed by the sort key field, as shown in Figure 3-30.

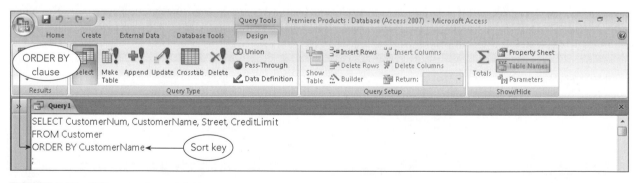

FIGURE 3-30   SQL query to sort data

The query results appear in Figure 3-31.

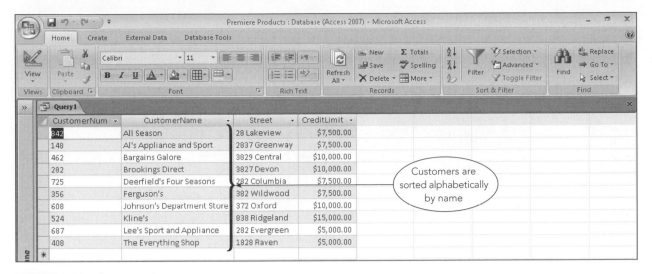

FIGURE 3-31 Query results

EXAMPLE 17

List the number, name, street, and credit limit of all customers. Order the customers by name within descending credit limit. (In other words, sort the customers by credit limit in descending order. Within each group of customers that have a common credit limit, sort the customers by name.)

When you need to sort data on two fields, the more important sort key is called the major sort key (also referred to as the primary sort key) and the less important sort key is called the minor sort key (also referred to as the secondary sort key). In this case, because you need to sort the output by name within credit limit, the CreditLimit field is the major sort key and the CustomerName field is the minor sort key. If there are two sort keys, as in Example 17, the major sort key will be listed first. You can specify to sort the output in descending (high-to-low) order by following the sort key with the word *DESC*, as shown in Figure 3-32.

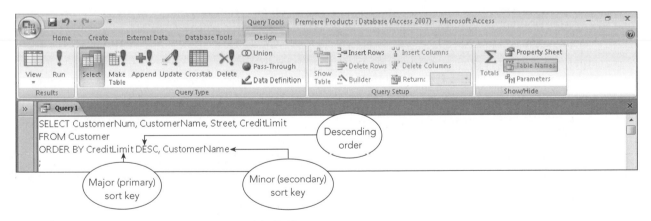

FIGURE 3-32 SQL query to sort data on multiple fields

The query results appear in Figure 3-33.

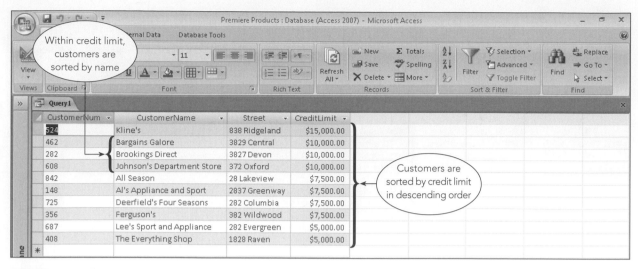

FIGURE 3-33 Query results

BUILT-IN FUNCTIONS

SQL has built-in functions (also called aggregate functions) to calculate the number of entries, the sum or average of all the entries in a given column, and the largest or smallest values in a given column. In SQL, these functions are called COUNT, SUM, AVG, MAX, and MIN, respectively.

EXAMPLE 18

How many parts are in item class HW?

In this query, you want to count the number of rows in the query results that have the value HW in the Class field. You could count the number of part numbers in the query results or the number of descriptions or the number of entries in any other field. It doesn't matter which column you choose because all columns will yield the correct answer. Rather than requiring you to pick a column arbitrarily, some versions of SQL allow you to use the * symbol to select any column. In SQL versions that support the * symbol, you could use the query design shown in Figure 3-34.

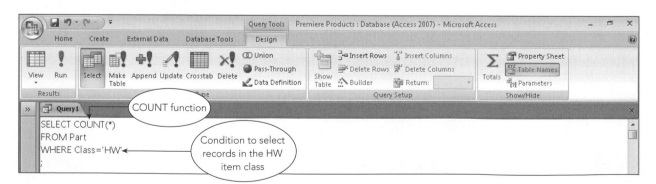

FIGURE 3-34 SQL query to count records

The query results appear in Figure 3-35.

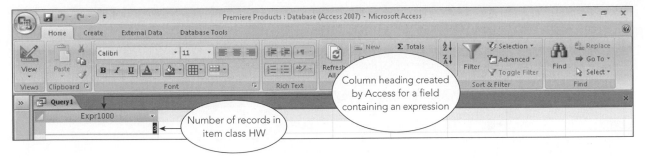

FIGURE 3-35 Query results

If your implementation of SQL doesn't permit the use of the * symbol, you could write the query as follows:

```
SELECT COUNT(PartNum)
FROM Part
WHERE Class='HW'
;
```

EXAMPLE 19

Find the number of customers and the total of their balances.

There are two differences between COUNT and SUM—other than the obvious fact that they are computing different statistics. In the case of SUM, you *must* specify the field for which you want a total and the field must be numeric. (How could you calculate a sum of names or addresses?) The query design appears in Figure 3-36.

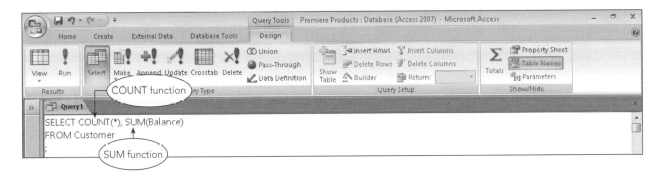

FIGURE 3-36 SQL query to count records and calculate a total

The query results appear in Figure 3-37.

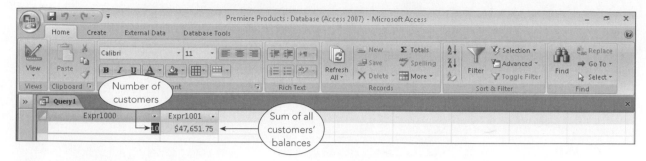

FIGURE 3-37 Query results

The use of AVG, MAX, and MIN is similar to the use of SUM. The only difference is that different statistics are calculated.

EXAMPLE 20

Find the total number of customers and the total of their balances. Change the column names for the number of customers and the total of their balances to CustomerCount and BalanceTotal, respectively.

As with computed fields, you can use the word *AS* to assign names to these computations, as shown in Figure 3-38.

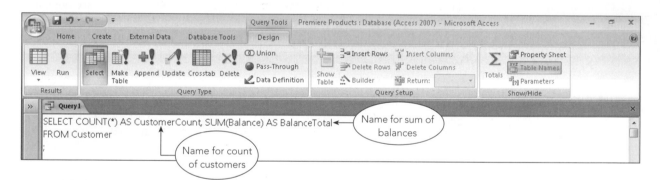

FIGURE 3-38 SQL query to perform calculations and rename columns

The query results appear in Figure 3-39.

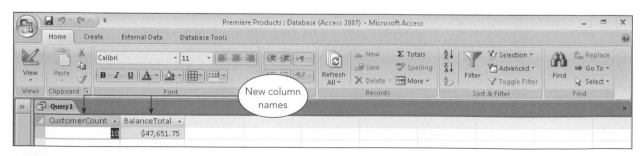

FIGURE 3-39 Query results

SUBQUERIES

In some cases, it is useful to obtain the results you want in two stages. You can do so by placing one query inside another. The inner query is called a **subquery** and is evaluated first. After the subquery has been evaluated, the outer query can be evaluated. Example 21 illustrates the process.

EXAMPLE 21

List the order number for each order that contains an order line for a part located in warehouse 3.

You can find the answer by using the Part table and creating a list of part numbers for those parts in warehouse 3. Then you can use the OrderLine table to find those order numbers present in any row on which the part number is in the results you created in the inner query. The corresponding query design appears in Figure 3-40.

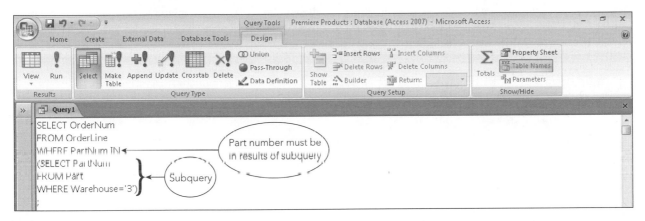

FIGURE 3-40 SQL query with a subquery

NOTE

Although not required, it is common to enclose subqueries in parentheses for readability.

The query results appear in Figure 3-41.

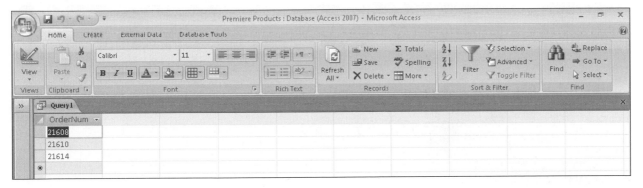

FIGURE 3-41 Query results

The subquery finds all the part numbers in the Part table with a warehouse number of 3. The subquery is evaluated first, producing a list of part numbers. After the subquery has been evaluated, the outer query is evaluated. Order numbers in the results appear in any row in the OrderLine table for which the part number in the row is in the subquery results.

GROUPING

Recall from Chapter 2 that grouping means creating groups of records that share some common characteristic. When grouping customers by sales rep number, for example, the customers of sales rep 20 would form one group, the customers of sales rep 35 would form a second group, and the customers of sales rep 65 would form a third group.

In Example 22, you need to group customers by rep number to perform the necessary calculations.

EXAMPLE 22

For each sales rep, list the rep number, the number of customers assigned to the rep, and the average balance of the rep's customers. Group the records by rep number and order the records by rep number.

This type of query requires grouping by rep number to make the correct calculations for each group. To indicate grouping in SQL, you use the **GROUP BY clause**, as shown in Figure 3-42. It is important to note that the GROUP BY clause does not mean that the query results will be sorted. To display the query results in a particular order, you must use the ORDER BY clause. The query design in Figure 3-42 uses the ORDER BY clause to sort the query results by rep number.

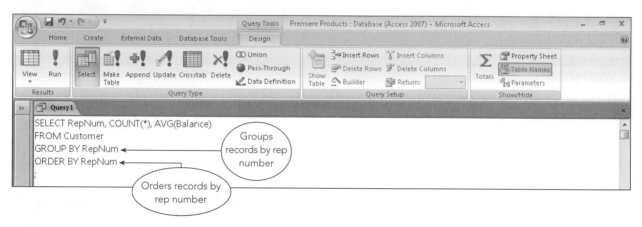

FIGURE 3-42   SQL query to group and sort records

The query results appear in Figure 3-43.

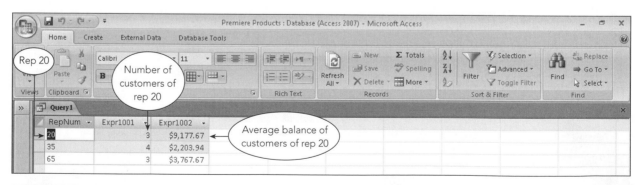

FIGURE 3-43   Query results

When rows are grouped, one line of output is produced for each group. Only statistics calculated for the group or fields whose values are the same for all rows in a group can be displayed in the grouped results.

Q & A

Question: Why is it appropriate to display the rep number?
Answer: Because the output is grouped by rep number, the rep number in one row in a group must be the same as the rep number in any other row in the group.

Q & A

Question: Would it be appropriate to display a customer number? Why or why not?
Answer: No, because the customer number will vary from one row in a group to another. (SQL could not determine which customer number to display for the group.)

EXAMPLE 23

For each sales rep with fewer than four customers, list the rep number, the number of customers assigned to the rep, and the average balance of the rep's customers. Rename the count of the number of customers and the average of the balances to NumCustomers and AverageBalance, respectively. Order the groups by rep number.

Examples 22 and 23 are similar, but there are two important differences: You need to rename the fields, and there is a restriction to display the calculations for only those reps having fewer than four customers. In other words, you want to display only those groups for which COUNT(*) is less than four. This restriction does not apply to individual rows, but to *groups*. Because the WHERE clause applies only to rows, it is not the appropriate clause to accomplish the kind of selection you need. Fortunately, the **HAVING clause** is to groups what the WHERE clause is to rows, as shown in Figure 3-44.

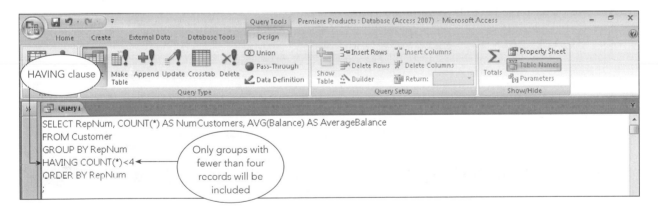

FIGURE 3-44 SQL query to restrict the groups that are included

The query results appear in Figure 3-45.

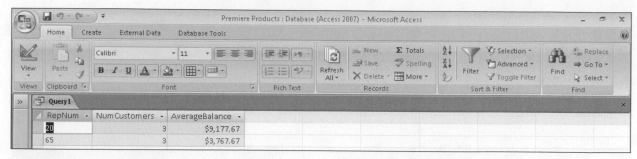

FIGURE 3-45 Query results

In this case, the row created for a group will be displayed only when the count of the number of records in the group is less than four.

You can include both a WHERE clause and a HAVING clause in the same query design, as shown in Figure 3-46.

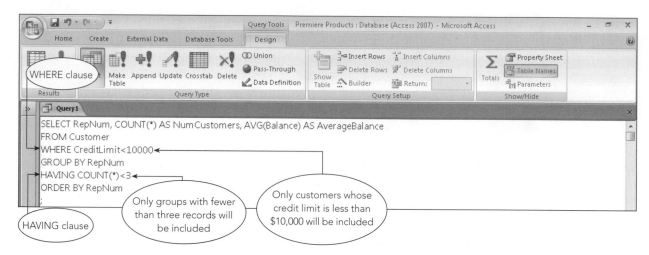

FIGURE 3-46 SQL query that includes WHERE and HAVING clauses

The query results appear in Figure 3-47.

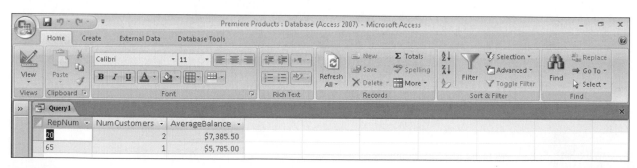

FIGURE 3-47 Query results

In this case, the WHERE clause will restrict the rows from the Customer table to those rows in which the credit limit is less than $10,000. These rows will be grouped by rep number. The HAVING clause then restricts the groups to be displayed to those for which the count of the rows in the group is less than three.

JOINING TABLES

Many queries require data from more than one table. As with QBE and relational algebra, it is necessary to be able to join tables so you can find rows in two or more tables that have identical values in matching fields. In SQL, this is accomplished by entering the appropriate conditions in the WHERE clause. (Appendix B includes information about an alternative way of joining tables in SQL that uses the FROM clause.)

EXAMPLE 24

List the number and name of each customer together with the number, last name, and first name of the sales rep who represents the customer. Order the records by customer number.

Because the numbers and names of customers are in the Customer table and the numbers and names of sales reps are in the Rep table, you need to include both tables in your SQL query. To join the tables, you'll construct the SQL command as follows:

1. In the SELECT clause, list all fields you want to display.
2. In the FROM clause, list all tables involved in the query.
3. In the WHERE clause, give the condition that will restrict the data to be retrieved to only those rows from the two tables that match; that is, you'll restrict it to the rows that have common values in matching fields.

As in relational algebra, it is often necessary to qualify a field name to specify the particular field you are referencing. To qualify a field name, precede the name of the field with the name of the table, followed by a period. For example, the RepNum field in the Rep table is written as Rep.RepNum and the RepNum field in the Customer table is written as Customer.RepNum. The query design appears in Figure 3-48.

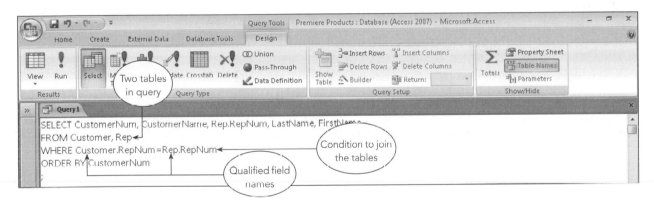

FIGURE 3-48 SQL query to join tables

The query results appear in Figure 3-49.

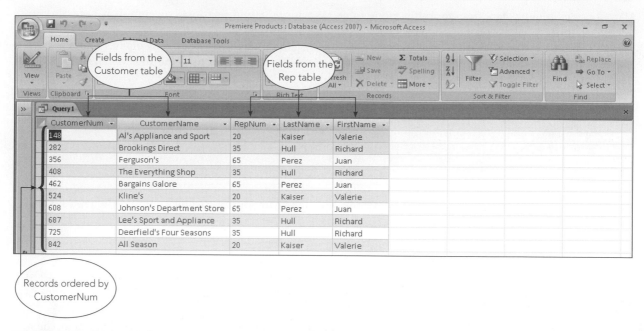

FIGURE 3-49 Query results

When there is potential ambiguity in listing field names, you *must* qualify the fields involved. It is permissible to qualify other fields as well, even if there is no possible confusion. Some people prefer to qualify all fields, which is certainly not a bad approach. In this text, however, you will qualify fields only when it is necessary to do so.

EXAMPLE 25

List the number and name of each customer whose credit limit is $10,000 together with the number, last name, and first name of the sales rep who represents the customer. Order the records by customer number.

In Example 24, the condition in the WHERE clause serves only to relate a customer to a sales rep. Although relating a customer to a sales rep is essential in this example as well, you also need to restrict the output to only those customers whose credit limit is $10,000. You can accomplish this goal by using the AND operator to create a compound condition, as shown in Figure 3-50.

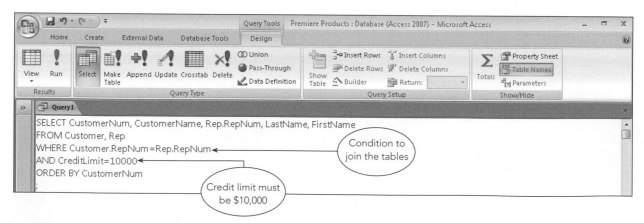

FIGURE 3-50 SQL query to restrict the records in a join

The query results appear in Figure 3-51.

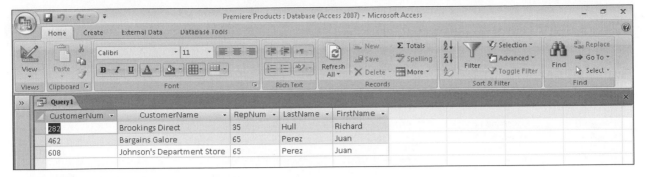

FIGURE 3-51 Query results

It is possible to join more than two tables, as illustrated in Example 26. For each pair of tables to join, you must include a condition indicating how the tables are related.

EXAMPLE 26

For every order, list the order number, order date, customer number, and customer name. In addition, for each order line within the order, list the part number, description, number ordered, and quoted price. Order the records by order number.

The order number and date are stored in the Orders table. The customer number and name are stored in the Customer table. The part number and description are stored in the Part table. The number ordered and quoted price are stored in the OrderLine table. Thus, you need to join *four* tables: Orders, Customer, Part, and OrderLine. The procedure for joining more than two tables is essentially the same as the one for joining two tables. The difference is that the condition in the WHERE clause will be a compound condition, as shown in Figure 3-52. The first condition relates an order to a customer, using the common CustomerNum columns. The second condition relates the order to an order line, using the common OrderNum columns. The final condition relates the order line to a part, using the common PartNum columns.

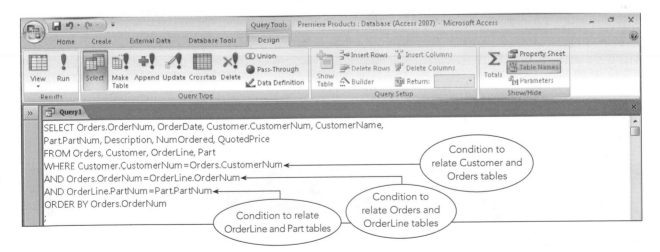

FIGURE 3-52 SQL query to join multiple tables

The query results appear in Figure 3-53.

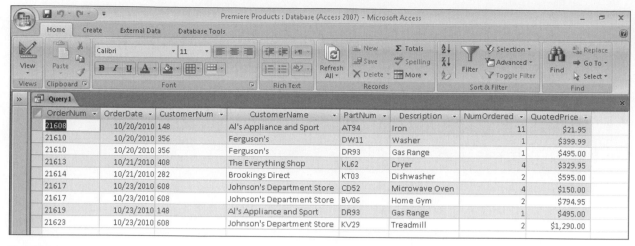

FIGURE 3-53 Query results

The query shown in Figure 3-52 is more complex than many of the previous ones. You might think that SQL is not such an easy language to use after all. If you take it one step at a time, however, you will find that the query in Example 26 isn't all that difficult. To construct a detailed query in a step-by-step fashion, do the following:

1. List in the SELECT clause all the columns you want to display. If the name of a column appears in more than one table, precede the column name with the table name (that is, qualify the column name).
2. List in the FROM clause all the tables involved in the query. Usually you include the tables that contain the columns listed in the SELECT clause. Occasionally, however, there might be a table that does not contain any columns used in the SELECT clause but that does contain columns used in the WHERE clause. In this case, you must also list the table. For example, if you do not need to list a customer number or name but you do need to list the sales rep name, you wouldn't include any columns from the Customer table in the SELECT clause. The Customer table is still required, however, because you must include columns from it in the WHERE clause.
3. Take one pair of related tables at a time and indicate in the WHERE clause the condition that relates the tables. Join these conditions with the AND operator. When there are other conditions, include them in the WHERE clause and connect them to the other conditions with the AND operator.

UNION

Recall from Chapter 2 that the union of two tables is a table containing all rows that are in the first table, the second table, or both tables. The two tables involved in a union *must* have the same structure, or be union compatible; in other words, they must have the same number of fields and their corresponding fields must have the same data types. If, for example, the first field in one table contains customer numbers, the first field in the other table also must contain customer numbers.

EXAMPLE 27

List the number and name of all customers that are represented by sales rep 35 or that currently have orders on file or both.

Because the two criteria are so different, you cannot use a simple OR criterion. Instead, you can create a table containing the number and name of all customers that are represented by sales rep 35 by selecting customer numbers and names from the Customer table in which the sales rep number is 35. Then you can create another table containing the number and name of every customer that currently has orders on file by joining the Customer and Orders tables. The two tables created by this process have the same structure—fields

Chapter 3

named CustomerNum and CustomerName. Because the tables are union compatible, it is possible to take the union of these two tables, which is the appropriate operation for this example, as shown in Figure 3-54.

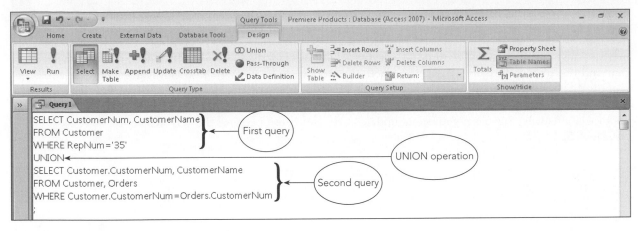

FIGURE 3-54 SQL query to perform a union

The query results appear in Figure 3-55.

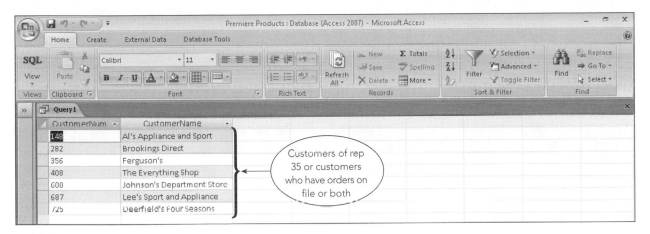

FIGURE 3-55 Query results

If an SQL implementation truly supports the union operation, it will remove any duplicate rows. For instance, any customers that are represented by sales rep 35 *and* that currently have orders on file will not appear twice. Some SQL implementations have a union operation but will not remove duplicate values.

UPDATING TABLES

There are more uses to SQL than simply retrieving data from a database and creating tables. SQL has several other capabilities, including the ability to update a database, as demonstrated in the following examples.

EXAMPLE 28

Change the street address of customer 524 to 1445 Rivard.

You can use the SQL **UPDATE** command to make changes to existing data. After the word *UPDATE*, you indicate the table to be updated. After the word *SET*, you indicate the field to be changed, followed by an equal sign and the new value. Finally, you can include a condition in the WHERE clause in which case only the records that satisfy the condition will be changed. The SQL command for this example appears in Figure 3-56. When you run this query in Access, a dialog box opens and indicates the number of records the UPDATE command will affect. In this case, you would update only one record because the WHERE clause selects customer 524.

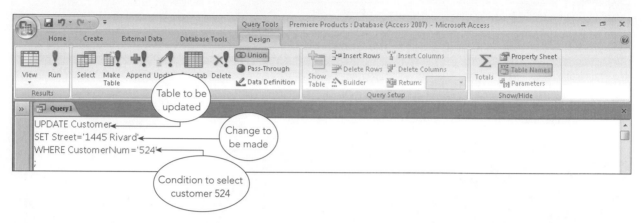

FIGURE 3-56   SQL query to update data

EXAMPLE 29

Add a new sales rep to the Rep table. Her number is 16; her name is Sharon Rands; and her address is 826 Raymond, Altonville, FL 32543. She has not yet earned any commission, but her commission rate is 5% (0.05).

To add new data to a table, you use the **INSERT** command. After the words *INSERT INTO*, you list the name of the table, followed by the word *VALUES*. Then you list the values in parentheses for each of the columns, as shown in Figure 3-57. Character values must be enclosed within single quotation marks. When you run this query in Access, a dialog box opens and indicates the number of records the INSERT command will append to the table. In this case, you would add one record to the Rep table.

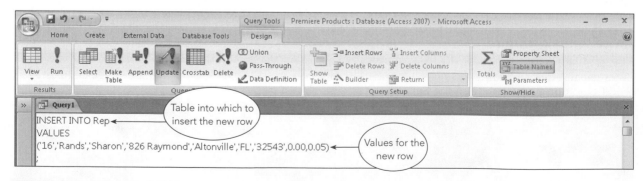

FIGURE 3-57   SQL query to insert a row

EXAMPLE 30

Delete any row in the OrderLine table in which the part number is BV06.

To delete data from the database, use the **DELETE** command, which consists of the word *DELETE* followed by a FROM clause identifying the table. Use a WHERE clause to specify a condition to select the records to delete. If you omit the condition for selecting the records to delete, when you run the query, it will delete all records from the table.

The DELETE command for this example is shown in Figure 3-58. When you run this query in Access, a dialog box opens and indicates the number of records the DELETE command will delete. In this case, you would delete only one record because the WHERE clause selects part number BV06.

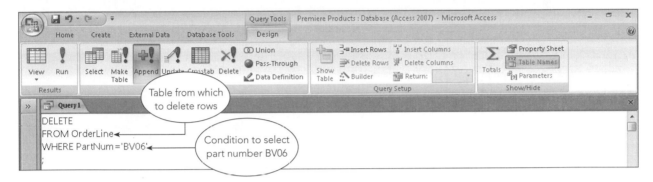

FIGURE 3-58 SQL query to delete rows

CREATING A TABLE FROM A QUERY

You can save the results of a query as a table by including the **INTO clause** in the query, as illustrated in Example 31.

EXAMPLE 31

Create a new table named SmallCust consisting of all fields from the Customer table and those rows in which the credit limit is less than or equal to $7,500.

To create the SmallCust table, create a query to select all fields from the Customer table, include a WHERE clause to restrict the rows to those in which CreditLimit <= 7500, and include an INTO clause. The INTO clause precedes the FROM clause and consists of the word *INTO* followed by the name of the table to be created. The query appears in Figure 3-59a. When you run this query in Access, a dialog box opens and indicates the number of records the INTO clause will paste into the new table. In this case, you would add six rows to the SmallCust table.

MySQL does not support the query shown in Figure 3-59a. To accomplish the same task, you would create the SmallCust table using a CREATE TABLE command. Then you would use an INSERT command to insert the appropriate data into the SmallCust table, as shown in Figure 3-59b.

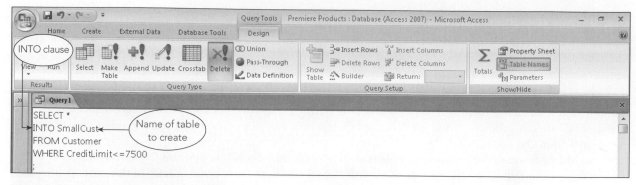

FIGURE 3-59a Query to create a new table (Access)

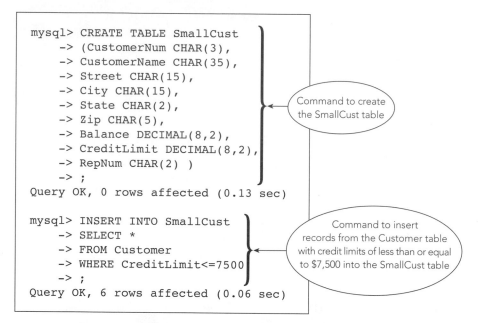

FIGURE 3-59b Query to create a new table (MySQL)

After you execute this query, you can use the SmallCust table shown in Figure 3-60, which is just like any other table you created using the CREATE TABLE command.

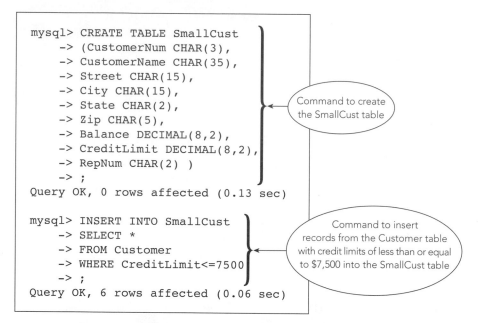

FIGURE 3-60 SmallCust table created by query

SUMMARY OF SQL COMMANDS

This section contains generic versions of SQL commands for every example presented in this chapter. (The example numbers match the ones used in the chapter, making it easy to return to the page in the chapter on which the example is described.) In most cases, commands in Access are identical to the generic versions. For those commands that differ in other SQL implementations, both the generic version and the Access version are included.

EXAMPLE 1

Use SQL to create the Rep table by describing its layout.

```
CREATE TABLE Rep
(RepNum  CHAR(2),
LastName  CHAR(15),
FirstName  CHAR(15),
Street  CHAR(15),
City  CHAR(15),
State  CHAR(2),
Zip  CHAR(5),
Commission  DECIMAL(7,2),
Rate  DECIMAL(3,2) )
;
```

Access:

```
CREATE TABLE Rep
(RepNum  CHAR(2),
LastName  CHAR(15),
FirstName  CHAR(15),
Street  CHAR(15),
City  CHAR(15),
State  CHAR(2),
Zip  CHAR(5),
Commission  CURRENCY,
Rate  NUMBER )
;
```

EXAMPLE 2

List the number, name, and balance of all customers.

```
SELECT CustomerNum, CustomerName, Balance
FROM Customer
;
```

EXAMPLE 3

List the complete Part table.

```
SELECT *
FROM Part
;
```

EXAMPLE 4

List the name of every customer with a $10,000 credit limit.

```
SELECT CustomerName
FROM Customer
WHERE CreditLimit=10000
;
```

EXAMPLE 5

Find the name of customer 148.

```
SELECT CustomerName
FROM Customer
WHERE CustomerNum='148'
;
```

EXAMPLE 6

Find the customer name for every customer located in the city of Grove.

```
SELECT CustomerName
FROM Customer
WHERE City='Grove'
;
```

EXAMPLE 7

List the number, name, credit limit, and balance for all customers with credit limits that exceed their balances.

```
SELECT CustomerNum, CustomerName, CreditLimit, Balance
FROM Customer
WHERE CreditLimit>Balance
;
```

EXAMPLE 8

List the descriptions of all parts that are located in warehouse 3 and for which there are more than 20 units on hand.

```
SELECT Description
FROM Part
WHERE Warehouse='3'
AND OnHand>20
;
```

EXAMPLE 9

List the descriptions of all parts that are located in warehouse 3 or for which there are more than 20 units on hand or both.

```
SELECT Description
FROM Part
WHERE Warehouse='3'
OR OnHand>20
;
```

EXAMPLE 10

List the descriptions of all parts that are not in warehouse 3.

```
SELECT Description
FROM Part
WHERE NOT Warehouse='3'
;
```

EXAMPLE 11

List the number, name, and balance of all customers with balances greater than or equal to $1,000 and less than or equal to $5,000.

```
SELECT CustomerNum, CustomerName, Balance
FROM Customer
WHERE Balance BETWEEN 1000 AND 5000
;
```

EXAMPLE 12

List the number, name, and available credit for all customers.

```
SELECT CustomerNum, CustomerName, CreditLimit-Balance
AS AvailableCredit
FROM Customer
;
```

EXAMPLE 13

List the number, name, and available credit for all customers with credit limits that exceed their balances.

```
SELECT CustomerNum, CustomerName, CreditLimit-Balance
AS AvailableCredit
FROM Customer
WHERE CreditLimit>Balance
;
```

EXAMPLE 14

List the number, name, and complete address of every customer located on a street that contains the letters *Oxford*.

```
SELECT CustomerNum, CustomerName, Street, City, State, Zip
FROM Customer
WHERE Street LIKE '%Oxford%'
;
```

Access:

```
SELECT CustomerNum, CustomerName, Street, City, State, Zip
FROM Customer
WHERE Street LIKE '*Oxford*'
;
```

EXAMPLE 15

List the number, name, street, and credit limit for every customer with a credit limit of $7,500, $10,000, or $15,000.

```
SELECT CustomerNum, CustomerName, Street, CreditLimit
FROM Customer
WHERE CreditLimit IN (7500, 10000, 15000)
;
```

EXAMPLE 16

List the number, name, street, and credit limit of all customers. Order (sort) the customers by name.

```
SELECT CustomerNum, CustomerName, Street, CreditLimit
FROM Customer
ORDER BY CustomerName
;
```

EXAMPLE 17

List the number, name, street, and credit limit of all customers. Order the customers by name within descending credit limit.

```
SELECT CustomerNum, CustomerName, Street, CreditLimit
FROM Customer
ORDER BY CreditLimit DESC, CustomerName
;
```

EXAMPLE 18

How many parts are in item class HW?

```
SELECT COUNT(*)
FROM Part
WHERE Class='HW'
;
```

EXAMPLE 19

Find the number of customers and the total of their balances.

```
SELECT COUNT(*), SUM(Balance)
FROM Customer
;
```

EXAMPLE 20

Find the total number of customers and the total of their balances. Change the column names for the number of customers and the total of their balances to CustomerCount and BalanceTotal, respectively.

```
SELECT COUNT(*) AS CustomerCount, SUM(Balance) AS BalanceTotal
From Customer
;
```

EXAMPLE 21

List the order number for each order that contains an order line for a part located in warehouse 3.

```
SELECT OrderNum
FROM OrderLine
WHERE PartNum IN
(SELECT PartNum
FROM Part
WHERE Warehouse='3')
;
```

EXAMPLE 22

For each sales rep, list the rep number, the number of customers assigned to the rep, and the average balance of the rep's customers. Group the records by rep number and order the records by rep number.

```
SELECT RepNum, COUNT(*), AVG(Balance)
FROM Customer
GROUP BY RepNum
ORDER BY RepNum
;
```

EXAMPLE 23

For each sales rep with fewer than four customers, list the rep number, the number of customers assigned to the rep, and the average balance of the rep's customers. Rename the count of the number of customers and the average of the balances to NumCustomers and AverageBalance, respectively. Order the groups by rep number.

```
SELECT RepNum, COUNT(*) AS NumCustomers, AVG(Balance)
AS AverageBalance
FROM Customer
GROUP BY RepNum
HAVING COUNT(*)<4
ORDER BY RepNum
;
```

EXAMPLE 24

List the number and name of each customer together with the number, last name, and first name of the sales rep who represents the customer. Order the records by customer number.

```
SELECT CustomerNum, CustomerName, Rep.RepNum, LastName, FirstName
FROM Customer, Rep
WHERE Customer.RepNum=Rep.RepNum
ORDER BY CustomerNum
;
```

EXAMPLE 25

List the number and name of each customer whose credit limit is $10,000 together with the number, last name, and first name of the sales rep who represents the customer. Order the records by customer number.

```
SELECT CustomerNum, CustomerName, Rep.RepNum, LastName, FirstName
FROM Customer, Rep
WHERE Customer.RepNum=Rep.RepNum
AND CreditLimit=10000
ORDER BY CustomerNum
;
```

EXAMPLE 26

For every order, list the order number, order date, customer number, and customer name. In addition, for each order line within the order, list the part number, description, number ordered, and quoted price. Order the records by order number.

```
SELECT Orders.OrderNum, OrderDate, Customer.CustomerNum,
    CustomerName, Part.PartNum, Description, NumOrdered, QuotedPrice
FROM Orders, Customer, OrderLine, Part
WHERE Customer.CustomerNum=Orders.CustomerNum
AND Orders.OrderNum=OrderLine.OrderNum
AND OrderLine.PartNum=Part.PartNum
ORDER BY Orders.OrderNum
;
```

EXAMPLE 27

List the number and name of all customers that are represented by sales rep 35 or that currently have orders on file or both.

```
SELECT CustomerNum, CustomerName
FROM Customer
WHERE RepNum='35'
UNION
SELECT Customer.CustomerNum, CustomerName
FROM Customer, Orders
WHERE Customer.CustomerNum=Orders.CustomerNum
;
```

EXAMPLE 28

Change the street address of customer 524 to 1445 Rivard.

```
UPDATE Customer
SET Street='1445 Rivard'
WHERE CustomerNum='524'
;
```

EXAMPLE 29

Add a new sales rep to the Rep table. Her number is 16; her name is Sharon Rands; and her address is 826 Raymond, Altonville, FL 32543. She has not yet earned any commission, but her commission rate is 5% (0.05).

```
INSERT INTO Rep
VALUES
('16','Rands','Sharon','826 Raymond','Altonville','FL','32543',0.00,0.05)
;
```

EXAMPLE 30

Delete any row in the OrderLine table in which the part number is BV06.

```
DELETE
FROM OrderLine
WHERE PartNum='BV06'
;
```

EXAMPLE 31

Create a new table named SmallCust consisting of all fields from the Customer table and those rows in which the credit limit is less than or equal to $7,500.

```
SELECT *
INTO SmallCust
FROM Customer
WHERE CreditLimit<=7500
;
```

MySQL:

```
CREATE TABLE SmallCust
(CustomerNum CHAR(3),
CustomerName CHAR(35),
Street CHAR(15),
City CHAR(15),
State CHAR(2),
Zip CHAR(5),
Balance DECIMAL(8,2),
CreditLimit DECIMAL(8,2),
RepNum CHAR(2) )
;

INSERT INTO SmallCust
SELECT *
FROM Customer
WHERE CreditLimit<=7500
;
```

Summary

- Structured Query Language (SQL) is a language that is used to manipulate relational databases.
- The basic form of an SQL query is SELECT-FROM-WHERE.
- Use the CREATE TABLE command to describe a table's layout to the DBMS, which creates the table in the databse.
- In SQL retrieval commands, fields are listed after SELECT, tables are listed after FROM, and conditions are listed after WHERE.
- In conditions, character values must be enclosed in single quotation marks.
- Compound conditions are formed by combining simple conditions using either or both of the following operators: AND and OR.
- Sorting is accomplished using the ORDER BY clause. The field on which the records are sorted is called the sort key. When the data is sorted in more than one field, the more important field is called the major sort key or primary sort key. The less important field is called the minor sort key or secondary sort key.
- Grouping is accomplished in SQL by using the GROUP BY clause. To restrict the rows to be displayed, use the HAVING clause.
- Joining tables is accomplished in SQL by using a condition that relates matching rows in the tables to be joined.
- SQL has the built-in (also called aggregate) functions COUNT, SUM, AVG, MAX, and MIN.
- One SQL query can be placed inside another. The subquery is evaluated first.
- The union of the results of two queries is specified by placing the UNION operator between the two queries.
- Calculated fields are specified in SQL queries by including the calculation, followed by the word *AS*, followed by the name of the calculated field.
- The INSERT command is used to add a new row to a table.
- The UPDATE command is used to change existing data.
- The DELETE command is used to delete records.
- The INTO clause is used in a SELECT command to create a table containing the results of the query.

Key Terms

CHAR(*n*)	INTO clause
command	ORDER BY clause
compound condition	reserved word
CREATE TABLE	SELECT clause
DATE	simple condition
DECIMAL(*p*,*q*)	SMALLINT
DELETE	SQL (Structured Query Language)
FROM clause	statement history
GROUP BY clause	subquery
HAVING clause	UPDATE
INSERT	WHERE clause
INTEGER	

Review Questions

1. Describe the process of creating a table in SQL and the different data types you can use for fields.
2. What is the purpose of the WHERE clause in SQL? Which comparison operators can you use in a WHERE clause?
3. How do you write a compound condition in an SQL query? When is a compound condition true?

4. What is a computed field? How can you use one in an SQL query? How do you assign a name to a computed field?

5. How do you use the LIKE and IN operators in an SQL query?

6. How do you sort data in SQL? When there is more than one sort key, how do you indicate which one is the major sort key? How do you sort data in descending order?

7. What are the SQL built-in functions? How do you use them in an SQL query?

8. What is a subquery? When is a subquery executed?

9. How do you group data in SQL? When you group data in SQL, are there any restrictions on the items that you can include in the SELECT clause? Explain.

10. How do you join tables in SQL?

11. How do you qualify the name of a field in an SQL query? When is it necessary to do so?

12. How do you take the union of two tables in SQL? What criteria must the tables meet to make a union possible?

13. Describe the three update commands in SQL.

14. How do you save the results of an SQL query as a table?

Premiere Products Exercises

In the following exercises, you will use the data in the Premiere Products database shown in Figure 2-1 in Chapter 2. (If you use a computer to complete these exercises, use a copy of the original Premiere Products database so you will still have the original data when you complete Chapter 4.) In each step, use SQL to obtain the desired results. You can use a DBMS to complete the exercises using a computer, or you can simply write the SQL command to complete each step. Check with your instructor if you are uncertain about which approach to take.

1. List the number and name of all customers.

2. List the complete Part table.

3. List the number and name of every customer represented by sales rep 35.

4. List the number and name of all customers that are represented by sales rep 35 and that have credit limits of $10,000.

5. List the number and name of all customers that are represented by sales rep 35 or that have credit limits of $10,000.

6. For each order, list the order number, order date, number of the customer that placed the order, and name of the customer that placed the order.

7. List the number and name of all customers represented by Juan Perez.

8. How many orders were placed on 10/20/2010?

9. Find the total of the balances for all customers represented by sales rep 35.

10. Give the part number, description, and on-hand value (OnHand * Price) for each part in item class HW.

11. List all columns and all rows in the Part table. Sort the results by part description.

12. List all columns and all rows in the Part table. Sort the results by part number within item class.

13. List the item class and the sum of the number of units on hand. Group the results by item class.

14. Create a new table named SportingGoods to contain the columns PartNum, Description, OnHand, Warehouse, and Price for all rows in which the item class is SG.

15. In the SportingGoods table, change the description of part BV06 to "Fitness Gym."

16. In the SportingGoods table, delete every row in which the price is greater than $1,000.

Henry Books Case

Ray Henry is very aware of the importance of the SQL language in database management. He realizes that he can use SQL to perform the same functions that you performed with queries in Chapter 2. In each of the following steps, use SQL to obtain the desired results using the data shown in Figures 1-17 through 1-20 in Chapter 1. (If you use a computer to complete these exercises, use a copy of the original Henry Books database so you will still have the original data when you complete Chapter 4.) You can use a DBMS to complete the exercises using a computer, or you can simply write the SQL command to complete each step. Check with your instructor if you are uncertain about which approach to take.

1. List the author number and last name for every author.

2. List the complete Branch table (all rows and all columns).

3. List the name of every publisher located in Boston.

4. List the name of every publisher not located in Boston.

5. List the name of every branch that has at least nine employees.

6. List the book code and title of every book that has the type SFI.

7. List the book code and title of every book that has the type SFI and that is a paperback.

8. List the book code and title of every book that has the type SFI or that has the publisher code PE.

9. List the book code, title, and price for each book with a price that is greater than or equal to $5 but less than or equal to $10.

10. List the book code and title of every book that has the type FIC and a price of less than $10.

11. Customers that are part of a special program get a 15% discount off regular book prices. To determine the discounted prices, list the book code, title, and discounted price of every book. (Your calculated column should calculate 85% of the current price, which is 100% less a 15% discount.)

12. List the book code and title of every book that has the type SFI, HOR, or ART.

13. List the book code, title, and publisher code for all books. Sort the results by title within publisher code.

14. How many books have the type SFI?

15. Calculate the average price for each type of book.

16. For every book, list the book code, book title, publisher code, and publisher name.

17. For every book published by Taunton Press, list the book title and book price.

18. List the book title and book code for every book published by Putnam Publishing Group that has a book price greater than $15.

19. Create a new table named Fiction using the data in the BookCode, Title, PublisherCode, and Price columns in the Book table for those books that have the type FIC.

20. Use an update query to change the price of any book in the Fiction table with a current price of $14.00 to $14.50.

21. Use a delete query to delete all books in the Fiction table that have the publisher code VB.

Alexamara Marina Group Case

In the following exercises, you will use the data in the Alexamara Marina Group database shown in Figures 1-20 through 1-24 in Chapter 1. (If you use a computer to complete these exercises, use a copy of the Alexamara Marina Group database so you will still have the original data when you complete Chapter 4.) In each step, use SQL to obtain the desired results. You can use the query feature in a DBMS to complete the exercises using a computer, or you can simply write the SQL command to complete each step. Check with your instructor if you are uncertain about which approach to take.

1. List the owner number, last name, and first name of every boat owner.

2. List the complete Marina table (all rows and all columns).

3. List the last name and first name of every owner located in Bowton.

4. List the last name and first name of every owner not located in Bowton.

5. List the marina number and slip number for every slip whose length is equal to or less than 30 feet.

6. List the marina number and slip number for every boat with the type Dolphin 28.

7. List the slip number for every boat with the type Dolphin 28 that is located in marina 1.

8. List the boat name for each boat located in a slip whose length is between 25 and 30 feet.

9. List the slip number for every slip in marina 1 whose annual rental fee is less than $3,000.

10. Labor is billed at the rate of $60 per hour. List the slip ID, category number, estimated hours, and estimated labor cost for every service request. To obtain the estimated labor cost, multiply the estimated hours by 60. Use the column name EstimatedCost for the estimated labor cost.

11. List the marina number and slip number for all slips containing a boat with the type Sprite 4000, Sprite 3000, or Ray 4025.

12. List the marina number, slip number, and boat name for all boats. Sort the results by boat name within the marina number.

13. How many Dolphin 28 boats are stored at both marinas?

14. Calculate the total rental fees Alexamara receives each year based on the length of the slip.

15. For every boat, list the marina number, slip number, boat name, owner number, owner's first name, and owner's last name.

16. For every completed or open service request for routine engine maintenance, list the slip ID, description, and status.

17. For every service request for routine engine maintenance, list the slip ID, marina number, slip number, estimated hours, spent hours, owner number, and owner's last name.

18. Create a new table named LargeSlip using the data in the MarinaNum, SlipNum, RentalFee, BoatName, and OwnerNum columns in the MarinaSlip table for slips with lengths of 40 feet.

19. Use an update query to change the rental fee of any slip in the LargeSlip table whose fee is currently $3,800 to $3,900.

20. Use a delete query to delete slips in the LargeSlip table whose rental fee is $3,600.

CHAPTER 4

THE RELATIONAL MODEL 3: ADVANCED TOPICS

LEARNING OBJECTIVES

Objectives

- Define, describe, and use views
- Use indexes to improve database performance
- Examine the security features of a DBMS
- Discuss entity, referential, and legal-values integrity
- Make changes to the structure of a relational database
- Define and use the system catalog
- Discuss stored procedures and triggers

INTRODUCTION

In Chapter 3, you used SQL to define and manipulate table data. In this chapter, you will investigate some other aspects of the relational model. You will learn about views, which represent a way of giving each user his or her own view of the data in a database. You will examine indexes and use them to improve database performance. You also will investigate the features of a DBMS that provide security. Then you will learn about important integrity rules and examine ways to change the structure of a database. You will use the system catalog found in many relational DBMSs to provide users with information about the structure of a database. Finally, you will examine the use of stored procedures and triggers.

> **NOTE**
>
> In this chapter, concepts are introduced using SQL and followed by the method you would use to accomplish the same task in Microsoft Access. Unless otherwise specified, the SQL commands in this chapter function in MySQL exactly as indicated.

> **NOTE**
>
> If you plan to work through the examples in this chapter using a computer, you should use a copy of the original Premiere Products database because the version of the database used in this chapter does not include the changes made in Chapter 3.

VIEWS

Most DBMSs support the creation of views. A **view** is an application program's or an individual user's picture of the database. An individual can use a view to create reports, charts, and other objects that show database data. In many cases, a user can use a view to examine table data as well. Because a view is usually less involved than the full database, its use can represent a great simplification. Views also provide a measure of security because omitting sensitive tables or fields from a view will render them unavailable to anyone who is accessing the database via that view.

To illustrate the idea of a view, suppose Juan is interested in the part number, part description, units on hand, and unit price for Premiere Products parts that are in class HW. He is not interested in any of the other fields in the Part table, nor is he interested in any of the rows that correspond to parts in other item classes. Viewing this data would be simpler for Juan if the other rows and fields were not even present.

Although you cannot change the structure of the Part table and omit some of its rows just for Juan, you can do the next best thing. You can provide him with a view that consists of precisely the rows and fields he needs to access. Using SQL, the following CREATE VIEW command creates the view that Juan can use to see the data he needs.

```
CREATE VIEW Housewares AS
SELECT PartNum, Description, OnHand, Price
FROM Part
WHERE Class='HW'
;
```

The SELECT command that creates the view, which is called the **defining query**, indicates what to include in the view. Conceptually, given the current data in the Premiere Products database, this view will contain the data shown in Figure 4-1. The data does not really exist in this form, however; nor will it *ever* exist in this form. It is tempting to think that when this view is used, the query is executed and will produce some sort of temporary table named Housewares, which Juan then could access; but that is *not* what happens.

Housewares

PartNum	Description	OnHand	Price
AT94	Iron	50	$24.95
DL71	Cordless Drill	21	$129.95
FD21	Stand Mixer	22	$159.95

FIGURE 4-1 Housewares view

Instead, the query acts as a sort of window into the database, as shown in Figure 4-2. As far as Juan is concerned, the entire database is just the darker shaded portion of the Part table. Juan can see any change that affects the darker portion of the Part table, but he is totally unaware of any other changes that are made in the database.

Rep

RepNum	LastName	FirstName	Street	City	State	Zip	Commission	Rate
20	Kaiser	Valerie	624 Randall	Grove	FL	33321	$20,542.50	0.05
35	Hull	Richard	532 Jackson	Sheldon	FL	33553	$39,216.00	0.07
65	Perez	Juan	1626 Taylor	Fillmore	FL	33336	$23,487.00	0.05

Customer

CustomerNum	CustomerName	Street	City	State	Zip	Balance	CreditLimit	RepNum
148	Al's Appliance and Sport	2837 Greenway	Fillmore	FL	33336	$6,550.00	$7,500.00	20
282	Brookings Direct	3827 Devon	Grove	FL	33321	$431.50	$10,000.00	35
356	Ferguson's	382 Wildwood	Northfield	FL	33146	$5,785.00	$7,500.00	65
408	The Everything Shop	1828 Raven	Crystal	FL	33503	$5,285.25	$5,000.00	35
462	Bargains Galore	3829 Central	Grove	FL	33321	$3,412.00	$10,000.00	65
524	Kline's	838 Ridgeland	Fillmore	FL	33336	$12,762.00	$15,000.00	20
608	Johnson's Department Store	372 Oxford	Sheldon	FL	33553	$2,106.00	$10,000.00	65
687	Lee's Sport and Appliance	282 Evergreen	Altonville	FL	32543	$2,851.00	$5,000.00	35
725	Deerfield's Four Seasons	282 Columbia	Sheldon	FL	33553	$248.00	$7,500.00	35
842	All Season	28 Lakeview	Grove	FL	33321	$8,221.00	$7,500.00	20

Orders

OrderNum	OrderDate	CustomerNum
21608	10/20/2010	148
21610	10/20/2010	356
21613	10/21/2010	408
21614	10/21/2010	282
21617	10/23/2010	608
21619	10/23/2010	148
21623	10/23/2010	608

OrderLine

OrderNum	PartNum	NumOrdered	QuotedPrice
21608	AT94	11	$21.95
21610	DR93	1	$495.00
21610	DW11	1	$399.99
21613	KL62	4	$329.95
21614	KT03	2	$595.00
21617	BV06	2	$794.95
21617	CD52	4	$150.00
21619	DR93	1	$495.00
21623	KV29	2	$1,290.00

Part

PartNum	Description	OnHand	Class	Warehouse	Price
AT94	Iron	50	HW	3	$24.95
BV06	Home Gym	45	SG	2	$794.95
CD52	Microwave Oven	32	AP	1	$165.00
DL71	Cordless Drill	21	HW	3	$129.95
DR93	Gas Range	8	AP	2	$495.00
DW11	Washer	12	AP	3	$399.99
FD21	Stand Mixer	22	HW	3	$159.95
KL62	Dryer	12	AP	1	$349.95
KT03	Dishwasher	8	AP	3	$595.00
KV29	Treadmill	9	SG	2	$1,390.00

FIGURE 4-2   Housewares view of the Premiere Products database

When you create a query that involves a view, the DBMS changes the query to one that selects data from the table(s) in the database that created the view. Suppose, for example, Juan creates the following query:

```
SELECT *
FROM Housewares
WHERE OnHand<25
;
```

The DBMS does *not* execute the query in this form. Instead, it merges the query Juan entered with the query that defines the view to form the query that is actually executed. When the DBMS merges the query that creates the view with the query to select rows where the OnHand value is less than 25, the query that the DBMS actually executes is as follows:

```
SELECT PartNum, Description, OnHand, Price
FROM Part
WHERE Class='HW'
AND OnHand<25
;
```

In the query that the DBMS executes, the FROM clause lists the Part table rather than the Housewares view, the SELECT clause lists fields from the Part table instead of * to select all fields from the Housewares view, and the WHERE clause contains a compound condition to select only those parts in the HW class (as Juan sees in the Housewares view) and only those parts with OnHand values of less than 25.

Juan, however, is unaware that this kind of activity is taking place. To Juan, it seems as though he is using a table named Housewares. One advantage of this approach is that because the Housewares view never exists in its own right, any update to the Part table is *immediately* available in Juan's Housewares view. If the Housewares view were really a table, that would not be the case.

To create a view in Access, you simply create and save a query. For example, to create the Housewares view, you would include the PartNum, Description, OnHand, and Price fields from the Part table. You would also include the Class field in the design grid and enter HW as the criterion. Because the Class field isn't included in the view, you would remove the check mark from the Class field's Show check box. Finally, you would save the query using the name Housewares, as shown in Figure 4-3.

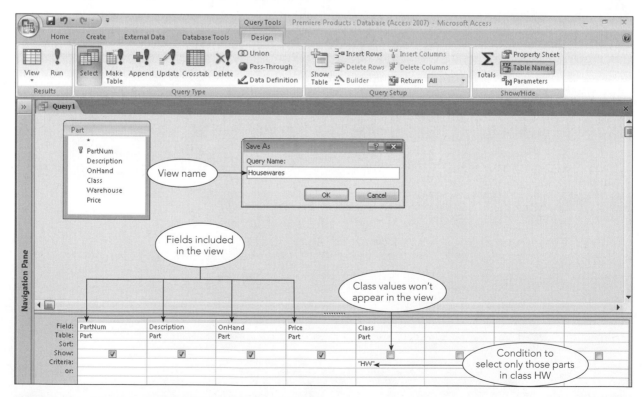

FIGURE 4-3   Access query design of the Housewares view

After creating the view, you can use it right away. Figure 4-4 shows the data in the Housewares view. You can create a form for the view, base a report on the view, and treat the view as though it were a table.

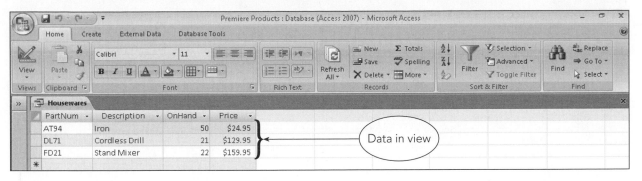

FIGURE 4-4   Housewares view datasheet

What if Juan wanted different names for the fields? You can use SQL to change the field names in a view by including the new field names in the CREATE VIEW command. For example, if Juan wanted the names of the PartNum, Description, OnHand, and Price fields to be PNum, PDesc, OnHd, and Price, respectively, the CREATE VIEW command would be as follows:

```
CREATE VIEW Housewares (PNum, PDesc, OnHd, Price) AS
SELECT PartNum, Description, OnHand, Price
FROM Part
WHERE Class='HW'
;
```

Now when Juan accesses the Housewares view, he uses the field names PNum, PDesc, OnHd, and Price rather than PartNum, Description, OnHand, and Price, respectively.

In Access, you can change the field names by preceding the name of the field with the desired name, followed by a colon, as shown in Figure 4-5.

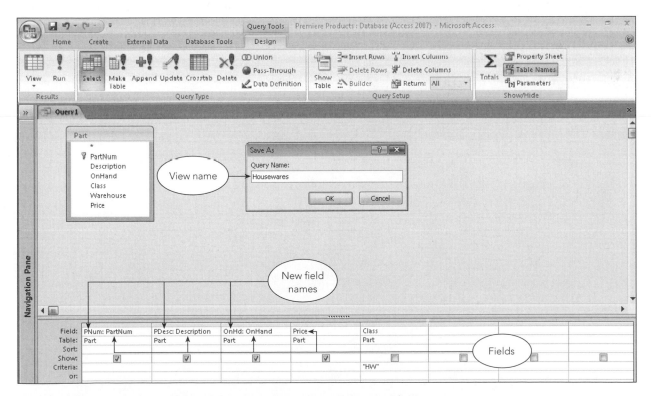

FIGURE 4-5   Access query design of the Housewares view with changed field names

In the query results shown in Figure 4-6, the column headings are PNum, PDesc, OnHd, and Price.

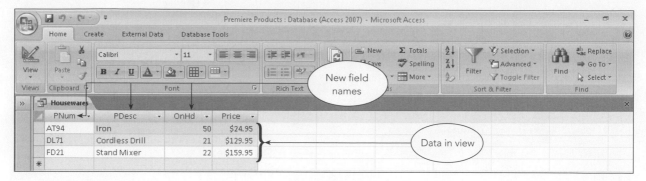

FIGURE 4-6 Datasheet for the Housewares view with changed field names

The Housewares view is an example of a **row-and-column subset view** beause it consists of a subset of the rows and columns in some individual table, which, in this case, is the Part table. Because the query can be any SQL query, a view can also join two or more tables.

Suppose, for example, Francesca needs to know the number and name of each sales rep, along with the number and name of the customers represented by each sales rep. It would be much simpler for her if this information were stored in a single table instead of in two tables that she would have to join together. She would like a single table that contains the sales rep number, sales rep name, customer number, and customer name. Suppose she would also like these fields to be named SNum, SLast, SFirst, CNum, and CName, respectively. She could use a join in the CREATE VIEW command as follows:

```
CREATE VIEW SalesCust (SNum, SLast, SFirst, CNum, CName) AS
SELECT Rep.RepNum, LastName, FirstName, CustomerNum, CustomerName
FROM Rep, Customer
WHERE Rep.RepNum=Customer.RepNum
;
```

Given the current data in the Premiere Products database, conceptually this view is the table shown in Figure 4-7.

SalesCust

SNum	SLast	SFirst	CNum	CName
20	Kaiser	Valerie	148	Al's Appliance and Sport
20	Kaiser	Valerie	524	Kline's
20	Kaiser	Valerie	842	All Season
35	Hull	Richard	282	Brookings Direct
35	Hull	Richard	408	The Everything Shop
35	Hull	Richard	687	Lee's Sport and Appliance
35	Hull	Richard	725	Deerfield's Four Seasons
65	Perez	Juan	356	Ferguson's
65	Perez	Juan	462	Bargains Galore
65	Perez	Juan	608	Johnson's Department Store

FIGURE 4-7 SalesCust view

To Francesca, the SalesCust view is a real table; she does not need to know what goes on behind the scenes in order to use it. She could find the number and name of the sales rep who represents customer 282, for example, by using the following query:

```
SELECT SNum, SLast, SFirst
FROM SalesCust
WHERE CNum='282'
;
```

Francesca is completely unaware that, behind the scenes, the DBMS converts her query as follows:

```
SELECT Rep.RepNum AS SNum, LastName AS SLast, FirstName AS SFirst
FROM Rep, Customer
WHERE Rep.RepNum=Customer.RepNum
AND CustomerNum='282'
;
```

In Access, the query for the SalesCust view appears in Figure 4-8.

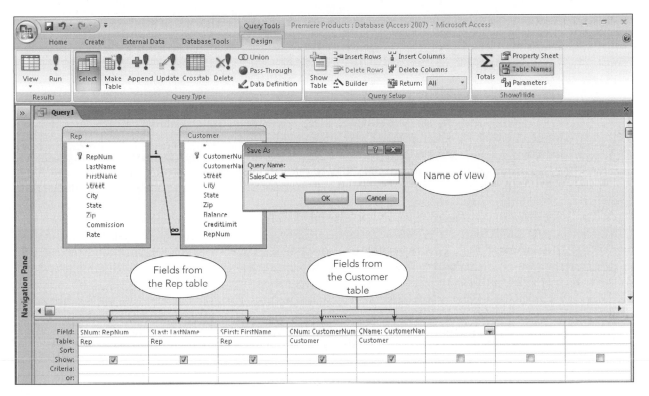

FIGURE 4-8   Access query design of the SalesCust view

The datasheet for the SalesCust view appears in Figure 4-9.

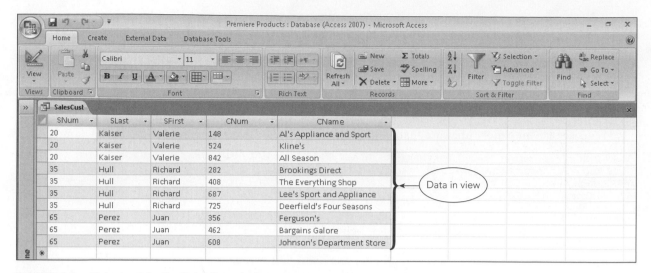

FIGURE 4-9 Datasheet for the SalesCust view

The use of views provides several advantages:

- Views provide data independence. If the database structure changes (by fields being added or relationships changing between tables, for example) in such a way that the view can still be derived from existing data, the user can still access and use the same view. If adding extra fields to tables in the database is the only change and these fields are not required by the view's user, the defining query may not even need to be changed for the user to continue using the view. If relationships are changed, the defining query may be different; but because users need not be aware of the defining query, this difference is unknown to them. They continue accessing the database through the same view as though nothing has changed.
- Because each user has his or her own view, different users can view the same data in different ways.
- A view should contain only those fields required by a given user. This practice has two advantages. First, because the view will, in all probability, contain fewer fields than the overall database and the view is conceptually a single table, rather than a collection of tables, it greatly simplifies the user's perception of the database. Second, views provide a measure of security. Fields that are not included in the view are not accessible to the view's user. For example, omitting the Balance field from a view will ensure that a user of the view cannot access any customer's balance. Likewise, rows that are not included in the view are not accessible. A user of the Housewares view, for example, cannot obtain any information about parts in the AP or SG item classes.

INDEXES

If you want to find a discussion of a given topic in a book, you could scan the entire book from start to finish, looking for references to the topic you had in mind. More than likely, however, you wouldn't have to resort to this technique. If the book had a good index, you could use it to quickly identify the pages on which your topic is discussed.

Within relational model systems on both mainframes and personal computers, the main mechanism for increasing the efficiency with which data is retrieved from the database is the **index**. Conceptually, these indexes are very much like the index in a book. Consider Figure 4-10, for example, which shows the Customer table for Premiere Products together with one extra field, RecordNum. This extra field gives the location of the record in the file. (Customer 148 is the first record in the table and is on record 1, customer 282 is on record 2, and so on.) These record numbers are automatically assigned and used by the DBMS, not by the users, which is why you do not normally see them. For illustrative purposes, Figure 4-10 includes a RecordNum column to show how an index works.

Customer

RecordNum	CustomerNum	CustomerName	...	Balance	CreditLimit	RepNum
1	148	Al's Appliance and Sport	...	$6,550.00	$7,500.00	20
2	282	Brookings Direct	...	$431.50	$10,000.00	35
3	356	Ferguson's	...	$5,785.00	$7,500.00	65
4	408	The Everything Shop	...	$5,285.25	$5,000.00	35
5	462	Bargains Galore	...	$3,412.00	$10,000.00	65
6	524	Kline's	...	$12,762.00	$15,000.00	20
7	608	Johnson's Department Store	...	$2,106.00	$10,000.00	65
8	687	Lee's Sport and Appliance	...	$2,851.00	$5,000.00	35
9	725	Deerfield's Four Seasons	...	$248.00	$7,500.00	35
10	842	All Season	...	$8,221.00	$7,500.00	20

FIGURE 4-10   Customer table with record numbers

To rapidly access a customer's record on the basis of his or her record number, you might choose to create and use an index, as shown in Figure 4-11.

CustomerNum Index

CustomerNum	RecordNum
148	1
282	2
356	3
408	4
462	5
524	6
608	7
687	8
725	9
842	10

FIGURE 4-11   Index for the Customer table on the CustomerNum field

The index has two fields. The first field contains a customer number, and the second field contains the number of the record on which the customer number is found. Because customer numbers are unique, there is only a single corresponding record number in this index. That is not always the case, however. Suppose, for example, you wanted to quickly access all customers with a specific credit limit or all customers that are represented by a specific sales rep. You might choose to create and use an index on credit limit as well as an index on sales rep number. These two indexes are shown in Figure 4-12.

CreditLimit Index

CreditLimit	RecordNum
$5,000	4, 8
$7,500	1, 3, 9, 10
$10,000	2, 5, 7
$15,000	6

RepNum Index

RepNum	RecordNum
20	1, 6, 10
35	2, 4, 8, 9
65	3, 5, 7

FIGURE 4-12 Indexes for the Customer table on the CreditLimit and RepNum fields

By examining the CreditLimit index in Figure 4-12, you can see that each credit limit occurs in the index along with the numbers of the records on which that credit limit occurs. Credit limit $7,500, for example, occurs on records 1, 3, 9, and 10. Further, the credit limits appear in the index in numerical order. If the DBMS uses this index to find those records on which the credit limit is $10,000, for example, it could scan the credit limits in the index to find $10,000. After doing that, it would determine the corresponding record numbers (2, 5, and 7) and then immediately go to those records in the Customer table, finding these customers more quickly than if it had to scan the entire Customer table one record at a time. Thus, indexes can make the process of retrieving records fast and efficient.

> **NOTE**
>
> With relatively small tables, the increased efficiency associated with indexes will not be readily apparent. In practice, it is common to encounter tables with thousands, tens of thousands, or even hundreds of thousands of records. In such cases, the increase in efficiency is dramatic. In fact, without indexes, many operations in such databases would simply not be practical—they would take too long to complete.

The field or combination of fields on which the index is built is called the **index key**. In the index shown in Figure 4-11, the index key is CustomerNum; in the indexes shown in Figure 4-12, the index keys are CreditLimit and RepNum. The index key for an index can be any field or combination of fields in any table.

After creating an index, you can use it to facilitate data retrieval. In powerful mainframe relational systems, the decision concerning which index(es) to use (if any) during a particular type of retrieval is a function of the DBMS.

As you would expect, the use of any index is not purely advantageous or disadvantageous. An advantage was already mentioned: An index makes certain types of retrieval more efficient. There are two disadvantages. First, the index occupies space on disk. Using this space for an index, however, is technically unnecessary because any retrieval that can be made using the index can also be made without the index, although less efficiently. The other disadvantage is that the DBMS must update the index whenever corresponding data in the database is updated. Without the index, the DBMS would not need to make these updates. The main question you must ask when considering whether to create a given index is this: Do the benefits derived during retrieval outweigh the additional storage required and the extra processing involved in update operations? The following guidelines should help you make this determination. Create an index on a field (or combination of fields) when one or more of the following conditions exist:

- The field is the primary key of the table. (In some systems, the DBMS might create this index automatically.)
- The field is the foreign key in a relationship you have created.
- You will frequently use the field as a sort field.
- You will frequently need to locate a record based on a value in this field.

You can add and delete indexes as necessary. You can create an index after the database is built—the index doesn't need to be created at the same time as the database. Likewise, when it appears that an existing index is unnecessary, you can delete it.

The exact process for creating an index varies from one DBMS to another. A common SQL command to create an index is as follows:

```
CREATE INDEX CustomerName
ON Customer (CustomerName)
;
```

This **CREATE INDEX** command creates an index named CustomerName. The index is for the Customer table, and the index key is the CustomerName field. In this example, the index name is the same as the index key. This format is not a requirement, but it is a good general practice.

Figure 4-13 shows the creation of an index on the CustomerName field in the Customer table using Access. As illustrated in the figure, there are three choices for index options: No, Yes (Duplicates OK), and Yes (No Duplicates).

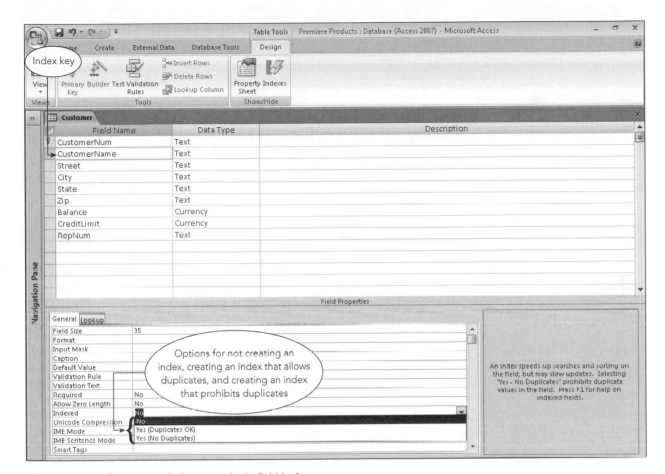

FIGURE 4-13   Creating an index on a single field in Access

The first Indexed option, No, is the default. You select No when you need to remove a previously created index. You select Yes (Duplicates OK) to create an index that allows duplicate values. In this case, Access allows more than one customer with the same name. When you select Yes (No Duplicates), Access creates the index, but you cannot add a customer with the same name as an existing customer in the database. The third option is used to enforce uniqueness when it is appropriate. For example, this would be a good choice for a Social Security number field.

When you create an index whose key is a single field, you have created a **single-field index** (also called a **single-column index**). A **multiple-field index** (also called a **multiple-column index**) is an index with more than one key field. When creating a multiple-field index, you list the more important key first. In addition, if data for either key appears in descending order, you must follow the field name with the word *DESC*.

To create an index named RepBal with the keys RepNum and Balance and with the balances listed in descending order, you could use the following SQL command:

```
CREATE INDEX RepBal
ON Customer (RepNum, Balance DESC)
;
```

Creating multiple-field indexes in Access involves a slightly different process than creating single-field indexes. To create multiple-field indexes, click the Indexes button in the Show/Hide group on the Table Tools Design tab, enter a name for the index, and then select the fields that make up the index key. If data for any of the fields is to appear in descending order, change the corresponding entry in the Sort Order column to Descending, as shown in Figure 4-14.

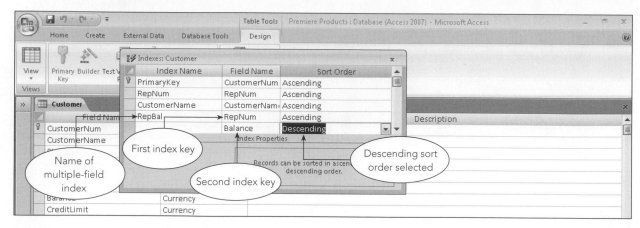

FIGURE 4-14   Creating a multiple-field index in Access

The SQL command used to drop (delete) an index that is no longer necessary is **DROP INDEX**, which consists of the words *DROP INDEX* followed by the name of the index to drop. To drop the RepBal index, for example, the command is as follows:

```
DROP INDEX RepBal
;
```

In MySQL, the DROP INDEX command must include an ON clause that specifies the name of the table. The corresponding command in MySQL is as follows:

```
DROP INDEX RepBal ON CUSTOMER
;
```

To delete an index in Access, select the index in the Indexes dialog box (see Figure 4-14), right-click it, and then click Delete Rows on the shortcut menu.

SECURITY

Security is the prevention of unauthorized access to the database. Within an organization, the database administrator determines the types of access various users can have to the database. Some users may be able to retrieve and update anything in the database. Other users may be able to retrieve any data from the database but not make any changes to it. Still other users may be able to access only a portion of the database. For example, Bill Kaiser may be able to retrieve and update sales rep and customer data, but not be permitted to retrieve data about parts and orders. Mary Smith may be able to retrieve part data and nothing else. Kyung Park may be able to retrieve and update data on parts in the HW class, but not in other classes.

After the database administrator has determined the access different users of the database will have, it is up to the DBMS to enforce it. In particular, it is up to whatever security mechanism the DBMS provides. In SQL systems, there are two security mechanisms. You have already seen that views furnish a certain amount of security. (When users are accessing the database through a view, they cannot access any data that is not included in the view.) The main mechanism for providing access to a database, however, is the GRANT statement.

The basic idea of the **GRANT** statement is that different types of privileges can be granted to users and, if necessary, later revoked. These privileges include such things as the right to select, insert, update, and delete table data. You can revoke user privileges using the **REVOKE** statement. Following are examples of these two statements.

The following command will enable user Jones to retrieve data from the Customer table, but not take any other action.

```
GRANT SELECT ON Customer TO Jones
;
```

The following command will enable users Smith and Park to add new records to the Part table.

```
GRANT INSERT ON Part TO Smith, Park
;
```

The following command will revoke the ability to retrieve Customer records from user Jones; that is, Jones will no longer have the privilege granted earlier.

```
REVOKE SELECT ON Customer FROM Jones
;
```

INTEGRITY RULES

A relational DBMS must enforce two important integrity rules that were defined by Dr. E. F. Codd (Codd, E. F. "Extending the Relational Database Model to Capture More Meaning." In *ACM TODS 4*, no. 4 (December 1979)). Both rules are related to two special types of keys: primary keys and foreign keys. The two integrity rules are called entity integrity and referential integrity.

Entity Integrity

In some DBMSs, when you describe a database, you can indicate that certain fields can accept a special value, called null. Essentially, setting the value in a given field to null is similar to not entering a value in the field at all. Nulls are used when a value is missing, unknown, or inapplicable. It is *not* the same as a blank or zero value, both of which are actual values. For example, a value of zero in the Balance field for a particular customer indicates that the customer has a zero balance. A value of null in a customer's Balance field, on the other hand, indicates that, for whatever reason, the customer's balance is unknown.

When you indicate that the Balance field can be null, you are saying that this situation (a customer with an unknown balance) is something you want to allow. If you don't want to allow unknown values, you indicate that Balance field values cannot be null.

The decision about allowing nulls is generally made on a field-by-field basis. There is one type of field for which you should *never* allow nulls, however, and that is the primary key. After all, the primary key is supposed to uniquely identify a given row, which would not happen if nulls were allowed. How, for example, could you tell two customers apart if both had null customer numbers? The restriction that the primary key cannot allow null values is called entity integrity.

Definition: **Entity integrity** is the rule that no field that is part of the primary key may accept null values.

Entity integrity guarantees that each record will indeed have its own identity. In other words, preventing the primary key from accepting null values ensures that you can distinguish one record from another. Typically, the DBMS handles this distinction automatically. All you need to do is specify which field or fields make up the primary key.

In SQL, you can specify the primary key by entering a **PRIMARY KEY** clause in either an ALTER TABLE (covered later in this chapter) or a CREATE TABLE command. For example, to use the PRIMARY KEY clause to indicate that CustomerNum is the primary key for the Customer table, the clause would be as follows:

```
PRIMARY KEY (CustomerNum)
```

In general, the PRIMARY KEY clause has the form PRIMARY KEY followed in parentheses by the field or fields that make up the primary key. When more than one field is included, the fields are separated by commas. Thus, the PRIMARY KEY clause for the OrderLine table is as follows:

```
PRIMARY KEY (OrderNum, PartNum)
```

In Access, you designate the primary key by selecting the primary key field in Table Design view and clicking the Primary Key button in the Tools group on the Table Tools Design tab. A key symbol will appear in the field's row selector to indicate that it is the primary key, as shown in Figure 4-15.

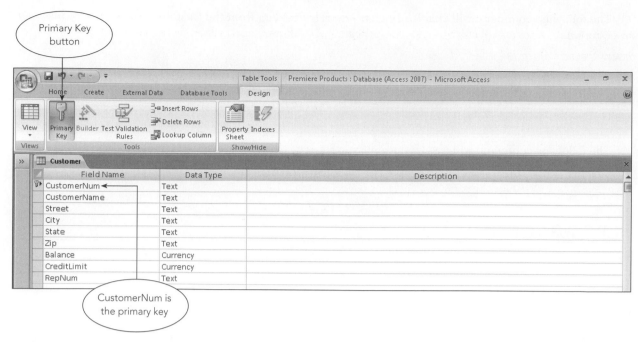

FIGURE 4-15 Specifying a primary key in Access

If the primary key consists of more than one field, select the first field, press and hold down the Ctrl key, and then click the other field or fields that make up the primary key. Clicking the Primary Key button adds the key symbol to the row selectors of the primary key fields, as shown in Figure 4-16.

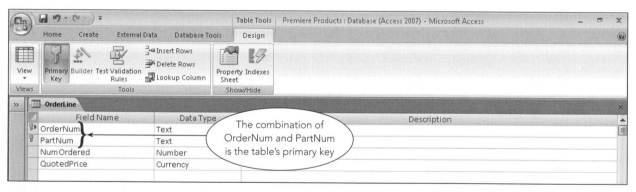

FIGURE 4-16 Specifying a primary key consisting of more than one field in Access

Referential Integrity

In the relational model you've examined thus far, you have created the relationships between tables by having common fields in two or more tables. The relationship between sales reps and customers, for example, is accomplished by including the primary key of the Rep table (RepNum) as a field in the Customer table.

This approach has several drawbacks. First, relationships are not very obvious. If you were not already familiar with the relationships in the Premiere Products database, you would have to find the matching fields in separate tables in order to locate the relationship. Even then, you couldn't be sure that the matching field names indicate a relationship. Two fields having the same name could be just a coincidence—the fields might have nothing to do with each other. Second, what if the primary key in the Rep table is named RepNum but the corresponding field in the Customer table is named SlsrNo? Unless you are aware that these two fields are identical, the relationship between customers and sales reps would not be clear. In a database having as few tables and fields as the Premiere Products database, these problems might be manageable. But picture a database that has 20 tables, each containing an average of 30 fields. As the number of tables and fields increases, so do the potential problems.

There is also another issue with the relational model. Nothing about the model itself would prevent a user from storing data about a customer whose sales rep number did not correspond to any sales rep already in the database. Clearly, this is not a desirable situation.

Fortunately, a solution exists for both issues. It involves using foreign keys.

Definition: A **foreign key** is a field (or collection of fields) in a table whose value is required to match the value of the primary key for a second table.

The RepNum field in the Customer table is a foreign key that must match the primary key of the Rep table. In practice, this means that the sales rep number for any customer must be the same as the number of a sales rep that is already in the database.

There is one possible exception to this rule. Perhaps Premiere Products does not require a customer to have a sales rep—it is strictly optional. This situation could be indicated in the Customer table by setting such a customer's sales rep number to null. Technically, however, a null sales rep number would violate the restrictions that you have indicated for a foreign key. Thus, if you were to use a null sales rep number, you would have to modify the definition of a foreign key to include the possibility of nulls. You would insist, though, that if the foreign key contained a value *other than null*, it would have to match the value of the primary key in some row in the other table. (In the example, for instance, a customer's sales rep number could be null. If it were not null, it would have to be the number of an actual sales rep.) This general property is called referential integrity.

Definition: **Referential integrity** is the rule that if table A contains a foreign key that matches the primary key of table B, the values of this foreign key must match the value of the primary key for some row in table B or be null.

Usually a foreign key is in a table that is different from the primary key it is required to match. In the Premiere Products database, for example, to be able to determine the rep for any customer, you include the rep number as a foreign key in the Customer table that must match the primary key in the Rep table. It is possible for the foreign key and the matching primary key to be in the same table, however. As an example of this situation, suppose one of the requirements in a particular database is that, given an employee, you must be able to determine the manager of that employee. You might have an Employee table with a primary key of EmployeeNum (the employee number). To determine the employee's manager, you would include the manager's employee number as a foreign key in the Employee table. Because the manager is also an employee, however, the manager will be in the same Employee table. Thus, this foreign key in the Employee table would need to match the primary key in the same Employee table. The only restriction is that the foreign key must have a name that is different from the primary key because the fields are in the same table. For example, you could name the foreign key ManagerEmployeeNum.

Using foreign keys solves the previously mentioned problems. Indicating that the RepNum field in the Customer table is a foreign key that must match the RepNum field in the Rep table explicitly specifies the relationship between customers and sales reps—you do not need to look for common fields in several tables. Further, with foreign keys, matching fields that have different names no longer pose a problem. For example, it would not matter if the name of the foreign key in the Customer table were SlsrNo and the primary key in the Rep table were RepNum; the only thing that *would* matter is that this field is a foreign key that matches the Rep table. Finally, through referential integrity, it is possible for a customer not to have a sales rep number, but it is not possible for a customer to have an *invalid* sales rep number; that is, a customer's sales rep number *must* be null or *must* be the number of a sales rep who is already in the database.

In SQL, you specify referential integrity using a **FOREIGN KEY** clause in either the CREATE TABLE or ALTER TABLE commands. To specify a foreign key, you need to specify both the field that is a foreign key and the table whose primary key it is to match. In the Customer table, for example, the RepNum field is a foreign key that must match the primary key in the Rep table as follows:

```
FOREIGN KEY (RepNum) REFERENCES Rep
```

The general form is FOREIGN KEY, followed by the field or combination of fields that make up the foreign key, which is followed by the word *REFERENCES* and the name of the table containing the primary key that the foreign key is supposed to match.

In Access, referential integrity is specified as part of the process of defining relationships, as shown in Figure 4-17.

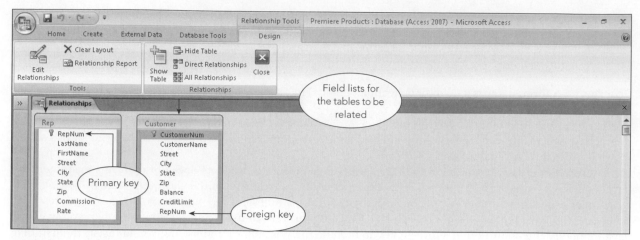

FIGURE 4-17   Using the Relationships window to relate tables in Access

You use the pointer to drag the primary key of the Rep table (RepNum) to the foreign key of the Customer table (RepNum). After releasing the mouse button, you can request Access to enforce referential integrity, as shown in Figure 4-18. You also can specify whether update or delete operations will "cascade." Selecting the **cascade delete** option ensures that the deletion of a sales rep record also deletes all customer records related to that sales rep; selecting the **cascade update** option ensures that changes made to the primary key of a sales rep record are also made in the related customer record. In Figure 4-18, the cascade delete and cascade update options are not selected.

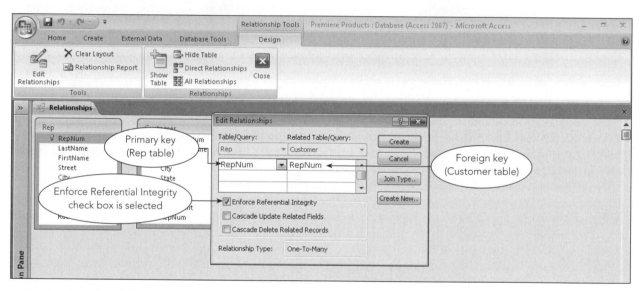

FIGURE 4-18   Specifying referential integrity in Access

With referential integrity enforced, users are not allowed to enter a customer record whose sales rep number does not match any sales rep currently in the Rep table. An error message, such as the one shown in Figure 4-19, appears when an attempt is made to enter an invalid sales rep number.

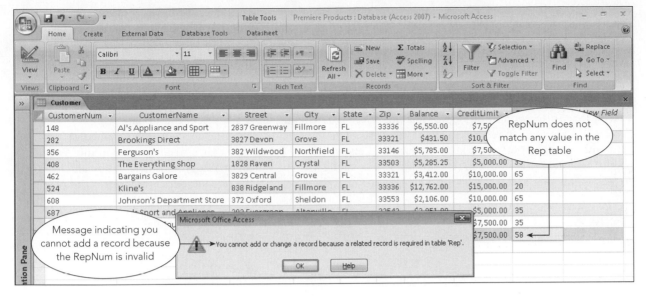

FIGURE 4-19 Referential integrity violation when attempting to add a record

Deleting a sales rep who currently has customers on file would also violate referential integrity because the sales rep's customers would no longer match any sales rep in the Rep table. The DBMS must refuse to carry out this type of deletion and then produce an error message, such as the one shown in Figure 4-20. If sales rep 20 leaves Premiere Products, all of her customers would need to be assigned to other sales reps before her record would be deleted from the Rep table.

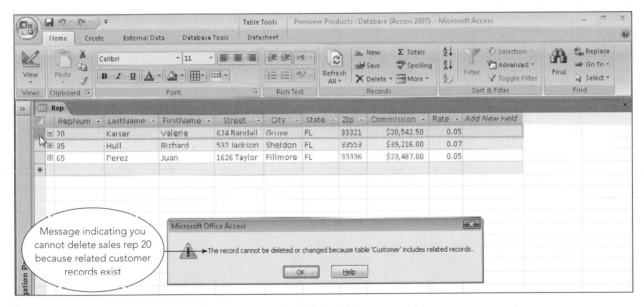

FIGURE 4-20 Referential integrity violation when attempting to delete a record

Legal-Values Integrity

In addition to the two integrity rules defined by Codd, there is a third type of integrity, called **legal-values integrity**. Often there is a particular set of values, called the legal values, that are allowable in a field. Legal-values integrity is the property that states that no record can exist in the database with a value in the field other than one of the legal values. For example, the only legal values for the CreditLimit field are $5,000, $7,500, $10,000, and $15,000. A record on which the credit limit is $12,500 would not be permitted.

In SQL, you use the **CHECK** clause to enforce legal-values integrity. For example, to ensure that the only legal values for credit limits are $5,000, $7,500, $10,000, or $15,000, the CHECK clause that you would include in a CREATE TABLE or ALTER TABLE command would be as follows:

```
CHECK (CreditLimit IN (5000, 7500, 10000, 15000))
```

The general form of the CHECK clause is the word *CHECK* followed by a condition. In the previous CHECK clause, the credit limit must be in the set consisting of 5000, 7500, 10000, or 15000. If any update to the database results in the condition being violated, the DBMS will automatically reject the update.

In Access, you can restrict the legal values accepted by a field by entering an appropriate **validation rule** that data entered in the field must follow. Figure 4-21 shows the validation rule that restricts entries in the CreditLimit field to 5000, 7500, 10000, and 15000. Along with the validation rule, you usually enter **validation text** to inform the user of the reason for the rejection when the user attempts to enter data that violates the rule.

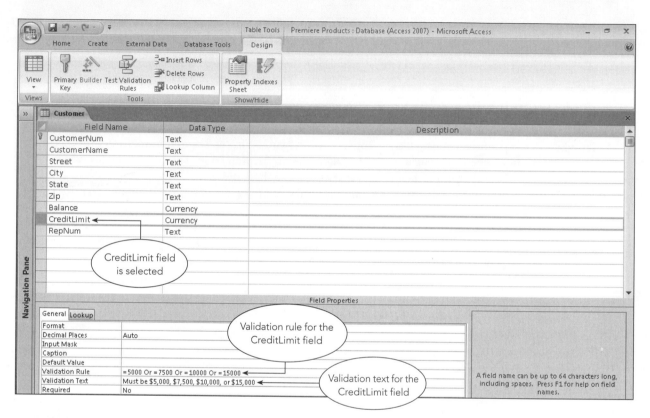

FIGURE 4-21   Specifying a validation rule in Access

STRUCTURE CHANGES

An important feature of relational DBMSs is the ease with which you can change the database structure by adding and removing tables and fields, by changing the characteristics of existing fields, or by creating and dropping indexes. Although the exact manner in which you accomplish these changes varies from one system to another, most systems allow you to make all of these changes quickly and easily.

Changes to a table's structure are made using the SQL **ALTER TABLE** command. Virtually every implementation of SQL allows the creation of new fields in existing tables. For example, suppose you need to maintain a customer type for each customer in the Premiere Products database. You can decide to assign regular customers type R, distributors type D, and special customers type S. To implement this change, you would add a new field to the Customer table as follows:

```
ALTER TABLE Customer
ADD CustType CHAR(1)
;
```

In Access, you can add a field in Table Design view at any time. Figure 4-22 shows the Customer table with the addition of the CustType field.

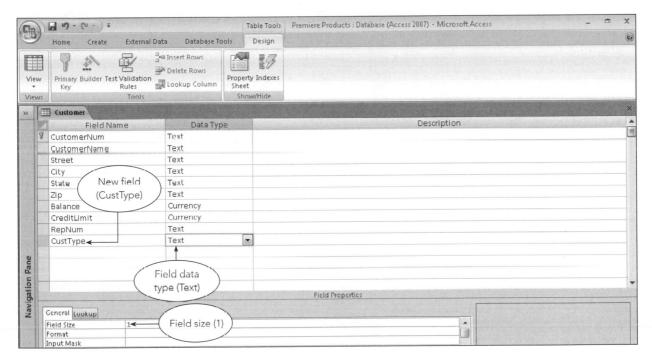

FIGURE 4-22   Adding a field in Access

At this point, the Customer table contains an extra field, CustType. For rows (customers) added from this point on, the value of CustType is entered just like any other field. For existing rows, CustType is typically assigned a null value by the DBMS automatically. The user then can change these values if desired.

Some systems allow changes to the properties of existing fields, such as increasing the length of a character field that was found to be inadequate. Assume the CustomerName field in the Customer table needed to be increased from 35 to 40 characters; in that case, the SQL ALTER TABLE command would be as follows:

```
ALTER TABLE Customer
CHANGE COLUMN CustomerName TO CHAR(40)
;
```

In Access, you can change field properties in Table Design view, as shown in Figure 4-23, in which the size of the CustomerName field, which was 35, was changed to 40.

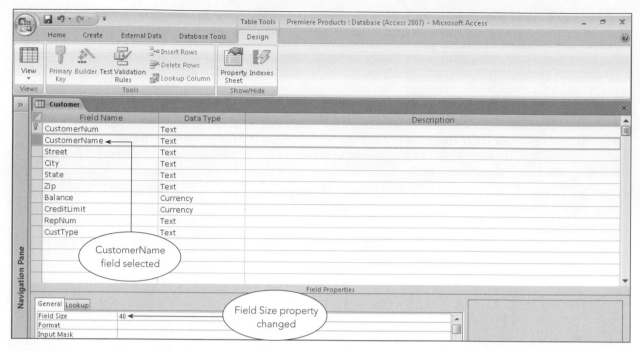

FIGURE 4-23 Changing a field property in Access

Some systems allow existing fields to be deleted. (Oracle is one system that does not allow existing fields to be deleted.) The SQL command for deleting the Warehouse field from the Part table is as follows:

```
ALTER TABLE Part
DELETE Warehouse
;
```

In Access, you can delete a field in Table Design view by selecting the field and pressing the Delete key. Access then will ask you to confirm the deletion of the field, as shown in Figure 4-24. Clicking the Yes button permanently deletes the field and the data it stores.

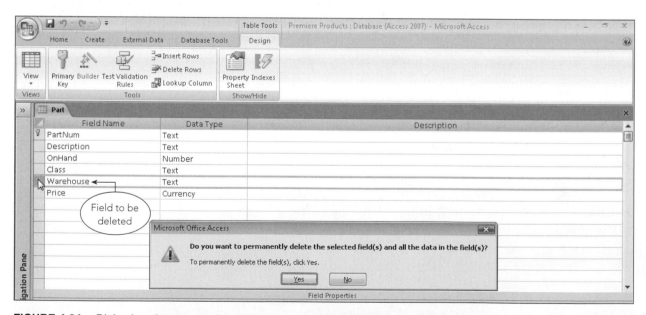

FIGURE 4-24 Dialog box that opens when a field in Access is deleted

With the SQL **DROP TABLE** command, you can delete a table that is no longer needed. If the SmallCust table (created in Chapter 3) was no longer needed in the Premiere Products database, you could remove it using the following command:

```
DROP TABLE SmallCust
;
```

The table and all indexes and views defined on the table would be deleted. The DROP TABLE command deletes the table structure as well as its data.

In Access, you can drop (delete) a table by right-clicking the table on the Navigation Pane and then clicking Delete on the shortcut menu, as shown in Figure 4-25.

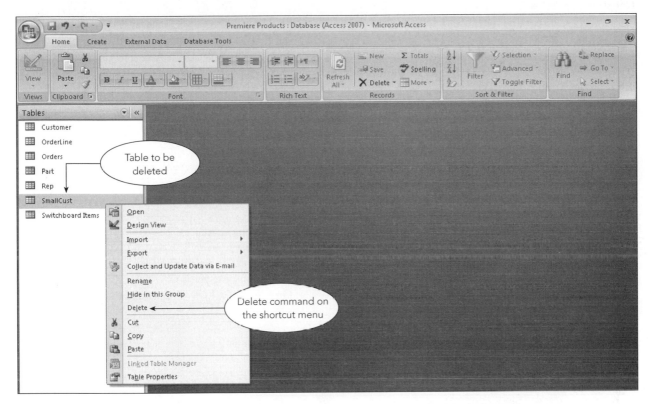

FIGURE 4-25   Deleting a table in Access

Making Complex Changes

In some cases, you might need to change a table's structure in ways that are beyond the capabilities of your DBMS. Perhaps you need to eliminate a field, change the field order, or combine data from two tables into one, but your system does not allow these types of changes. For example, some systems, including Oracle, do not allow you to reduce the size of a field or change its data type. In these situations, you can use the CREATE TABLE command to describe the new table, and then insert values into it using the INSERT command combined with an appropriate SELECT clause, as you learned in Chapter 3. If you are using a version of SQL that supports the SELECT INTO command, as Access does, you can use it to create the new table in a single operation.

SYSTEM CATALOG

Information about tables in the database is kept in the **system catalog** (or the **catalog**). The catalog is maintained automatically by the DBMS. When a user adds a new table, changes the structure of an existing table, or deletes a table, the DBMS updates the catalog to reflect these changes.

This section describes the types of items kept in the catalog and the way in which you can query it to determine information about the database structure. (This description represents the way catalogs are used in a typical SQL implementation.) Although catalogs in individual relational DBMSs will vary from the examples shown here, the general ideas apply to most relational systems.

The catalog you'll consider contains two tables: **Systables** (information about the tables known to SQL) and **Syscolumns** (information about the columns or fields within these tables). An actual catalog contains other tables as well, such as **Sysindexes** (information about the indexes that are defined on these tables) and **Sysviews** (information about the views that have been created). Although these tables have many fields, only a few are of concern here.

As shown in Figure 4-26, the Systables table contains the Name, Creator, and Colcount fields. The Name field identifies the name of a table, the Creator field identifies the person or group that created the table, and the Colcount field contains the number of fields in the table being described. If, for example, the user named Brown created the Rep table and the Rep table has nine fields, there would be a row in the Systables table in which the Name is Rep, the Creator is Brown, and the Colcount is 9. Similar rows would exist for all tables known to the system.

Systables

Name	Creator	Colcount
Customer	Brown	10
Part	Brown	6
Orders	Brown	3
Order Line	Brown	4
Rep	Brown	9

FIGURE 4-26 Systables table

The Syscolumns table contains the Colname, Tbname, and Coltype fields, as shown in Figure 4-27. The Colname field identifies the name of a field in one of the tables. The table in which the field is found is stored in the Tbname field, and the data type for the field is found in the Coltype field. There is a row in the Syscolumns table for each field in the Rep table, for example. On each of these rows, Tbname is Rep. On one of these rows, Colname is RepNum and Coltype is CHAR(2). On another row, Colname is LastName and Coltype is CHAR(15).

Syscolumns

Colname	Tbname	Coltype
Balance	Customer	DECIMAL(8,2)
City	Customer	CHAR(15)
City	Rep	CHAR(15)
Class	Part	CHAR(2)
Commission	Rep	DECIMAL(7,2)
CreditLimit	Customer	DECIMAL(8,2)
CustomerName	Customer	CHAR(35)
CustomerNum	Customer	CHAR(3)
CustomerNum	Orders	CHAR(3)
Description	Part	CHAR(15)
FirstName	Rep	CHAR(15)
LastName	Rep	CHAR(15)
NumOrdered	OrderLine	DECIMAL(3,0)
OnHand	Part	DECIMAL(4,0)
OrderDate	Orders	DATE
OrderNum	OrderLine	CHAR(5)
OrderNum	Orders	CHAR(5)
PartNum	OrderLine	CHAR(4)
PartNum	Part	CHAR(4)
Price	Part	DECIMAL(6,2)
QuotedPrice	OrderLine	DECIMAL(6,2)
Rate	Rep	DECIMAL(3,2)
RepNum	Customer	CHAR(2)
RepNum	Rep	CHAR(2)
State	Customer	CHAR(2)
State	Rep	CHAR(2)
Street	Customer	CHAR(15)
Street	Rep	CHAR(15)
Warehouse	Part	CHAR(2)
Zip	Customer	CHAR(5)
Zip	Rep	CHAR(5)

FIGURE 4-27   Syscolumns table

A DBMS furnishes ways of using the catalog to determine information about the structure of the database. In some cases, this simply involves using SQL to query the tables in the catalog. For example, to list the name and type of all fields (columns) in the Part table, you could use the following SQL command:

```
SELECT Colname, Coltype
FROM Syscolumns
WHERE Tbname='Part'
;
```

NOTE

In MySQL, you use the SHOW TABLES command to produce a list of all tables in the current database. The SHOW INDEX command produces a list of all indexes. The SHOW COLUMNS command, which consists of the words *SHOW COLUMNS FROM* followed by the name of a table, produces details concerning all columns in the indicated table. Running the command SHOW COLUMNS FROM CUSTOMER, for example, lists details concerning all the columns in the Customer table.

In other cases, special tools provide the desired documentation. For example, Access has a tool called the **Documenter**, which allows you to print detailed documentation about any table, query, report, form, or other object in the database. To document the objects in an Access database, click the Database Tools tab, and then click the Database Documenter button in the Analyze group.

STORED PROCEDURES

In a **client/server system**, the database resides on a computer called the **server** and users access the database through clients. A **client** is a computer that is connected to a network and has access through the server to the database. Every time a user executes a query, the DBMS must determine the best way to process the query and provide the results. For example, the DBMS must determine which indexes are available and whether it can use those indexes to make the processing of the query more efficient.

If you anticipate running a particular query often, for example, you can improve overall performance by saving the query in a special file called a **stored procedure**. The stored procedure is placed on the server. The DBMS compiles the stored procedure (translating it into machine code) and creates an execution plan, which is the most efficient way of obtaining the results. From that point on, users execute the compiled, optimized code in the stored procedure.

Another reason for saving a query as a stored procedure, even when you are not working in a client/server system, is convenience. Rather than retyping the entire query each time you need it, you can use the stored procedure. For example, suppose you frequently execute a query to change a customer's credit limit. You can use the same query to select the record using the customer's number and to change the credit limit. Instead of running the query each time and changing the customer number and the credit limit, it would be simpler to store the query in a stored procedure. When you run the stored procedure, you need to enter only the appropriate customer number and the new credit limit.

Although it is easy to create and use a stored procedure, you need to be aware of one issue you will face when creating a stored procedure in MySQL. The semicolon in a MySQL command marks the end of the command. In this context, the semicolon is called a **delimiter**. When MySQL encounters a semicolon in a command, it executes the command. Because a semicolon also indicates the end of a stored procedure, you need to change the delimiter temporarily for the stored procedure. In the following example, the first line temporarily changes the delimiter to a double dollar sign (*$$*). The last line in the command changes the delimiter back to a semicolon. (Notice that there is a space between the word *DELIMITER* and the semicolon.)

```
DELIMITER $$
CREATE PROCEDURE Change_Credit
(CNum CHAR(4), CLimit DECIMAL (8,2))
BEGIN
UPDATE Customer
SET CreditLimit = CLimit
WHERE CustomerNum = CNum ;
END
$$
DELIMITER ;
```

The CREATE PROCEDURE line in the stored procedure causes MySQL to create a procedure named Change_Credit. The second line indicates that there are two arguments, CNum and CLimit, with the appropriate data types. These values are used in the SQL statement that will be stored. (Although it is not required, creating aliases for the field names simplifies the command.) When users run this stored procedure, they furnish values for CNum and CLimit. The stored procedure then updates the customer whose number is stored in CNum by changing the customer's credit limit to the value stored in CLimit.

The next line contains the word *BEGIN* to mark the beginning of the SQL command to be stored. The next three lines contain the command, including the semicolon. The line that contains the word *END* marks the end of the command. The *$$* delimiter indicates that the CREATE PROCEDURE command is complete. MySQL then executes this command and creates the stored procedure. The final line changes the delimiter back to the semicolon.

To use this stored procedure, a user enters the word *CALL* followed by the procedure name. After the procedure name, the user enters the values of the two arguments in parentheses. For example, changing the credit limit of customer number 356 to $10,000, requires the following command:

```
CALL Change_Credit ('356', 10000);
```

Although Access does not support stored procedures, you can achieve some of the same convenience by creating a parameter query that prompts the user for the arguments you would otherwise use in a stored procedure.

TRIGGERS

A **trigger** is an action that occurs automatically in response to an associated database operation such as an INSERT, UPDATE, or DELETE command. Like a stored procedure, a trigger is stored and compiled on the server. Unlike a stored procedure, which is executed in response to a user request, a trigger is executed in response to a command that causes the associated database operation to occur.

The examples in this section assume there is a new column named OnOrder in the Part table. This column represents the number of units of a part currently on order. For example, if there are two separate order lines for a part and the number ordered on one order line is 3 and the number ordered on the other order line is 2, the OnOrder value for that part will be 5. Adding, changing, or deleting order lines affects the value in the OnOrder column for the part. To ensure that the value is updated appropriately, you can use a trigger.

When a user adds an order line, the following MySQL trigger, which is named AddOrderLine, is executed. When a user adds an order line, the trigger must update the OnOrder value for the corresponding part to reflect the order line. For example, if the value in the OnOrder column for part CD52 is 4 and the user adds an order line on which the part number is CD52 and the number of units ordered is 2, 6 units of part CD52 will be on order. When a record is added to the OrderLine table, the AddOrderLine trigger updates the Part table by adding the number of units ordered on the order line to the previous value in the OnOrder column.

```
DELIMITER $$
CREATE TRIGGER AddOrderLine
AFTER INSERT ON OrderLine
FOR EACH ROW
BEGIN
UPDATE PART
SET OnOrder = OnOrder + New.NumOrdered
WHERE PartNum = New.PartNum ;
END
$$
DELIMITER ;
```

Because semicolons cause the same issues with triggers as with stored procedures, the command starts by changing the delimiter to $$. The second line indicates that the command is creating a trigger named AddOrderLine. The third line indicates that this trigger will be executed after an order line is inserted. The fourth line indicates that the SQL command is to occur for each row that is added. Like stored procedures, the SQL command is enclosed between the words *BEGIN* and *END*. In this case, the SQL command is an UPDATE command. The command uses the New qualifier. The New qualifier refers to the row that is added to the OrderLine table. If an order line is added on which the part number is CD52 and the number ordered is 2, New.PartNum will be CD52 and New.NumOrdered will be 2.

The following UpdateOrderLine trigger is executed when a user attempts to update an order line. There are two differences between the UpdateOrderLine trigger and the AddOrderLine trigger. First, the third line of the UpdateOrderLine trigger indicates that this trigger is executed after an UPDATE of an order line, rather than an INSERT. Second, the computation to update the OnOrder column includes both New.NumOrdered and Old.NumOrdered. As with the AddOrderLine trigger, New.NumOrdered refers to the new value. In an UPDATE command, however, there is also an old value, which is the value before the update takes place. If an update changes the value for NumOrdered from 1 to 3, Old.NumOrdered is 1 and New.NumOrdered is 3. Adding New.NumOrdered and subtracting Old.NumOrdered results in a net change of an increase of 2. (The net change could also be negative, in which case the OnOrder value decreases.)

```
DELIMITER $$
CREATE TRIGGER UpdateOrderLine
AFTER UPDATE ON OrderLine
FOR EACH ROW
BEGIN
UPDATE PART
SET OnOrder = OnOrder + New.NumOrdered - Old.NumOrdered
WHERE Part.PartNum = New.PartNum ;
END
$$
DELIMITER ;
```

The following DeleteOrderLine trigger performs a function similar to the other two. When an order line is deleted, the OnOrder value for the corresponding part is updated by subtracting Old.NumOrdered from the current OnOrder value. (In a delete operation, there is no New.NumOrdered.)

```
DELIMITER $$
CREATE TRIGGER DeleteOrderLine
AFTER DELETE ON OrderLine
FOR EACH ROW
BEGIN
UPDATE PART
SET OnOrder = OnOrder - Old.NumOrdered
WHERE PartNum = Old.PartNum ;
END
$$
DELIMITER ;
```

Access does not support triggers. When using a form to update table data, you can achieve some of the same funtionality by creating VBA code to be executed after the insertion, update, or deletion of records.

Summary

- Views are used to give each user his or her own view of the data in a database. In SQL, a defining query creates a view. When you enter a query that references a view, it is merged with the defining query to produce the query that is actually executed. In Access, views are created by saving queries that select the data to use in the view.

- Indexes are often used to facilitate data retrieval from the database. You can create an index on any field or combination of fields.

- Security is provided in SQL systems using the GRANT and REVOKE commands.

- Entity integrity is the property that states that no field that is part of the primary key can accept null values.

- Referential integrity is the property that states that the value in any foreign key field must be null or must match an actual value in the primary key field of another table. Referential integrity is specified in SQL using the FOREIGN KEY clause. In Access, foreign keys are specified by creating relationships.

- Legal-values integrity is the property that states that the value entered in a field must be one of the legal values that satisfies some particular condition. Legal-values integrity is specified in SQL using the CHECK clause. In Access, legal-values integrity is specified using validation rules.

- The ALTER TABLE command allows you to add fields to a table, delete fields, or change the characteristics of fields. In Access, you can change the structure of a table by making the desired changes in the table design.

- The DROP TABLE command lets you delete a table from a database. In Access, you can delete a table by selecting the Delete command on the table's shortcut menu in the Navigation Pane.

- The system catalog is a feature of many relational DBMSs that stores information about the structure of a database. The system updates the catalog automatically. Each DBMS includes facilities to produce documentation of the database structure using the information in the catalog.

- A stored procedure is a query saved in a file that users can execute later.

- A trigger is an action that occurs automatically in response to an associated database operation such as an INSERT, UPDATE, or DELETE command. Like a stored procedure, a trigger is stored and compiled on the server. Unlike a stored procedure, which is executed in response to a user request, a trigger is executed in response to a command that causes the associated database operation to occur.

Key Terms

ALTER TABLE	multiple-column index
cascade delete	multiple-field index
cascade update	PRIMARY KEY
catalog	referential integrity
CHECK	REVOKE
client	row-and-column subset view
client/server system	security
CREATE INDEX	server
defining query	single-column index
delimiter	single-field index
Documenter	stored procedure
DROP INDEX	Syscolumns
DROP TABLE	Sysindexes
entity integrity	Systables
FOREIGN KEY	system catalog
foreign key	Sysviews
GRANT	trigger
index	validation rule
index key	validation text
legal-values integrity	view

Review Questions

1. What is a view? How is it defined? Do the data described in a view definition ever exist in that form? What happens when a user accesses a database through a view?

2. Using data from the Premiere Products database, define a view named TopLevelCust. It consists of the number, name, address, balance, and credit limit of all customers with credit limits that are greater than or equal to $10,000.

    a. Using SQL, write the view definition for TopLevelCust.

    b. Write an SQL query to retrieve the number and name of all customers in the TopLevelCust view with balances that exceed their credit limits.

    c. Convert the query you wrote in Question 2b to the query that the DBMS will actually execute.

3. Define a view named PartOrder. It consists of the part number, description, price, order number, order date, number ordered, and quoted price for all order lines currently on file.

    a. Using SQL, write the view definition for PartOrder.

    b. Write an SQL query to retrieve the part number, description, order number, and quoted price for all orders in the PartOrder view for parts with quoted prices that exceed $100.

    c. Convert the query you wrote in Question 3b to the query that the DBMS will actually execute.

4. What is an index? What are the advantages and disadvantages of using indexes? How do you use SQL to create an index?

5. Describe the GRANT statement and explain how it relates to security. What types of privileges may be granted? How are they revoked?

6. Write the SQL commands to grant the following privileges:

    a. User Stillwell must be able to retrieve data from the Part table.

    b. Users Webb and Bradley must be able to add new orders and order lines.

7. Write the SQL command to revoke user Stillwell's privilege.

8. What is the system catalog? Name three items about which the catalog maintains information.

9. Write the SQL commands to obtain the following information from the system catalog:

    a. List every table that you created.

    b. List every field in the Customer table and its associated data type.

    c. List every table that contains a field named PartNum.

10. Why is it a good idea for the DBMS to update the catalog automatically when a change is made in the database structure? Could users cause problems by updating the catalog themselves? Explain.

11. What are nulls? Which field cannot accept null values? Why?

12. State the three integrity rules. Indicate the reasons for enforcing each rule.

13. The Orders table contains a foreign key, CustomerNum, that must match the primary key of the Customer table. What type of update to the Orders table would violate referential integrity? If deletes do not cascade, what type of update to the Customer table would violate referential integrity? If deletes do cascade, what would happen when a customer is deleted?

14. How would you use SQL to change a table's structure? What general types of changes are possible? Which commands are used to implement these changes?

15. What are stored procedures? What purpose do they serve?

16. What are triggers? What purpose do they serve?

Premiere Products Exercises

In the following exercises, you will use the data in the Premiere Products database shown in Figure 2-1 in Chapter 2. (If you use a computer to complete these exercises, use a copy of the original Premiere Products database so your data will not reflect the changes you made in Chapter 3.) If you have access to a DBMS, use the DBMS to perform the tasks and explain the steps you used in the process. If not, explain how you would use SQL to obtain the desired results. Check with your instructor if you are not certain about which approach to take.

1. Create the TopLevelCust view described in Review Question 2. Display the data in the view.

2. Create the PartOrder view described in Review Question 3. Display the data in the view.

3. Create a view named OrdTot. It consists of the order number and order total for each order currently on file. (The order total is the sum of the number ordered multiplied by the quoted price on each order line for each order.) Display the data in the view.

4. Create the following indexes. If it is necessary to name the index in your DBMS, use the indicated name.

    a. Create an index named PartIndex1 on the PartNum field in the OrderLine table.

    b. Create an index named PartIndex2 on the Warehouse field in the Part table.

    c. Create an index named PartIndex3 on the Warehouse and Class fields in the Part table.

    d. Create an index named PartIndex4 on the Warehouse and OnHand fields in the Part table and list units on hand in descending order.

5. Drop the PartIndex3 index.

6. Assume the Part table has been created but there are no integrity constraints. Create the necessary integrity constraint to ensure that the only allowable values for the Class field are AP, HW, and SG. Ensure that the PartNum field is the primary key and that the PartNum field in the OrderLine table is a foreign key that must match the primary key of the Part table.

7. Add a field named Allocation to the Part table. The allocation is a number representing the number of units of each part that have been allocated to each customer. Set all values of Allocation to zero. Calculate the number of units of part number KV29 currently on order. Change the value of Allocation for part number KV29 to this number. Display all the data in the Part table.

8. Increase the length of the Warehouse field in the Part table to two characters. Change the warehouse number for warehouse 1 to 1a. Display all the data in the Part table.

9. Delete the Allocation field from the Part table. Display all the data in the Part table.

10. What command would delete the Part table from the Premiere Products database? (Do not delete the Part table.)

11. Write a stored procedure that will change the price of a part with a given part number. How would you use this stored procedure to change the price of part AT94 to $26.95?

12. Write the code for the following triggers following the style shown in the text.

    a. When adding a customer, add the customer's balance times the sales rep's commission rate to the commission for the corresponding sales rep.

    b. When updating a customer, add the difference between the new balance and the old balance multipled by the sales rep's commission rate to the commission for the corresponding sales rep.

    c. When deleting a customer, subtract the balance multiplied by the sales rep's commission rate from the commission for the corresponding sales rep.

Henry Books Case

Ray Henry would like you to complete the following tasks to help him maintain his database. In the following exercises, you will use the data in the Henry Books database shown in Figures 1-17 through 1-20 in Chapter 1. (If you use a computer to complete these exercises, use a copy of the original Henry Books database so your data will not reflect the changes you made in Chapter 3.) If you have access to a DBMS, use the DBMS to perform the tasks and explain the steps you used in the process. If not, explain how you would use SQL to obtain the desired results. Check with your instructor if you are uncertain about which approach to take.

1. Create a view named PenguinBooks. It consists of the book code, book title, book type, and book price for every book published by Penguin USA. Display the data in the view.

2. Create a view named Paperback. It consists of the book code, book title, publisher name, and book price for every book that is available in paperback. Display the data in the view.

3. Create a view named BookInventory. It consists of the book code and the total number of units of the book on hand at any branch. Display the data in the view.

4. Create the following indexes. If it is necessary to name the index in your DBMS, use the indicated name.

    a. Create an index named BookIndex1 on the PublisherName field in the Publisher table.

b. Create an index named BookIndex2 on the Type field in the Book table.

c. Create an index named BookIndex3 on the Type and Price fields in the Book table and list the prices in descending order.

5. Drop the BookIndex3 index.

6. Specify the integrity constraint that the price of any book must be less than $90.

7. Ensure that the following are foreign keys (that is, specify referential integrity) within the Henry Books database.

a. PublisherCode is a foreign key in the Book table.

b. BranchNum is a foreign key in the Inventory table.

c. AuthorNum is a foreign key in the Wrote table.

8. Add to the Book table a new character field named Classic that is one character in length.

9. Change the Classic field in the Book table to Y for the book titled *The Grapes of Wrath*.

10. Change the length of the Title field in the Book table to 60.

11. What command would delete the Books table from the Henry Books database? (Do not delete the Book table.)

12. Write a stored procedure that will change the price of a book with a given book code. How would you use this stored procedure to change the price of book 0189 to $8.49?

13. Assume the Book table contains a column called TotalOnHand that represents the total units on hand in all branches for that book. Following the style shown in the text, write the code for the following triggers:

a. When inserting a row in the Inventory table, add the OnHand value to the TotalOnHand value for the appropriate book.

b. When updating a row in the Inventory table, add the difference between the new OnHand value and the old OnHand value to the TotalOnHand value for the appropriate book.

c. When deleting a row in the Inventory table, subtract the OnHand value from the TotalOnHand value for the appropriate book.

Alexamara Marina Group Case

In the following exercises, you will use the data in the Alexamara Marina Group database shown in Figures 1-20 through 1-24. (If you use a computer to complete these exercises, use a copy of the original Alexamara database so your data will not reflect the changes you made in Chapter 3.) If you have access to a DBMS, use the DBMS to perform the tasks and explain the steps you used in the process. If not, explain how you would use SQL to obtain the desired results. Check with your instructor if you are not certain about which approach to take.

1. Create a view named LargeSlip using the data in the MarinaNum, SlipNum, RentalFee, BoatName, and OwnerNum columns in the MarinaSlip table for those slips with lengths of 40 feet. Display the data in the view.

2. Create a view named InitialService using the slip ID, category number, category description, and estimated hours for every service request for which the spent hours are zero. Display the data in the view.

3. Create a view named TypesOfBoats using the boat type and a count of all boats of each type. Display the data in the view.

4. Create the following indexes. If it is necessary to name the index in your DBMS, use the indicated name.

a. Create an index named BoatIndex on the BoatName field in the MarinaSlip table.

b. Create an index named BoatIndex2 on the OwnerNum field in the MarinaSlip table.

c. Create an index named BoatIndex3 on the Length and BoatName fields in the MarinaSlip table and list the lengths in descending order.

5. Drop the BoatIndex3 index.

6. Assume the MarinaSlip table has been created but there are no integrity constraints. Create the necessary integrity constraints so the rental fee must be less than $5,000 and the slip length must be 25, 30, or 40.

7. Ensure that the following are foreign keys (that is, specify referential integrity) in the Alexamara database.

   a. MarinaNum is a foreign key in the MarinaSlip table.

   b. OwnerNum is a foreign key in the MarinaSlip table.

   c. CategoryNum is a foreign key in the ServiceRequest table.

   d. SlipID is a foreign key in the SerivceRequest table.

8. Add to the MarinaSlip table a new character field named FeePaid that is one character in length. On all records, change the value for the FeePaid field to Y.

9. Change the FeePaid field in the MarinaSlip table to N for the slip whose slip ID is 4.

10. Change the length of the BoatName field in the MarinaSlip table to 60.

11. Write a stored procedure that will change the rental fee of a slip with a given slip ID. How would you use this stored procedure to change the rental fee of slip ID 3 to 3,700.00?

12. Assume the Owner table contains a column called TotalRental that represents the total rental fee for all slips rented by that owner. Write the code for the following triggers following the style shown in the text.

   a. When inserting a row in the MarinaSlip table, add the rental fee to the total rental for the appropriate owner.

   b. When updating a row in the MarinaSlip table, add the difference between the new rental fee and the old rental fee to the total rental for the appropriate owner.

   c. When deleting a row in the MarinaSlip table, subtract the rental fee from the total rental for the appropriate owner.

DATABASE DESIGN 1: NORMALIZATION

INTRODUCTION

You have examined the basic relational model, its structure, and the various ways of manipulating data within a relational database. In this chapter, you will learn about the normalization process and its underlying concepts and features. The **normalization process** enables you to identify the existence of potential problems, called **update anomalies**, in the design of a relational database. This process also supplies methods for correcting these problems.

To correct update anomalies in a database, you must convert tables to various types of **normal forms**. A table in a particular normal form possesses a certain desirable collection of properties. The most common normal forms are first normal form (1NF), second normal form (2NF), third normal form (3NF), and fourth normal form (4NF). Normalization is a progression in which a table that is in first normal form is better (freer from problems) than a table that is not in first normal form, a table that is in second normal form is better than one that is in first normal form, and so on. The goal of normalization is to take a table or collection of tables and produce a new collection of tables that represents the same information but that is free of update anomalies.

In this chapter, you will learn about two crucial concepts that are fundamental to understanding the normalization process: functional dependence and keys. You will also learn about first, second, third, and fourth normal form.

Many of the examples in this chapter use data from the Premiere Products database, which is shown in Figure 5-1.

Rep

RepNum	LastName	FirstName	Street	City	State	Zip	Commission	Rate
20	Kaiser	Valerie	624 Randall	Grove	FL	33321	$20,542.50	0.05
35	Hull	Richard	532 Jackson	Sheldon	FL	33553	$39,216.00	0.07
65	Perez	Juan	1626 Taylor	Fillmore	FL	33336	$23,487.00	0.05

Customer

CustomerNum	CustomerName	Street	City	State	Zip	Balance	CreditLimit	RepNum
148	Al's Appliance and Sport	2837 Greenway	Fillmore	FL	33336	$6,550.00	$7,500.00	20
282	Brookings Direct	3827 Devon	Grove	FL	33321	$431.50	$10,000.00	35
356	Ferguson's	382 Wildwood	Northfield	FL	33146	$5,785.00	$7,500.00	65
408	The Everything Shop	1828 Raven	Crystal	FL	33503	$5,285.25	$5,000.00	35
462	Bargains Galore	3829 Central	Grove	FL	33321	$3,412.00	$10,000.00	65
524	Kline's	838 Ridgeland	Fillmore	FL	33336	$12,762.00	$15,000.00	20
608	Johnson's Department Store	372 Oxford	Sheldon	FL	33553	$2,106.00	$10,000.00	65
687	Lee's Sport and Appliance	282 Evergreen	Altonville	FL	32543	$2,851.00	$5,000.00	35
725	Deerfield's Four Seasons	282 Columbia	Sheldon	FL	33553	$248.00	$7,500.00	35
842	All Season	28 Lakeview	Grove	FL	33321	$8,221.00	$7,500.00	20

Orders

OrderNum	OrderDate	CustomerNum
21608	10/20/2010	148
21610	10/20/2010	356
21613	10/21/2010	408
21614	10/21/2010	282
21617	10/23/2010	608
21619	10/23/2010	148
21623	10/23/2010	608

OrderLine

OrderNum	PartNum	NumOrdered	QuotedPrice
21608	AT94	11	$21.95
21610	DR93	1	$495.00
21610	DW11	1	$399.99
21613	KL62	4	$329.95
21614	KT03	2	$595.00
21617	BV06	2	$794.95
21617	CD52	4	$150.00
21619	DR93	1	$495.00
21623	KV29	2	$1,290.00

Part

PartNum	Description	OnHand	Class	Warehouse	Price
AT94	Iron	50	HW	3	$24.95
BV06	Home Gym	45	SG	2	$794.95
CD52	Microwave Oven	32	AP	1	$165.00
DL71	Cordless Drill	21	HW	3	$129.95
DR93	Gas Range	8	AP	2	$495.00
DW11	Washer	12	AP	3	$399.99
FD21	Stand Mixer	22	HW	3	$159.95
KL62	Dryer	12	AP	1	$349.95
KT03	Dishwasher	8	AP	3	$595.00
KV29	Treadmill	9	SG	2	$1,390.00

FIGURE 5-1 Premiere Products data

FUNCTIONAL DEPENDENCE

Understanding functional dependence is crucial to learning the material in the rest of this chapter. **Functional dependence** is a formal name for what is basically a simple idea. To understand functional dependence, suppose the Rep table for Premiere Products contains an additional column named PayClass, as shown in Figure 5-2.

Rep

RepNum	LastName	FirstName	Street	City	State	Zip	Commission	PayClass	Rate
20	Kaiser	Valerie	624 Randall	Grove	FL	33321	$20,542.50	1	0.05
35	Hull	Richard	532 Jackson	Sheldon	FL	33553	$39,216.00	2	0.07
65	Perez	Juan	1626 Taylor	Fillmore	FL	33336	$23,487.00	1	0.05

FIGURE 5-2   Rep table with additional column, PayClass

Assume one of the policies at Premiere Products is that all sales reps in any given pay class earn the same commission rate. How might you convey this fact to someone else? You might say that a sales rep's pay class *determines* his or her commission rate. Another way to convey this fact is to say that a sales rep's commission rate *depends on* his or her pay class. This phrasing uses the words *determines* and *depends on* exactly the way you will use them in connection with database design. If you wanted to be more formal, you would precede either expression with the word *functionally*. Thus, you might say, "A sales rep's pay class *functionally determines* his or her commission rate" or "A sales rep's commission rate *functionally depends on* his or her pay class."

The formal definition of functional dependence is as follows:

Definition: A column (attribute) B is **functionally dependent** on another column A (or possibly a collection of columns) when each value for A in the database is associated with exactly one value of B.

You can think of functional dependence as follows: If you are given a value for A in the database, do you know whether it will be associated with exactly one value of B? If so, B is functionally dependent on A (written as A → B). If B is functionally dependent on A, you can also say that A **functionally determines** B.

In the Rep table, LastName is functionally dependent on RepNum. If you are given a value of 20 for RepNum, for example, you know that you will find a *single* LastName (in this case, Kaiser) associated with it. (**Note:** You need to be concerned only with actual values of RepNum in the database. If you are given a value of 21 for RepNum, for example, you will not find any names associated with it because there is no row in the Rep table on which the rep number is 21.)

Q & A

Question: In the Customer table, is CustomerName functionally dependent on RepNum?
Answer: No. Rep number 20, for example, occurs on a row in which the customer name is Al's Appliance and Sport, on a row in which the customer name is Kline's, and on a row in which the customer name is All Season. Thus, a rep number can be associated with more than one customer name.

Q & A

Question: In the OrderLine table, is QuotedPrice functionally dependent on OrderNum?
Answer: No. Order number 21617, for example, occurs on a row in which the quoted price is $794.95 and on a row in which the quoted price is $150.00. Thus, an order number can be associated with more than one quoted price.

Q & A

Question: Is QuotedPrice functionally dependent on PartNum?
Answer: No. A given part number such as DR93 can occur on more than one row in the OrderLine table. The quoted price can be different on each row. With current data, the two rows on which the part number is DR93 actually have the same price, $495.00. What is important, however, is not that they happen to be the same now, but whether there is a guarantee that they will always be the same. That is certainly not the case. In fact, the reason for storing quoted prices in the OrderLine table is to allow two rows with the same part number to have different prices.

Q & A

Question: On which columns is QuotedPrice functionally dependent?
Answer: For any combination of an order number and a part number, there can be only one row in the OrderLine table. Thus, any combination of an order number and part number in the OrderLine table is associated with exactly one quoted price. Consequently, QuotedPrice is functionally dependent on the combination (formally called the **concatenation**) of OrderNum and PartNum.

At this point, a question naturally arises: How do you determine functional dependencies? Can you determine them by looking at sample data, for example? The answer is no.

Consider the Rep table shown in Figure 5-3, in which all last names are unique. It is very tempting to say that LastName functionally determines Street, City, State, and Zip (or equivalently that Street, City, State, and Zip are all functionally dependent on LastName). After all, given the last name of a rep, you can find his or her address.

Rep

RepNum	LastName	FirstName	Street	City	State	Zip	Commission	Rate
20	Kaiser	Valerie	624 Randall	Grove	FL	33321	$20,542.50	0.05
35	Hull	Richard	532 Jackson	Sheldon	FL	33553	$39,216.00	0.07
65	Perez	Juan	1626 Taylor	Fillmore	FL	33336	$23,487.00	0.05

FIGURE 5-3 Rep table

What happens when you add rep 85, whose last name also is Kaiser, to the database? Now you have the situation illustrated in Figure 5-4. If the last name you are given is Kaiser, you no longer can find a single address. Thus, you were misled by the original sample data. The only way to determine the functional dependencies that exist is to examine users' policies through discussions with users, an examination of user documentation, and so on.

Rep

RepNum	LastName	FirstName	Street	City	State	Zip	Commission	Rate
20	Kaiser	Valerie	624 Randall	Grove	FL	33321	$20,542.50	0.05
35	Hull	Richard	532 Jackson	Sheldon	FL	33553	$39,216.00	0.07
65	Perez	Juan	1626 Taylor	Fillmore	FL	33336	$23,487.00	0.05
85	Kaiser	Wil	172 Bahia	Norton	FL	39281	$0.00	0.05

FIGURE 5-4 Rep table with second rep named Kaiser added

Q & A

Question: Assume the following columns exist in a relation named Student:

- StudentNum (student number)
- StudentLast (student last name)
- StudentFirst (student first name)
- HighSchoolNum (number of the high school from which the student graduated)
- HighSchoolName (name of the high school from which the student graduated)
- AdvisorNum (number of the student's advisor)
- AdvisorLast (last name of the student's advisor)
- AdvisorFirst (first name of the student's advisor)

Student numbers, high school numbers, and advisor numbers are unique; no two students have the same number, no two high schools have the same number, and no two advisors have the same number. Use this information to determine the functional dependencies in the Student relation.

Answer: Because student numbers are unique, any given student number in the database is associated with a single last name, first name, high school number, high school name, advisor number, advisor last name, and advisor first name. Thus, all the other columns in the Student relation are functionally dependent on StudentNum, which is represented as follows:

```
StudentNum → StudentLast, StudentFirst, HighSchoolNum, HighSchoolName,
      AdvisorNum, AdvisorLast, AdvisorFirst
```

Because two students can have the same first and last names, StudentFirst and StudentLast do not determine anything else. Because high school numbers are unique, any given high school number is associated with exactly one high school name. If high school 128 is Robbins High, for example, any student whose high school number is 128 *must* have the high school name Robbins High. Thus, HighSchoolName is functionally dependent on HighSchoolNum, which is represented as follows:

```
HighSchoolNum → HighSchoolName
```

Because advisor numbers are unique, any given advisor number is associated with exactly one advisor first name and exactly one advisor last name. If advisor 20 is Mary Webb, for example, any student whose advisor number is 20 *must* have the advisor's first name Mary and the advisor's last name Webb. Thus, AdvisorFirst and AdvisorLast are functionally dependent on AdvisorNum, which is represented as follows:

```
AdvisorNum → AdvisorLast, AdvisorFirst
```

As with students, an advisor's first and last names are not necessarily unique, so AdvisorFirst and AdvisorLast do not determine anything. The complete collection of functional dependencies is as follows:

```
StudentNum → StudentLast, StudentFirst, HighSchoolNum, HighSchoolName,
      AdvisorNum, AdvisorLast, AdvisorFirst
HighSchoolNum → HighSchoolName
AdvisorNum → AdvisorLast, AdvisorFirst
```

KEYS

A second underlying concept of the normalization process is that of the primary key. You already encountered the basic concept of the primary key in earlier chapters. In this chapter, however, you need a more precise definition.

Definition: Column A (or a collection of columns) is the **primary key** for a relation (table) R, if:

Property 1. *All* columns in R are functionally dependent on A.

Property 2. No subcollection of the columns in A (assuming A is a collection of columns and not just a single column) also has Property 1.

Q & A

Question: Is Class the primary key for the Part table?

Answer: No, because the other columns are not functionally dependent on the class. The item class HW, for example, appears on a row in the Part table in which the part number is AT94, a row in which the part number is DL71, and a row in which the part number is FD21. The item class HW is associated with three part numbers, so the part number is not functionally dependent on the class.

Q & A

Question: Is CustomerNum the primary key for the Customer table?

Answer: Yes, because customer numbers are unique. A given customer number cannot appear on more than one row. Thus, each customer number is associated with a single name, a single street, a single city, a single state, a single zip code, a single balance, a single credit limit, and a single rep number. In other words, all columns in the Customer table are functionally dependent on CustomerNum.

Q & A

Question: Is OrderNum the primary key for the OrderLine table?

Answer: No, because it does not uniquely determine NumOrdered or QuotedPrice. The order number 21617, for example, appears on a row in the OrderLine table in which the number ordered is 2 and the quoted price is $794.95 and on a row in which the number ordered is 4 and the quoted price is $150.00.

Q & A

Question: Is the combination of OrderNum and PartNum the primary key for the OrderLine table?

Answer: Yes, because all columns are functionally dependent on this combination. Any combination of an order number and a part number occurs on only one row in the OrderLine table and is associated with only one value for NumOrdered and only one value for QuotedPrice. Further, neither OrderNum nor PartNum alone has this property. For example, order number 21617 appears on more than one row, as does part DR93.

Q & A

Question: Is the combination of PartNum and Description the primary key for the Part table?

Answer: No. It is true that this combination functionally determines all columns in the Part table. PartNum alone, however, also has this property.

Q & A

Question: You already determined the functional dependencies in a Student relation containing the following columns: StudentNum, StudentLast, StudentFirst, HighSchoolNum, HighSchoolName, AdvisorNum, AdvisorLast, and AdvisorFirst. The functional dependencies you determined were as follows:

```
StudentNum → StudentLast, StudentFirst, HighSchoolNum, HighSchoolName, AdvisorNum,
     AdvisorLast, AdvisorFirst
HighSchoolNum → HighSchoolName
AdvisorNum → AdvisorLast, AdvisorFirst
```

What is the primary key for the Student relation?

Answer: The only column that determines all the other columns is StudentNum, so it is the primary key for the Student relation.

Occasionally (but not often), there might be more than one possibility for the primary key. For example, if the Premiere Products database included an Employee table to store employee numbers and Social Security numbers, either the employee number or the Social Security number could serve as the primary key. In this case, both columns are referred to as candidate keys. Like a primary key, a **candidate key** is a column or a collection of columns on which all columns in the table are functionally dependent; the definition for primary key also defines a candidate key. From all the candidate keys, one is chosen to be the primary key. The candidate keys that are not chosen as the primary key are often referred to as **alternate keys**.

NOTE

The primary key is often called the *key* in other studies on database management and the relational model. This text will continue to use the term *primary key* to distinguish between the different concepts of a key that you will encounter throughout this text.

FIRST NORMAL FORM

A relation (table) that contains a **repeating group** (or multiple entries for a single record) is called an **unnormalized relation**. Removing repeating groups is the starting point in the quest to create tables that are as free of problems as possible. Tables without repeating groups are said to be in first normal form.

Definition: A table (relation) is in **first normal form (1NF)** when it does not contain repeating groups.

As an example, consider the Orders table shown in Figure 5-5, in which there is a repeating group consisting of PartNum and NumOrdered.

Orders

OrderNum	OrderDate	PartNum	NumOrdered
21608	10/20/2010	AT94	11
21610	10/20/2010	DR93	1
		DW11	1
21613	10/21/2010	KL62	4
21614	10/21/2010	KT03	2
21617	10/23/2010	BV06	2
		CD52	4
21619	10/23/2010	DR93	1
21623	10/23/2010	KV29	2

FIGURE 5-5 Sample unnormalized table

The notation for describing the Orders table is as follows:

```
Orders (OrderNum, OrderDate, (PartNum, NumOrdered) )
```

This notation indicates a table named Orders consisting of a primary key (OrderNum) and a column named OrderDate. The inner parentheses indicate that there is a repeating group. The repeating group contains two columns, PartNum and NumOrdered. This means that for a single order, there can be multiple combinations of a part number and a corresponding number of units ordered, as illustrated in Figure 5-5. The row for order 21617, for example, contains two such combinations. In the first combination, the part number is BV06 and the number ordered is 2. In the second combination, the part number is CD52 and the number ordered is 4.

To convert the Orders table to first normal form, you remove the repeating group as follows:

Orders (<u>OrderNum</u>, OrderDate, <u>PartNum</u>, NumOrdered)

Figure 5-6 shows the new table, which is now in first normal form.

Orders

OrderNum	OrderDate	PartNum	NumOrdered
21608	10/20/2010	AT94	11
21610	10/20/2010	DR93	1
21610	10/20/2010	DW11	1
21613	10/20/2010	KL62	4
21614	10/20/2010	KT03	2
21617	10/20/2010	BV06	2
21617	10/20/2010	CD52	4
21619	10/20/2010	DR93	1
21623	10/20/2010	KV29	2

FIGURE 5-6   Result of normalization (conversion to first normal form)

Note that the fifth row of the unnormalized table (see Figure 5-5) indicates that part BV06 and part CD52 are both present for order 21617. In the normalized table (see Figure 5-6), this information is represented by *two* rows, the sixth and seventh. The primary key to the unnormalized Orders table was OrderNum alone. The primary key to the normalized table is now the combination of OrderNum and PartNum.

In general, when converting a table that is not in first normal form to first normal form, the primary key will usually include the original primary key concatenated with the key to the repeating group, which is the column that distinguishes one occurrence of the repeating group from another on a given row in the table. In this case, PartNum is the key to the repeating group; thus, PartNum becomes part of the primary key of the first normal form table.

SECOND NORMAL FORM

A table that is in first normal form still might contain problems that will require you to restructure it. Consider the following table:

Orders (<u>OrderNum</u>, OrderDate, <u>PartNum</u>, Description, NumOrdered, QuotedPrice)

This table has the following functional dependencies:

OrderNum $\rightarrow$ OrderDate
PartNum $\rightarrow$ Description
OrderNum, PartNum $\rightarrow$ NumOrdered, QuotedPrice, OrderDate, Description

This notation indicates that OrderNum alone determines OrderDate and that PartNum alone determines Description but that *both* an OrderNum *and* a PartNum are required to determine either NumOrdered or QuotedPrice. (The combination of OrderNum and PartNum also determines both OrderDate and Description because OrderNum determines OrderDate and PartNum determines Description.) Consider the sample of this table shown in Figure 5-7.

Orders

OrderNum	OrderDate	PartNum	Description	NumOrdered	QuotedPrice
21608	10/20/2010	AT94	Iron	11	$21.95
21610	10/20/2010	DR93	Gas Range	1	$495.00
21610	10/20/2010	DW11	Washer	1	$399.99
21613	10/21/2010	KL62	Dryer	4	$329.95
21614	10/21/2010	KT03	Dishwasher	2	$595.00
21617	10/23/2010	BV06	Home Gym	2	$794.95
21617	10/23/2010	CD52	Microwave Oven	4	$150.00
21619	10/23/2010	DR93	Gas Range	1	$495.00
21623	10/23/2010	KV29	Treadmill	2	$1290.00

FIGURE 5-7   Sample Orders table

The description of a specific part (DR93, for example) occurs twice in the table. This redundancy causes several problems. It is wasteful of space, but that is not nearly as serious as some of the other problems. These other problems are called update anomalies, and they fall into four categories:

1. **Update.** A change to the description of part DR93 requires not one change to the table, but two changes—you have to change each row on which part DR93 appears. Changing multiple rows makes the update process more cumbersome; it also is more complicated logically and takes more time to update.
2. **Inconsistent data.** There is nothing about the design that would prohibit part DR93 from having two different descriptions in the database. In fact, if part DR93 were to occur on 20 rows, it could potentially have 20 *different* descriptions in the database!
3. **Additions.** You have a real problem when you try to add a new part and its description to the database. Because the primary key for the table consists of both OrderNum and PartNum, you need values for both columns when you want to add a new row. If you have a part to add but there are no orders for it yet, what do you use for an order number? The only solution is to make up a dummy order number and then replace it with a real order number after an order for the new part is received. Certainly, this is not an acceptable solution.
4. **Deletions.** If you deleted order 21608 from the database, you would *lose* all information about part AT94. For example, you would no longer know that part AT94 is an iron.

These problems occur because you have a column, Description, that is dependent on only a portion of the primary key (PartNum) and *not* on the complete primary key. This problem leads to the definition of second normal form. Second normal form represents an improvement over first normal form because it eliminates update anomalies in these situations. To understand second normal form, you need to understand the term *nonkey column*.

Definition: A column is a **nonkey column** (also called a **nonkey attribute**) when it is not a part of the primary key.

Definition: A table (relation) is in **second normal form (2NF)** when it is in first normal form and no nonkey column is dependent on only a portion of the primary key.

For another perspective on second normal form, consider Figure 5-8. This type of diagram, sometimes called a **dependency diagram**, uses arrows to indicate all the functional dependencies present in the Orders table. The arrows above the boxes indicate the normal dependencies that should be present; in other words, the primary key functionally determines all other columns. (In the Orders table, the concatenation of OrderNum and PartNum determines all other columns.) The arrows below the boxes prevent the table from being in second normal form. These arrows represent types of dependencies that are often called **partial dependencies**, which are dependencies on only a portion of the primary key. In fact, another definition for second normal form is a table that is in first normal form but that contains no partial dependencies.

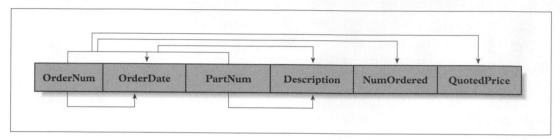

FIGURE 5-8   Dependences in the Orders table

Regardless of which definition of second normal form you use, you now can identify the fundamental problem with the Orders table: It is *not* in second normal form. Although it may be pleasing to have a name for the problem, what you really need is a method to *correct* it; you need a way to convert tables to second normal form. To do so, first take each subset of the set of columns that makes up the primary key; then begin a new table with this subset as the primary key. For the Orders table, this would give the following:

```
(OrderNum,
(PartNum,
(OrderNum,  PartNum,
```

Next, place each of the other columns with its appropriate primary key; that is, place each primary key with the minimal collection of columns on which it depends. For the Orders table, this would yield the following:

```
(OrderNum,  OrderDate)
(PartNum,  Description)
(OrderNum,  PartNum,  NumOrdered,  QuotedPrice)
```

Now you can give each new table a name that is descriptive of the table's contents, such as Orders, Part, or OrderLine. Figure 5-9 shows the original Orders table on top; the resulting Orders, Part, and OrderLine tables created after the Orders table was converted to second normal form appear below it.

Orders

OrderNum	OrderDate	PartNum	Description	NumOrdered	QuotedPrice
21608	10/20/2010	AT94	Iron	11	$21.95
21610	10/20/2010	DR93	Gas Range	1	$495.00
21610	10/20/2010	DW11	Washer	1	$399.99
21613	10/21/2010	KL62	Dryer	4	$329.95
21614	10/21/2010	KT03	Dishwasher	2	$595.00
21617	10/23/2010	BV06	Home Gym	2	$794.95
21617	10/23/2010	CD52	Microwave Oven	4	$150.00
21619	10/23/2010	DR93	Gas Range	1	$495.00
21623	10/23/2010	KV29	Treadmill	2	$1290.00

Orders

OrderNum	OrderDate
21608	10/20/2010
21610	10/20/2010
21613	10/21/2010
21614	10/21/2010
21617	10/23/2010
21619	10/23/2010
21623	10/23/2010

Part

PartNum	Description
AT94	Iron
BV06	Home Gym
CD52	Microwave Oven
DL71	Cordless Drill
DR93	Gas Range
DW11	Washer
FD21	Stand Mixer
KL62	Dryer
KT03	Dishwasher
KV29	Treadmill

OrderLine

OrderNum	PartNum	NumOrdered	QuotedPrice
21608	AT94	11	$21.95
21610	DR93	1	$495.00
21610	DW11	1	$399.99
21613	KL62	4	$329.95
21614	KT03	2	$595.00
21617	BV06	2	$794.95
21617	CD52	4	$150.00
21619	DR93	1	$495.00
21623	KV29	2	$1290.00

FIGURE 5-9　Conversion to second normal form

Now you have eliminated the update anomalies. A description appears only once for each part, so you do not have the redundancy that you did in the previous design. Changing the description for part DR93 from Gas Range to Deluxe Range, for example, now is a simple process involving a single change. Because the description for a part occurs in a single place, it is not possible to have multiple descriptions for a single part in the database at the same time.

To add a new part and its description, you create a new row in the Part table; there is no need to have an existing order for that part. Also, deleting order 21608 does not delete part AT94 from the Part table, and you still have its description (Iron) in the database. Finally, you have not lost any information in the process—you can reconstruct the data in the original design from the data in the new design.

THIRD NORMAL FORM

Problems can still exist with tables that are in second normal form. Consider the following Customer table:

Customer (<u>CustomerNum</u>, CustomerName, Balance, CreditLimit, RepNum, LastName, FirstName)

The functional dependencies in this table are as follows:

CustomerNum $\rightarrow$ CustomerName, Balance, CreditLimit, RepNum, LastName, FirstName
RepNum $\rightarrow$ LastName, FirstName

CustomerNum determines all the other columns. In addition, RepNum determines LastName and FirstName.

When the primary key of a table is a single column, the table is automatically in second normal form. (If the table were not in second normal form, some columns would be dependent on only a *portion* of the primary key, which is impossible when the primary key is just one column.) Thus, the Customer table is in second normal form.

The sample Customer table shown in Figure 5-10 illustrates that this table possesses problems similar to those encountered earlier even though it is in second normal form. In this case, the name of a sales rep can occur many times in the table; see sales rep 65 (Juan Perez), for example.

Customer

CustomerNum	CustomerName	Balance	CreditLimit	RepNum	LastName	FirstName
148	Al's Appliance and Sport	$6,550.00	$7,500.00	20	Kaiser	Valerie
282	Brookings Direct	$431.50	$10,000.00	35	Hull	Richard
356	Ferguson's	$5,785.00	$7,500.00	65	Perez	Juan
408	The Everything Shop	$5,285.25	$5,000.00	35	Hull	Richard
462	Bargains Galore	$3,412.00	$10,000.00	65	Perez	Juan
524	Kline's	$12,762.00	$15,000.00	20	Kaiser	Valerie
608	Johnson's Department Store	$2,106.00	$10,000.00	65	Perez	Juan
687	Lee's Sport and Appliance	$2,851.00	$5,000.00	35	Hull	Richard
725	Deerfield's Four Seasons	$248.00	$7,500.00	35	Hull	Richard
842	All Season	$8,221.00	$7,500.00	20	Kaiser	Valerie

FIGURE 5-10   Sample Customer table

This redundancy creates the same set of problems that you examined in the first normal form Orders table. In addition to the problem of wasted space, you have similar update anomalies as follows:

1. *Updates.* A change to the name of a sales rep requires not one change to the table, but several, making the update process cumbersome.
2. *Inconsistent data.* There is nothing about the design that would prohibit a sales rep from having two different names in the database. In fact, if the same sales rep represents 20 customers (and thus would be found on 20 different rows), he or she could have 20 *different* names in the database.
3. *Additions.* In order to add sales rep 87 (Mary Daniels) to the database, she must already represent at least one customer. If she has not yet been assigned any customers, you must add her record and create a fictitious customer for her to represent. Again, this is not a desirable solution to the problem.
4. *Deletions.* If you deleted all the customers of sales rep 35 from the database, you would lose all information concerning sales rep 35.

These update anomalies are due to the fact that RepNum determines LastName and FirstName, but RepNum is not the primary key. As a result, the same RepNum and consequently the same LastName and FirstName can appear on many different rows.

You've seen that second normal form is an improvement over first normal form; but to eliminate second normal form problems, you need an even better strategy for creating tables in the database. Third normal form provides that strategy. Before looking at third normal form, however, you need to become familiar with the special name that is given to any column that determines another column (as RepNum does in the Customer table).

Definition: Any column (or collection of columns) that determines another column is called a **determinant**.
Certainly, the primary key in a table will be a determinant. In fact, by definition, any candidate key will be a determinant. (Remember that a candidate key is a column or a collection of columns that could function as the primary key.) In this case, RepNum is a determinant; but it is not a candidate key, and that is the problem.

Definition: A table (relation) is in **third normal form (3NF)** when it is in second normal form and the only determinants it contains are candidate keys.

NOTE

The previous definition is not the original definition of third normal form. This more recent definition, which is preferable to the original, is often referred to as **Boyce-Codd normal form (BCNF)** when it is important to make a distinction between this definition and the original. This text will not make such a distinction, but will take this to be the definition of third normal form.

Again, for an additional perspective, you can use a dependency diagram, as shown in Figure 5-11. The arrows above the boxes represent the normal dependencies of all columns on the primary key. The arrows below the boxes represent the problem—these arrows make RepNum a determinant. If there were arrows from RepNum to all the columns, RepNum would be a candidate key and you would not have a problem. The absence of these arrows indicates that this table contains a determinant that is not a candidate key. Thus, the table is not in third normal form.

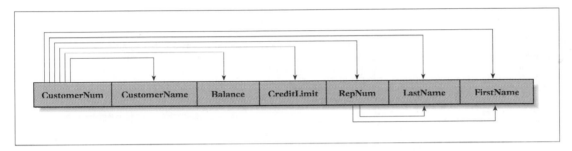

FIGURE 5-11   Dependencies in the Customer table

You now have identified the problem with the Customer table: It is not in third normal form. The following method corrects the deficiency in the Customer table and in all tables having similar deficiencies.

First, for each determinant that is not a candidate key, remove from the table the columns that depend on this determinant (but don't remove the determinant). Next, create a new table containing all the columns from the original table that depend on this determinant. Finally, make the determinant the primary key of this new table. In the Customer table, for example, you would remove LastName and FirstName because they depend on the determinant RepNum, which is not a candidate key. A new table is formed, consisting of RepNum (as the primary key), LastName, and FirstName.

```
Customer (CustomerNum, CustomerName, Balance, CreditLimit, RepNum)
Rep (RepNum, LastName, FirstName)
```

Figure 5-12 shows samples of the revised Customer table and the new Rep table.

Customer

CustomerNum	CustomerName	Balance	CreditLimit	RepNum	LastName	FirstName
148	Al's Appliance and Sport	$6,550.00	$7,500.00	20	Kaiser	Valerie
282	Brookings Direct	$431.50	$10,000.00	35	Hull	Richard
356	Ferguson's	$5,785.00	$7,500.00	65	Perez	Juan
408	The Everything Shop	$5,285.25	$5,000.00	35	Hull	Richard
462	Bargains Galore	$3,412.00	$10,000.00	65	Perez	Juan
524	Kline's	$12,762.00	$15,000.00	20	Kaiser	Valerie
608	Johnson's Department Store	$2,106.00	$10,000.00	65	Perez	Juan
687	Lee's Sport and Appliance	$2,851.00	$5,000.00	35	Hull	Richard
725	Deerfield's Four Seasons	$248.00	$7,500.00	35	Hull	Richard
842	All Season	$8,221.00	$7,500.00	20	Kaiser	Valerie

Customer

CustomerNum	CustomerName	Balance	CreditLimit	RepNum
148	Al's Appliance and Sport	$6,550.00	$7,500.00	20
282	Brookings Direct	$431.50	$10,000.00	35
356	Ferguson's	$5,785.00	$7,500.00	65
408	The Everything Shop	$5,285.25	$5,000.00	35
462	Bargains Galore	$3,412.00	$10,000.00	65
524	Kline's	$12,762.00	$15,000.00	20
608	Johnson's Department Store	$2,106.00	$10,000.00	65
687	Lee's Sport and Appliance	$2,851.00	$5,000.00	35
725	Deerfield's Four Seasons	$248.00	$7,500.00	35
842	All Season	$8,221.00	$7,500.00	20

Rep

RepNum	LastName	FirstName
20	Kaiser	Valerie
35	Hull	Richard
65	Perez	Juan

FIGURE 5-12 Conversion to third normal form

Have you now corrected all previously identified problems? A sales rep's name appears only once, thus avoiding redundancy and simplifying the process of changing a sales rep's name. With this design, it is not possible for the same sales rep to have different names in the database. To add a new sales rep to the database, you can add a row in the Rep table without requiring the rep to have at least one assigned customer. Finally, deleting all the customers of a given sales rep will not remove the sales rep's record from the Rep table, retaining the sales rep's name. In addition, you can reconstruct all the data in the original table from the data in the new collection of tables. All previously mentioned problems have indeed been solved.

INCORRECT DECOMPOSITIONS

It is important to note that the decomposition of a table into two or more third normal form tables *must* be accomplished by the method indicated in the previous section, even though there are other possibilities that at first glance might seem to be legitimate. For example, you can examine two other decompositions of the Customer table into third normal form tables to understand the difficulties they pose. Assume in the decomposition process that

Customer (<u>CustomerNum</u>, CustomerName, Balance, CreditLimit, RepNum, LastName, FirstName)

    is replaced by

Customer (<u>CustomerNum</u>, CustomerName, Balance, CreditLimit, RepNum)
Rep (<u>CustomerNum</u>, LastName, FirstName)

    Samples of these tables appear in Figure 5-13. Both new tables are in third normal form. In addition, by joining these two tables on CustomerNum, you can reconstruct the original Customer table. The result, however, still suffers from some of the same kinds of problems as the original Customer table.

Customer

CustomerNum	CustomerName	Balance	CreditLimit	RepNum	LastName	FirstName
148	Al's Appliance and Sport	$6,550.00	$7,500.00	20	Kaiser	Valerie
282	Brookings Direct	$431.50	$10,000.00	35	Hull	Richard
356	Ferguson's	$5,785.00	$7,500.00	65	Perez	Juan
408	The Everything Shop	$5,285.25	$5,000.00	35	Hull	Richard
462	Bargains Galore	$3,412.00	$10,000.00	65	Perez	Juan
524	Kline's	$12,762.00	$15,000.00	20	Kaiser	Valerie
608	Johnson's Department Store	$2,106.00	$10,000.00	65	Perez	Juan
687	Lee's Sport and Appliance	$2,851.00	$5,000.00	35	Hull	Richard
725	Deerfield's Four Seasons	$248.00	$7,500.00	35	Hull	Richard
842	All Season	$8,221.00	$7,500.00	20	Kaiser	Valerie

Customer

CustomerNum	CustomerName	Balance	CreditLimit	RepNum
148	Al's Appliance and Sport	$6,550.00	$7,500.00	20
282	Brookings Direct	$431.50	$10,000.00	35
356	Ferguson's	$5,785.00	$7,500.00	65
408	The Everything Shop	$5,285.25	$5,000.00	35
462	Bargains Galore	$3,412.00	$10,000.00	65
524	Kline's	$12,762.00	$15,000.00	20
608	Johnson's Department Store	$2,106.00	$10,000.00	65
687	Lee's Sport and Appliance	$2,851.00	$5,000.00	35
725	Deerfield's Four Seasons	$248.00	$7,500.00	35
842	All Season	$8,221.00	$7,500.00	20

Rep

CustomerNum	LastName	FirstName
148	Kaiser	Valerie
282	Hull	Richard
356	Perez	Juan
408	Hull	Richard
462	Perez	Juan
524	Kaiser	Valerie
608	Perez	Juan
687	Hull	Richard
725	Hull	Richard
842	Kaiser	Valerie

FIGURE 5-13 Incorrect decomposition of the Customer table

Consider, for example, the redundancy in the storage of sales reps' names, the problem encountered in changing the name of a sales rep, and the difficulty of adding a new sales rep who represents no customers. In addition, because the rep number and names are in different tables, you have actually *split a functional dependence across two different tables*. Thus, this seemingly valid decomposition is definitely not a desirable way to create third normal form tables.

There is another decomposition that you might choose, and that is to replace

Customer (<u>CustomerNum</u>, CustomerName, Balance, CreditLimit, RepNum, LastName, FirstName)

with

Customer (<u>CustomerNum</u>, CustomerName, Balance, CreditLimit, LastName, FirstName)
Rep (<u>RepNum</u>, LastName, FirstName)

Samples of these tables appear in Figure 5-14.

Customer

CustomerNum	CustomerName	Balance	CreditLimit	RepNum	LastName	FirstName
148	Al's Appliance and Sport	$6,550.00	$7,500.00	20	Kaiser	Valerie
282	Brookings Direct	$431.50	$10,000.00	35	Hull	Richard
356	Ferguson's	$5,785.00	$7,500.00	65	Perez	Juan
408	The Everything Shop	$5,285.25	$5,000.00	35	Hull	Richard
462	Bargains Galore	$3,412.00	$10,000.00	65	Perez	Juan
524	Kline's	$12,762.00	$15,000.00	20	Kaiser	Valerie
608	Johnson's Department Store	$2,106.00	$10,000.00	65	Perez	Juan
687	Lee's Sport and Appliance	$2,851.00	$5,000.00	35	Hull	Richard
725	Deerfield's Four Seasons	$248.00	$7,500.00	35	Hull	Richard
842	All Season	$8,221.00	$7,500.00	20	Kaiser	Valerie

Customer

CustomerNum	CustomerName	Balance	CreditLimit	LastName	FirstName
148	Al's Appliance and Sport	$6,550.00	$7,500.00	Kaiser	Valerie
282	Brookings Direct	$431.50	$10,000.00	Hull	Richard
356	Ferguson's	$5,785.00	$7,500.00	Perez	Juan
408	The Everything Shop	$5,285.25	$5,000.00	Hull	Richard
462	Bargains Galore	$3,412.00	$10,000.00	Perez	Juan
524	Kline's	$12,762.00	$15,000.00	Kaiser	Valerie
608	Johnson's Department Store	$2,106.00	$10,000.00	Perez	Juan
687	Lee's Sport and Appliance	$2,851.00	$5,000.00	Hull	Richard
725	Deerfield's Four Seasons	$248.00	$7,500.00	Hull	Richard
842	All Season	$8,221.00	$7,500.00	Kaiser	Valerie

Rep

RepNum	LastName	FirstName
20	Kaiser	Valerie
35	Hull	Richard
65	Perez	Juan

FIGURE 5-14   Second incorrect decomposition of the Customer table

This new design seems to be a possibility. Not only are both tables in third normal form, but joining them together based on LastName and FirstName seems to reconstruct the data in the original table. Or does it? Suppose the name of sales rep 65 is also Valerie Kaiser. In that case, when you join the two new tables, there would be no way to correctly identify which Valerie Kaiser represents which customers. Thus, you would get a row on which customer 148 (Al's Appliance and Sport) is associated with sales rep 20 (Valerie Kaiser) and *another* row on which customer 148 is associated with sales rep 65 (the other Valerie Kaiser). Because you obviously want decompositions that preserve the original information, this design is not appropriate.

Q & A

Question: Using the types of entities found in a college environment (faculty, students, departments, courses, and so on), create an example of a table that is in first normal form but not in second normal form and an example of a table that is in second normal form but not in third normal form. In each case, justify your solutions and show how to convert to the higher forms.

Answer: There are many possible solutions. Your answer may differ from the following solution, but that does not mean it is an unsatisfactory solution.

To create a first normal form table that is not in second normal form, you need a table that has no repeating groups and that has at least one column that is dependent on only a portion of the primary key. For a column to be dependent on a portion of the primary key, the primary key must contain at least two columns. Following is a picture of what you need:

(<u>1</u> , <u>2</u> , 3 , 4)

This table contains four columns, numbered 1, 2, 3, and 4, in which the combination of columns 1 and 2 functionally determines both columns 3 and 4. In addition, neither column 1 nor column 2 can determine *all* other columns; if either one could, the primary key would contain only this one column. Finally, you want part of the primary key (say, column 2) to determine another column (say, column 4).

Now that you know the pattern you need, you would like to find columns from within the college environment to fit it. One example would be as follows:

(<u>StudentNum</u>, <u>CourseNum</u>, Grade, CourseDescription)

In this example, the concatenation of StudentNum and CourseNum determines both Grade and CourseDescription. Both columns are required to determine Grade; thus, the primary key consists of their concatenation. CourseDescription, however, is dependent only on CourseNum, which violates second normal form. To convert this table to second normal form, you would replace it with two tables:

(<u>StudentNum</u>, <u>CourseNum</u>, Grade)

(<u>CourseNum</u>, CourseDescription)

You would, of course, now give these tables appropriate names.

To create a table that is in second normal form but not in third normal form, you need a second normal form table in which there is a determinant that is *not* a candidate key. If you choose a table that has a single column as the primary key, it is automatically in second normal form, so the real problem is the determinant. You need a table like the following:

(<u>1</u> , 2 , 3)

This table contains three columns, numbered 1, 2, and 3, in which column 1 determines each of the others and thus is the primary key. When column 2 also determines column 3, column 2 is a determinant. When column 2 does not also determine column 1, column 2 is not a candidate key. One example that fits this pattern would be as follows:

(<u>StudentNum</u>, AdvisorNum, AdvisorName)

Here the StudentNum determines both the student's AdvisorNum and AdvisorName. AdvisorNum determines AdvisorName, but AdvisorNum does not determine StudentNum because one advisor can have many advisees. This table is in second normal form but not in third normal form. To convert it to third normal form, you would replace it with the following:

(<u>StudentNum</u>, AdvisorNum)

(<u>AdvisorNum</u>, AdvisorName)

Q & A

Question: Convert the following table to third normal form. In this table, StudentNum determines StudentName, NumCredits, AdvisorNum, and AdvisorName. AdvisorNum determines AdvisorName. CourseNum determines Description. The combination of StudentNum and CourseNum determines Grade.

```
Student (StudentNum, StudentName, NumCredits, AdvisorNum,
AdvisorName, (CourseNum, Description, Grade) )
```

Answer: Step 1. Remove the repeating group to convert it to first normal form, yielding the following:

```
Student (StudentNum, StudentName, NumCredits, AdvisorNum, AdvisorName, CourseNum,
    Description, Grade)
```

This table is now in first normal form because it has no repeating groups. It is not, however, in second normal form because StudentName, for example, is dependent only on StudentNum, which is only a portion of the primary key.

Step 2. Convert the first normal form table to second normal form. First, for each subset of the primary key, start a table with that subset as its key, yielding the following:

```
(StudentNum,
(CourseNum,
(StudentNum, CourseNum,
```

Next, place the rest of the columns with the smallest collection of columns on which they depend, giving the following:

```
(StudentNum, StudentName, NumCredits, AdvisorNum, AdvisorName)
(CourseNum, Description)
(StudentNum, CourseNume, Grade)
```

Finally, assign names to each of the newly created tables as follows:

```
Student (StudentNum, StudentName, NumCredits, AdvisorNum, AdvisorName)
Course (CourseNum, Description)
StudentCourse (StudentNum, CourseNum, Grade)
```

Although these tables are all in second normal form, Course and StudentCourse are also in third normal form. The Student table is not in third normal form, however, because it contains a determinant (AdvisorNum) that is not a candidate key.

Step 3. Convert the second normal form Student table to third normal form by removing the column that depends on the determinant AdvisorNum and placing it in a separate table.

```
(StudentNum, StudentName, NumCredits, AdvisorNum)
(AdvisorNum, AdvisorName)
```

Step 4. Name these tables and put the entire collection together, giving the following:

```
Student (StudentNum, StudentName, NumCredits, AdvisorNum)
Advisor (AdvisorNum, AdvisorName)
Course (CourseNum, Description)
StudentCourse (StudentNum, CourseNum, Grade)
```

MULTIVALUED DEPENDENCIES AND FOURTH NORMAL FORM

By converting a given collection of tables to an equivalent third normal form collection of tables, you remove any problems arising from functional dependencies. Usually this means that you eliminate the types of previously discussed update anomalies. Converting to third normal form doesn't avoid all problems related to dependencies, however. A different kind of dependency also can lead to the same types of problems.

To illustrate the problem, suppose you are interested in faculty members at Marvel College. In addition to faculty members, you are interested in the students they advise and the committees on which the faculty members serve. A faculty member can advise many students. Because students can have more than one major, a student can have more than one faculty member as an advisor. A faculty member can serve on zero, one, or more committees. As an initial relational design for this situation, suppose you chose the following unnormalized table:

```
Faculty (FacultyNum, (StudentNum), (CommitteeCode) )
```

The single Faculty table has a primary key of FacultyNum (the number that identifies the faculty member) and two separate repeating groups, StudentNum (the number that identifies the student) and CommitteeCode (the code that identifies the committee, such as ADV for Advisory committee, PER for Personnel committee, and CUR for Curriculum committee). To convert this table to first normal form, you might be tempted to remove the two repeating groups and expand the primary key to include both StudentNum and CommitteeCode. That solution would give the following table:

Faculty (FacultyNum, StudentNum, CommitteeCode)

Samples of the table with repeating groups and with the repeating groups removed appear in Figure 5-15.

Faculty

FacultyNum	StudentNum	CommitteeCode
123	12805 24139	ADV HSG PER
444	57384	HSG
456	24139 36273 37573	CUR

Faculty

FacultyNum	StudentNum	CommitteeCode
123	12805	ADV
123	12805	HSG
123	12805	PER
123	24139	ADV
123	24139	HSG
123	24139	PER
444	57384	HSG
456	24139	CUR
456	36273	CUR
456	37573	CUR

FIGURE 5-15   Incorrect way to remove repeating groups—relation is not in fourth normal form

You already may have suspected that this approach has some problems. If so, you are correct. It is a strange way to normalize the original table. Yet it is precisely this approach for removing repeating groups that leads to the problems concerning multivalued dependencies. You will see how this table should have been normalized to avoid the problems altogether. For now, however, you'll examine this table to see what kinds of problems are present.

The first thing you should observe about this table is that it is in third normal form because no groups repeat, no column is dependent on only a portion of the primary key, and no determinants exist that are not candidate keys. There are several problems, however, with this third normal form table.

1.   *Update.* Changing the CommitteeCode for faculty member 123 requires more than one change. If this faculty member changes from an Advisory committee member to a Curriculum committee member, you would need to change the CommitteeCode from ADV to CUR in rows 1 and 4 of the table. After all, it doesn't make sense to say that the committee is ADV when associated with student 12805 and CUR when associated with student 24139. The same committee is served on by the same faculty member. The faculty member does not serve on one committee when advising one student and a different committee when advising another student.

2. ***Additions.*** Suppose faculty member 555 joins the faculty at Marvel College. Also suppose that this faculty member does not yet serve on a committee. When this faculty member begins advising student 44332, you have a problem because CommitteeCode is part of the primary key. You need to enter a fictitious CommitteeCode in this situation.

3. ***Deletions.*** If faculty member 444 no longer advises student 57384 and you delete the appropriate row from the table, you lose the information that faculty member 444 serves on the Housing committee (HSG).

These problems are similar to those encountered in the discussions of both second normal form and third normal form, but there are no functional dependencies among the columns in this table. A given faculty member is not associated with one student, as he or she would be if this were a functional dependence. Each faculty member, however, is associated with a specific collection of students. More importantly, this association is *independent* of any association with committees. This independence is what causes the problem. This type of dependency is called a multivalued dependency.

Definition: In a table with columns A, B, and C, there is a **multivalued dependence** of column B on column A (also read as "B is **multidependent** on A" or "A **multidetermines** B") when each value for A is associated with a specific collection of values for B and, further, this collection is independent of any values for C. This is usually written as follows:

$$A \rightarrow\rightarrow B$$

Definition: A table (relation) is in **fourth normal form (4NF)** when it is in third normal form and there are no multivalued dependencies.

As you might expect, converting a table to fourth normal form is similar to the normalization process encountered in the treatments of second normal form and third normal form. You split the third normal form table into separate tables, each containing the column that multidetermines the others, which, in this case, is FacultyNum. This means you replace

Faculty (<u>FacultyNum</u>, <u>StudentNum</u>, <u>CommitteeCode</u>)

with

FacStudent (<u>FacultyNum</u>, <u>StudentNum</u>)

FacCommittee (<u>FacultyNum</u>, <u>CommitteeCode</u>)

Figure 5-16 shows samples of these tables. As before, the problems have disappeared. There is no problem with changing the CommitteeCode ADV to CUR for faculty member 123 because the committee code occurs in only one place. To add the information that faculty member 555 advises student 44332, you need to add a row to the FacStudent table—it doesn't matter whether this faculty member serves on a committee. Finally, to delete the information that faculty member 444 advises student 57384, you need to remove a row from the FacStudent table. In this case, you do not lose the information that this faculty member serves on the Housing committee.

Faculty

FacultyNum	StudentNum	CommitteeCode
123	12805	ADV
123	12805	HSG
123	12805	PER
123	24139	ADV
123	24139	HSG
123	24139	PER
444	57384	HSG
456	24139	CUR
456	36273	CUR
456	37573	CUR

FacStudent

FacultyNum	StudentNum
123	12805
123	24139
444	57384
456	24139
456	36273
456	37573

FacCommittee

FacultyNum	CommitteeCode
123	ADV
123	HSG
123	PER
444	HSG
456	CUR

FIGURE 5-16   Conversion to fourth normal form

Figure 5-17 summarizes the four normal forms.

Normal Form	Meaning/Required Conditions	Notes
First normal form	No repeating groups exist	
Second normal form	First normal form and no non-key column is dependent on only a portion of the primary key	Automatically second normal form if the primary key contains only a single column
Third normal form	Second normal form and the only determinants are candidate keys	Actually Boyce-Codd normal form (BCNF)
Fourth normal form	Third normal form and no multivalued dependencies exist	

FIGURE 5-17   Normal forms

AVOIDING THE PROBLEM WITH MULTIVALUED DEPENDENCIES

Any table that is not in fourth normal form suffers some serious problems, but there is a way to avoid dealing with the issue. You should have a design methodology for normalizing tables that prevents this situation from occurring in the first place. You already have most of such a methodology in place from the discussion of the first normal form, second normal form, and third normal form normalization processes. All you need is a slightly more sophisticated method for converting an unnormalized table to first normal form.

The conversion of an unnormalized table to first normal form requires the removal of repeating groups. When this was first demonstrated, you merely removed the repeating group symbol and expanded the primary key. You will recall, for example, that

```
Orders (OrderNum, OrderDate, (PartNum, NumOrdered) )
```

became

```
Orders (OrderNum, OrderDate, PartNum, NumOrdered)
```

The primary key was expanded to include the primary key of the original table together with the key to the repeating group.

What if there are two or more repeating groups? The method you used earlier is inadequate for such situations. Instead, you must place each repeating group in a separate table. Each table will contain all the columns that make up the given repeating group, as well as the primary key of the original unnormalized table. The primary key to each new table will be the concatenation of the primary key of the original table and the primary key of the repeating group. For example, consider the following unnormalized table that contains two repeating groups.

```
Faculty (FacultyNum, FacultyName, (StudentNum, StudentName), (CommitteeCode,
    CommitteeName) )
```

In this example, FacultyName is the name of the faculty member and StudentName is the name of the student. The columns CommitteeCode and CommitteeName refer to the committee's code and name. (For example, one row in this table would have PER in the CommitteeCode column and Personnel Committee in the CommitteeName column.) Applying this new method to create first normal form tables would produce the following:

```
Faculty (FacultyNum, FacultyName)
FacStudent (FacultyNum, StudentNum, StudentName)
FacCommittee (FacultyNum, CommitteeCode, CommitteeName)
```

As you can see, this collection of tables avoids the problems with multivalued dependencies. At this point, you have a collection of first normal form tables that you still need to convert to third normal form. By using the above process, however, you are guaranteed that the result will also be in fourth normal form.

APPLICATION TO DATABASE DESIGN

The normalization process used to convert a relation or collection of relations to an equivalent collection of third normal form tables is a crucial part of the database design process. By following a careful and appropriate normalization methodology, you need not worry about normal forms higher than third normal form. There are three aspects concerning normalization that you need to keep in mind, however.

First, you should carefully convert tables to third normal form. Suppose the following columns exist in a Coach relation. (The ellipsis (...) represents additional columns that exist but are not included in this example.)

```
Coach (CoachNum, LastName, FirstName, Street, City, State, Zip, ...)
```

In addition to the functional dependencies that all the columns have on CoachNum, there are two other functional dependencies. As originally designed by the United States Postal Service, Zip determines both State and City.

Does this mean that you should replace the Coach relation with the following?

```
Coach (CoachNum, LastName, FirstName, Street, Zip, ...)
ZipCode (Zip, City, State)
```

If you are determined to ensure that every relation is in third normal form, you would replace the Coach relation with the revised Coach relation and the new ZipCode relation; but this approach is probably unnecessary. If you review the list of problems normally associated with relations that are not in third normal form, you will see that they don't apply here. Are you likely to need to change the state in which the zip code 49428 is located? Do you need to add the fact that zip code 49401 corresponds to Allendale, Michigan, if you have no customers who live in Allendale? In this case, the design of the original Coach relation is sufficient.

Second, there are currently situations where the same zip code corresponds to more than one city or even to more than one state. This situation illustrates the wisdom in not making the change and the fact that requirements and, consequently, the functional dependencies can change over time. It is critical to review assumptions and dependencies periodically to see if any changes to the design are warranted.

Third, by splitting relations to achieve third normal form tables, you create the need to express an **interrelation constraint**, a condition that involves two or more relations. In the example given earlier for converting to third normal form, you split the Customer relation in the Premiere Products database from

```
Customer (CustomerNum, CustomerName, Balance, CreditLimit, RepNum, LastName, FirstName)
```

to

```
Customer (CustomerNum, CustomerName, Balance, CreditLimit, RepNum)
Rep (RepNum, LastName, FirstName)
```

Nothing about these two relations by themselves would force the RepNum on a row in the Customer relation to match a value of RepNum in the Rep relation. Requiring this to take place is an example of an interrelation constraint. Foreign keys handle this type of interrelation constraint. You will learn more about and specify foreign keys during the database design process, which is covered in Chapter 6.

Summary

- A column (attribute) B is functionally dependent on another column A (or possibly a collection of columns) when each value for A in the database is associated with exactly one value of B.

- The primary key is a column (or a collection of columns) A such that all other columns are functionally dependent on A and no subcollection of the columns in A also has this property.

- When there is more than one choice for the primary key, one of the possibilities is chosen to be *the* primary key. The others are referred to as candidate keys.

- A table (relation) is in first normal form (1NF) when it does not contain repeating groups.

- A column is a nonkey column (also called a nonkey attribute) when it is not a part of the primary key.

- A table (relation) is in the second normal form (2NF) when it is in first normal form and no nonkey column is dependent on only a portion of the primary key.

- A determinant is any column that functionally determines another column.

- A table (relation) is in third normal form (3NF) when it is in second normal form and the only determinants it contains are candidate keys.

- A collection of tables (relations) that is not in third normal form has inherent problems called update anomalies. Replacing this collection with an equivalent collection of tables (relations) that is in third normal form removes these anomalies. This replacement must be done carefully, following a method like the one proposed in this text. If not, other problems, such as those discussed in this chapter, may be introduced.

- A table (relation) is in fourth normal form (4NF) when it is in third normal form and there are no multivalued dependencies.

Key Terms

alternate key	multidetermine
Boyce-Codd normal form (BCNF)	multivalued dependence
candidate key	nonkey attribute
concatenation	nonkey column
dependency diagram	normal form
determinant	normalization process
first normal form (1NF)	partial dependency
fourth normal form (4NF)	primary key
functional dependence	repeating group
functionally dependent	second normal form (2NF)
functionally determines	third normal form (3NF)
interrelation constraint	unnormalized relation
multidependent	update anomaly

Review Questions

1. Define functional dependence.
2. Give an example of a column A and a column B such that B is functionally dependent on A. Give an example of a column C and a column D such that D is *not* functionally dependent on C.
3. Define primary key.
4. Define candidate key.
5. Define first normal form.
6. Define second normal form. What types of problems would you find in tables that are not in second normal form?
7. Define third normal form. What types of problems would you find in tables that are not in third normal form?
8. Define fourth normal form. What types of problems would you find in tables that are not in fourth normal form?

9. Define interrelation constraint and give one example of such a constraint. How are interrelation constraints addressed?

10. Consider a Student table containing StudentNum, StudentName, student's StudentMajor, student's AdvisorNum, student's AdvisorName, student's AdvisorOfficeNum, student's AdvisorPhone, student's NumCredits, and student's Class (freshman, sophomore, and so on). List the functional dependencies that exist, along with the assumptions that would support those dependencies.

11. Convert the following table to an equivalent collection of tables that are in third normal form. This table contains information about patients of a dentist. Each patient belongs to a household.

```
Patient (HouseholdNum, HouseholdName, Street, City, State, Zip,
        Balance, PatientNum, PatientName, (ServiceCode, Description,
        Fee, Date) )
```

The following dependencies exist in the Patient table:

```
PatientNum →  HouseholdNum, HouseholdName, Street, City, State,
        Zip, Balance, PatientName
HouseholdNum →  HouseholdName, Street, City, State, Zip, Balance
ServiceCode →  Description, Fee
PatientNum, ServiceCode →  Date
```

12. Using your knowledge of the college environment, determine the functional dependencies that exist in the following table. After determining the functional dependencies, convert this table to an equivalent collection of tables that are in third normal form.

```
Student (StudentNum, StudentName, NumCredits, AdvisorNum,
        AdvisorName, DeptNum, DeptName, (CourseNum, Description,
        Term, Grade) )
```

13. Again, using your knowledge of the college environment, determine the functional or multivalued dependencies that exist in the following table. After determining the functional dependencies, convert this table to an equivalent collection of tables that are in fourth normal form. ActivityNum and ActivityName refer to activities in which a student can choose to participate. For example, activity number 1 might be soccer, activity 2 might be band, and activity 3 might be the debate team. A student can choose to participate in multiple activities. CourseNum and Description refer to courses the student is taking.

```
Student (StudentNum, StudentName, ActivityNum, ActivityName,
        CourseNum, Description)
```

Premiere Products Exercises

The following exercises are based on the Premiere Products database.

1. Using your knowledge of Premiere Products, determine the functional dependencies that exist in the following table. After determining the functional dependencies, convert this table to an equivalent collection of tables that are in third normal form.

```
Part (PartNum, Description, OnHand, Class, Warehouse,
        Price, (OrderNum, OrderDate, CustomerNum,
        CustomerName, RepNum, LastName, FirstName,
        NumOrdered, QuotedPrice) )
```

2. List the functional dependencies in the following table that concerns invoicing (an application Premiere Products is considering adding to its database), subject to the specified conditions. For a given invoice (identified by the InvoiceNum), there will be a single customer. The customer's number, name, and complete address appear on the invoice, as does the date. Also, there may be several different parts appearing on the invoice. For each part that appears, display the part number, description, price, and number shipped. Each customer that orders a particular part pays the same price. Convert this table to an equivalent collection of tables that are in third normal form.

```
Invoice (InvoiceNum, CustomerNum, LastName, FirstName,
        Street, City, State, Zip, Date, (PartNum,
        Description, Price, NumShipped) )
```

3. The requirements for Premiere Products have changed. A number *and* a name now identify each warehouse. Units of each part may be stored in multiple warehouses, and it is important to know precisely how many parts are stored in each warehouse. In addition, Premiere Products now wants to manage information about the suppliers from which it purchases parts. For each part, Premiere Products needs to know the number and name of each supplier as well as the expected lead time for delivering each part. (Lead time is the amount of time a supplier is expected to take to deliver the part after Premiere Products has ordered it.) Each part can have many

suppliers, and each supplier can supply many parts. Using this information, convert the following unnormalized relation to fourth normal form:

```
Part (PartNum, Description, Class, Price,
      (WarehouseNum, WarehouseName, OnHand),
      (SupplierNum, SupplierName, LeadTime) )
```

Henry Books Case

The following exercises are based on the Henry Books database.

1. Using the types of entities found in the Henry Books database (books, authors, and publishers), create an example of a table that is in first normal form but not in second normal form and an example of a table that is in second normal form but not in third normal form. In each case, justify your answers and show how to convert to the higher forms.

2. Henry Books is considering selling textbooks to a local college. To do so, it must maintain information about courses, textbooks, and instructors. Determine the multivalued dependencies in the following table; then convert this table to an equivalent collection of tables that are in fourth normal form. Each course is associated with a specific set of textbooks independent of the instructors who are teaching the course. In other words, although many instructors may be teaching the course, they all will use the same set of textbooks.

```
Course (CourseNum, Textbook, InstructorNum, InstructorName)
```

3. The following unnormalized table is similar in content to the table in Exercise 2. Note that this table has two separate repeating groups: one listing the textbooks used for the course and the other listing the instructors who teach the course. Convert it to fourth normal form. Did you encounter the table from Exercise 2 along the way?

```
Course (CourseNum, Description, NumCredits,
        (Textbook), (InstructorNum, InstructorName) )
```

4. Identify the functional dependencies in the following unnormalized table. Convert the table to third normal form. Is the result also in fourth normal form? Why or why not?

```
Book (BookCode, Title, PublisherCode, PublisherName,
      (AuthorNum, AuthorLast, AuthorFirst) )
```

Alexamara Marina Group Case

The following exercises are based on the Alexamara Marina Group database.

1. Using the types of entities found in the Alexamara Marina Group database (marinas, owners, boat slips, categories, and service requests), create an example of a table that is in first normal form but not in second normal form and an example of a table that is in second normal form but not in third normal form. In each case, justify your answer and show how to convert to the higher forms.

2. Determine the functional dependencies that exist in the following table; then convert this table to an equivalent collection of tables that are in third normal form:

```
Marina (MarinaNum, Name, (SlipNum, Length, RentalFee, BoatName) )
```

3. Determine the functional dependencies that exist in the following table; then convert this table to an equivalent collection of tables that are in third normal form:

```
MarinaSlip (SlipID, MarinaNum, SlipNum, Length, RentalFee, BoatName, BoatType,
            OwnerNum, LastName, FirstName)
```

CHAPTER 6

DATABASE DESIGN 2: DESIGN METHOD

LEARNING OBJECTIVES

Objectives

- Discuss the general process and goals of database design
- Define user views and explain their function
- Define Database Design Language (DBDL) and use it to document database designs
- Create an entity-relationship (E-R) diagram to visually represent a database design
- Present a method for database design at the information level and view examples illustrating this method
- Explain the physical-level design process
- Discuss top-down and bottom-up approaches to database design and examine the advantages and disadvantages of both methods
- Use a survey form to obtain information from users prior to beginning the database design process
- Review existing documents to obtain information prior to beginning the database design
- Discuss special issues related to implementing one-to-one relationships and many-to-many relationships involving more than two entities
- Discuss entity subtypes and their relationships to nulls
- Learn how to avoid potential problems when merging third normal form relations
- Examine the entity-relationship model for representing and designing databases

INTRODUCTION

Now that you have learned how to identify and correct poor table designs, you will turn your attention to the design process by determining the tables (relations) and columns (attributes) that make up the database. In addition, you will determine the relationships between the various tables.

Most designers tackle database design using a two-step process. In the first step, the database designers design a database that satisfies the organization's requirements as cleanly as possible. This step is called **information-level design**, and it is completed *independently* of any particular DBMS that the organization will ultimately use. In the second step, which is called the **physical-level design**, designers adapt the information-level design for the specific DBMS that the organization will use. During the physical-level design, designers must consider the characteristics of the particular DBMS that the organization will use.

After examining the information-level design process, you will explore the general database design method and view examples illustrating this method. You will construct entity-relationship diagrams to represent the database design visually. You'll then learn about the physical-level design process. You will also compare top-down and bottom-up approaches to database design.

You will explore special issues related to database design, including survey forms and their use in database design and the way to obtain important information from existing documents. You will examine issues related to the implementation of some special types of relationships. You will learn about entity subtypes and their relationship to nulls. You will examine issues related to merging third normal form relations. Finally, you will learn about the entity-relationship model.

USER VIEWS

Regardless of which approach an organization adopts to implement its database design, a complete database design that will satisfy all the organization's requirements is rarely a one-step process. Unless the requirements are simple, an organization will usually divide the overall job of database design into many smaller tasks by identifying the individual pieces of the design problem, called user views. A **user view** is the set of requirements that is necessary to support the operations of a particular database user. For example, at Premiere Products, the database must be capable of storing each part's number, description, units on hand, item class, number of the warehouse in which the part is located, and price. It is critical to analyze and determine these user views carefully before beginning the design process.

For each user view, designers must design the database structure to support the view and then merge it into a **cumulative design** that supports all the user views encountered during the design process. Each user view is generally much simpler than the total collection of requirements, so working on individual user views is usually more manageable than attempting to turn the design of the entire database into one large task.

INFORMATION-LEVEL DESIGN METHOD

The information-level design method in this text involves representing individual user views, refining them to eliminate any problems, and then merging them into a cumulative design. After you have represented and merged all user views, you can complete the cumulative design for the entire database.

When creating user views, a "user" can be a person or a group that will use the system, a report that the system must produce, or a type of transaction that the system must support. In the last two instances, you might think of the user as the person who will use the report or enter the transaction. In fact, if the same user requires three separate reports, for example, it is more efficient to consider each report as a separate user view, even though only one "user" is involved, because smaller user views are easier to construct.

For each user view, the information-level design method requires you to complete the following steps:

1. Represent the user view as a collection of tables.
2. Normalize these tables.
3. Identify all keys in these tables.
4. Merge the result of Steps 1 through 3 into the cumulative design.

In the following sections, you will examine each of these steps in detail.

REPRESENT THE USER VIEW AS A COLLECTION OF TABLES

When provided with a user view or some sort of stated requirement, you must develop a collection of tables that will support it. In some cases, the collection of tables may be obvious to you. For example, suppose a requested user view involves departments and employees, each department can hire many employees, and each employee can work in only one department (a typical restriction). A design similar to the following may have naturally occurred to you and is appropriate to represent the circumstances.

```
Department (DepartmentNum, Name, Location)
Employee (EmployeeNum, LastName, FirstName, Street, City,
    State, Zip, WageRate, SocSecNum, DepartmentNum)
```

You will undoubtedly find that the more designs you complete, the easier it will be for you to develop such a collection without resorting to any special procedure. The real question is this: What procedure should you follow when the correct design is not so obvious? In this case, you can complete the following four steps.

Step 1. Determine the entities involved and create a separate table for each type of entity. At this point, you do not need to do anything more than name the tables. For example, if a user view involves departments and employees, you can create a Department table and an Employee table. At this point, you will write something like this:

```
Department (
Employee (
```

That is, you will write the name of a table and an opening parenthesis, *and that is all*. You will assign columns to these tables in later steps.

Step 2. Determine the primary key for each table. In this step, you can add one or more columns depending on how many columns are required for the primary key. You'll add additional columns later. Even though you have yet to determine the columns in the table, you can usually determine the primary key. For example, the primary key in an Employee table will probably be EmployeeNum, and the primary key in a Department table will probably be DepartmentNum.

The primary key is the unique identifier, so the essential question is this: What does it take to uniquely identify an employee or a department? Even if you are trying to automate a previously designed manual system, you usually can find a unique identifier in that system. If no unique identifier is available, you'll need to assign one. For example, in a manual system, customers may not have been assigned numbers because the customer base was small and the organization did not require or use customer numbers. Because the organization is computerizing its records, however, now is a good time to assign customer numbers to become the unique identifiers you are seeking.

After creating unique identifiers, you add these primary keys to what you have written already. At this point, you will have something like the following:

```
Department (DepartmentNum,
Employee (EmployeeNum,
```

Now you have the name of the table and the primary key, but that is all. In later steps, you will add the other columns.

Step 3. Determine the properties for each entity. You can look at the user requirements and then determine the other required properties of each entity. These properties, along with the primary key identified in Step 2, will become columns in the appropriate tables. For example, an Employee entity may require columns for LastName, FirstName, Street, City, State, Zip, WageRate, and SocSecNum (Social Security number). The Department entity may require columns for Name (department name) and Location (department location). Adding these columns to what is already in place produces the following:

```
Department (DepartmentNum, Name, Location
Employee (EmployeeNum, LastName, FirstName, Street, City,
    State, Zip, WageRate, SocSecNum
```

Step 4. Determine relationships between the entities. The basic relationships are one-to-many, many-to-many, and one-to-one. You'll see how to handle each type of relationship next.

To create a one-to-many relationship, include the primary key of the "one" table as a foreign key in the "many" table. For example, assume each employee works in a single department but a department can have many employees. Thus, *one* department is related to *many* employees. In this case, you would include the primary key of the Department table (the "one" part) as a foreign key in the Employee table (the "many" part). The tables would now look like this:

```
Department (DepartmentNum, Name, Location)
Employee (EmployeeNum, LastName, FirstName, Street, City,
    State, Zip, WageRate, SocSecNum, DepartmentNum)
```

You create a **many-to-many relationship** by creating a new table whose primary key is the combination of the primary keys of the original tables. Assume each employee can work in multiple departments and each department can have many employees. In this case, you would create a new table whose primary key is the combination of EmployeeNum and DepartmentNum. Because the new table represents the fact that an employee *works in* a department, you might choose to call it WorksIn. Another method is to use a name that combines the names of the two tables being related. Using the second approach, the new table's name could be DepartmentEmployee or EmployeeDepartment. After creating the new table, the collection of tables is as follows:

```
Department (DepartmentNum, Name, Location)
Employee (EmployeeNum, LastName, FirstName, Street, City,
    State, Zip, WageRate, SocSecNum)
WorksIn (EmployeeNum, DepartmentNum)
```

In this design, there is a one-to-many relationship between the Department and WorksIn tables and a one-to-many relationship between the Employee and WorksIn tables. By creating the WorksIn table, which includes foreign keys from the Department and Employee tables, you have created a new table to implement a many-to-many relationship. The one-to-many relationship between each of the original tables with the new table creates the many-to-many relationship between the two original tables.

In some situations, no other columns will be required in the new table. The other columns in the WorksIn table would be those columns that depend on both the employee and the department, if such columns existed. One possibility, for example, would be the date the department hired the employee because it depends on *both* the employee *and* the department.

If each employee works in a single department and each department has only one employee, the relationship between employees and departments is **one-to-one**. (In practice, such relationships are rare.) The simplest way to implement a one-to-one relationship is to treat it as a one-to-many relationship. Which table is the "one" part of the relationship, and which table is the "many" part? Sometimes looking ahead helps. For example, you might ask this question: If the relationship changes in the future, is it more likely that one employee will work in many departments or that one department will hire several employees rather than just one? If your research determines that it is more likely that a department will hire more than one employee, you would make the Employee table the "many" part of the relationship. If both situations might happen, you could treat the relationship as many-to-many. If neither situation is likely to occur, you could arbitrarily choose the "many" part of the relationship.

NORMALIZE THE TABLES

After establishing the relationships between the entities, the next task is to normalize each table, with the target being third normal form. (The target is actually fourth normal form, but careful planning in the early phases of the normalization process usually rules out the need to consider fourth normal form.)

IDENTIFY ALL KEYS

For each table, you must identify the primary key and any alternate keys, secondary keys, and foreign keys. In the database containing information about employees and departments, you already determined the primary keys for each table in an earlier step.

Recall that an alternate key is a column or collection of columns that could have been chosen as a primary key but was not. It is not common to have alternate keys; if they do exist and the system must enforce their uniqueness, however, you should note them. You usually implement this restriction by creating a unique index on the field. If there are any **secondary keys** (columns that are of interest strictly for the purpose of retrieval), you should represent them at this point. If a user were to indicate, for example, that rapidly retrieving an employee record on the basis of his or her last name was important, you would designate the LastName column as a secondary key. You usually create a nonunique index for each secondary key.

In many ways, the foreign key is the most important key because it is through foreign keys that you create relationships between tables and enforce certain types of integrity constraints in a database. Remember that a foreign key is a column (or collection of columns) in one table that is required to match the value of the primary key for some row in another table or is required to be null. (This property is called referential integrity.) Consider, for example, the following tables:

```
Department (DepartmentNum, Name, Location)
Employee (EmployeeNum, LastName, FirstName, Street, City,
    State, Zip, WageRate, SocSecNum, DepartmentNum)
```

As before, the DepartmentNum column in the Employee table indicates the department in which the employee works. In this case, you say that the DepartmentNum column in the Employee table is a foreign key that *identifies* Department. Thus, the number in this column on any row in the Employee table must be a department number that is already in the database or the value must be set to null. (Null indicates that, for whatever reason, the employee is not assigned to a department.)

Types of Primary Keys

There are three types of primary keys that you can use in your database design. A **natural key** (also called a **logical key** or an **intelligent key**) is a primary key that consists of a column that uniquely identifies an entitiy, such as a person's Social Security number, a book's ISBN (International Standard Book Number), a product's UPC (Universal Product Code), or a vehicle's VIN (Vehicle Identification Number). These characteristics are inherent to the entity and visible to users. If a natural key exists for an entity, you usually can select it as the primary key.

If a natural key does not exist for an entity, it is common to create a primary key column that will be unique and accessible to users. The primary keys in the Premiere Products database (RepNum, CustomerNum, OrderNum, and PartNum) were created to serve as the primary keys. A column that you create for an entity to serve solely as the primary key and that is visible to users is called an **artificial key**.

The final type of primary key, which is called a **surrogate key** (or a **synthetic key**), is a system-generated primary key that is usually hidden from users. When a DBMS creates a surrogate key, it is usually an automatic numbering data type, such as the Access AutoNumber data type. For example, suppose you have the following relation for Customer payments.

```
Payment (CustomerNum, PaymentDate, PaymentAmount)
```

Because a customer can make multiple payments, CustomerNum cannot be the primary key. Assuming it is possible for a customer to make more than one payment on a particular day, the combination of CustomerNum and PaymentDate also cannot be the primary key. Adding an artificial key, such as PaymentNum, means you would have to assign a PaymentNum every time the customer makes a payment. Adding a surrogate key, such as PaymentId, would make more sense because the DBMS will automatically assign a unique value to each payment. Users do not need to be aware of the PaymentId value, however.

DATABASE DESIGN LANGUAGE (DBDL)

To carry out the design process, you must have a mechanism for representing tables and keys. The standard notation you have used thus far for representing tables is fine, but it does not go far enough—there is no way to represent alternate, secondary, or foreign keys. Because the information-level design method is based on the relational model, it is desirable to represent tables with the standard notation. To do so, you will add additional features capable of representing additional information. One approach to doing this is called **Database Design Language (DBDL)**. Figure 6-1 shows sample DBDL documentation for the Employee table.

```
Employee (EmployeeNum, LastName, FirstName, Street, City, State, Zip,
    WageRate, SocSecNum, DepartmentNum)
    AK    SocSecNum
    SK    LastName
    FK    DepartmentNum → Department
```

FIGURE 6-1   DBDL for the Employee table

In DBDL, you represent a table by listing all columns and then underlining the primary key. Below the table definition, you list any alternate keys, secondary keys, and foreign keys using the abbreviations AK, SK, and FK, respectively. For alternate and secondary keys, you can list the column or collection of columns by name. In the case of foreign keys, however, you must also represent the table whose primary key the foreign key must match. In DBDL, you write the foreign key followed by an arrow pointing to the table that the foreign key identifies.

The rules for defining tables and their keys using DBDL are as follows:

- Tables (relations), columns (attributes), and primary keys are written by first listing the table name and then in parentheses listing the columns that make up the table. The column(s) that make up the primary key are underlined.

- Alternate keys are identified by the abbreviation AK, followed by the column(s) that make up the alternate key.

- Secondary keys are identified by the abbreviation SK, followed by the column(s) that make up the secondary key.

- Foreign keys are identified by the abbreviation FK, followed by the column(s) that make up the foreign key. Foreign keys are followed by an arrow pointing to the table identified by the foreign key. When several tables are listed, a common practice places the table containing the foreign key below the table that the foreign key identifies, if possible.

Figure 6-1 shows that there is a table named Employee, containing the columns EmployeeNum, LastName, FirstName, Street, City, State, Zip, WageRate, SocSecNum, and DepartmentNum. The primary key is EmployeeNum. Another possible primary key is SocSecNum, which is listed as an alternate key. The LastName column is a secondary key, which allows you to retrieve data more efficiently based on an employee's last name. (You can add additional secondary key designations later as necessary.) The DepartmentNum column is a foreign key that identifies the department number in the Department table in which the employee works.

ENTITY-RELATIONSHIP (E-R) DIAGRAMS

A popular type of diagram that visually represents the structure of a database is the entity-relationship (E-R) diagram. In an E-R diagram, rectangles represent the entities (tables). Foreign key restrictions determine relationships between the tables, and these relationships are represented as lines joining the corresponding rectangles.

There are several different styles of E-R diagrams currently in use. In this text, the style you will use is called **IDEF1X**.

NOTE

IDEF stands for "Integrated Definition" and is the name for a family of modeling languages that began with a project of the U.S. Air Force called Integrated Computer Aided Manufacturing. There are languages for such areas as activity modeling (IDEF0), conceptual data modeling (IDEF1), simulation modeling (IDEF2), process modeling (IDEF3), and object-oriented software design (IDEF4). The language in this family for logical data modeling is IDEF1X.

Consider the following database design written in DBDL:

```
Department (DepartmentNum, Name, Location)
Employee (EmployeeNum, LastName, FirstName, Street, City,
    State, Zip, WageRate, SocSecNum, DepartmentNum)
    AK    SocSecNum
    SK    LastName, FirstName
    FK    DepartmentNum → Department
```

The E-R diagram for the preceding database design appears in Figure 6-2.

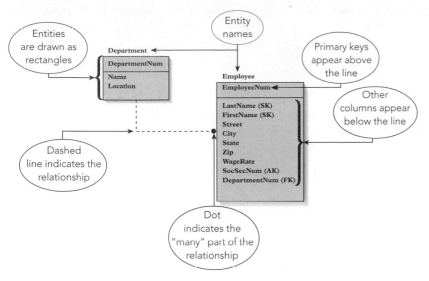

FIGURE 6-2 E-R diagram

The E-R diagram shown in Figure 6-2 has the following characteristics:

- A rectangle represents each entity in the E-R diagram—there is one rectangle for the Department entity and a second rectangle for the Employee entity. The name of each entity appears above the rectangle.
- The primary key for each entity appears above the line in the rectangle for each entity. DepartmentNum is the primary key of the Department entity, and EmployeeNum is the primary key of the Employee entity.
- The other columns in each entity appear below the line within each rectangle.
- The letters AK, SK, and FK appear in parentheses following the alternate key, secondary key, and foreign key, respectively, in the Employee entity. (The Department entity doesn't have an alternate, secondary, or foreign key.)
- For each foreign key, there is a line leading from the rectangle that corresponds to the table being identified to the rectangle that corresponds to the table containing the foreign key. The dot at the end of the line indicates the "many" part of the one-to-many relationship between the Department and Employee entities. (In Figure 6-2, *one* department is related to *many* employees, so the dot is at the end of the line connected to the Employee entity.)

When you use an E-R diagram to represent a database, it visually illustrates all the information listed in the DBDL. Thus, you would not also need to include the DBDL version of the design. There are other styles, however, that do not include such information within the diagram. In that case, you should represent the design with *both* the diagram *and* the DBDL.

MERGE THE RESULT INTO THE DESIGN

As soon as you have completed Steps 1 through 3 for a given user view, you can merge the results into the cumulative design. If the view on which you have been working is the first user view, the cumulative design will be identical to the design for the first user. Otherwise, you merge all the tables for this user with those tables that are currently in the cumulative design.

Next, you combine tables that have the same primary key to form a new table. The new table has the same primary key as those tables you have combined. The new table also contains all the columns from both tables. In the case of duplicate columns, you remove all but one copy of the column. For example, if the cumulative design already contains the following table

```
Employee (EmployeeNum, LastName, FirstName, WageRate, SocSecNum, DepartmentNum)
```

and the user view you just completed contains the following table

```
Employee (EmployeeNum, LastName, FirstName, Street, City, State, Zip)
```

you would combine the two tables because they have the same primary key. All the columns from both tables are in the new table, but without any duplicate columns. Thus, LastName and FirstName appear only once, even though they are in each table. The end result is as follows:

```
Employee (EmployeeNum, LastName, FirstName, WageRate,
     SocSecNum, DepartmentNum, Street, City, State, Zip)
```

If necessary, you could reorder the columns at this point. For example, you might move the Street, City, State, and Zip columns to follow the FirstName column, which is the more traditional arrangement of this type of data. This change would give the following:

```
Employee (EmployeeNum, LastName, FirstName, Street, City,
     State, Zip, WageRate, SocSecNum, DepartmentNum)
```

At this point, you need to check the new design to ensure that it is still in third normal form. If it is not, you should convert it to third normal form before proceeding.

Figure 6-3 summarizes the process that is repeated for each user view until all user views have been examined. At that point, the design is reviewed to resolve any problems that may remain and to ensure that it can meet the needs of all individual users. After all user view requirements are satisfied, the information-level design is considered to be complete.

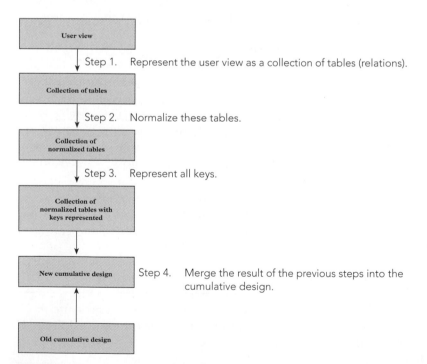

FIGURE 6-3   Information-level design method

DATABASE DESIGN EXAMPLES

Now that you understand how to represent a database in DBDL and in an E-R diagram, you can examine the requirements of another database, the Premiere Products database. In the process, you will see how a set of requirements led to the database with which you have been working throughout this text.

EXAMPLE 1

Complete an information-level design for a database that satisfies the following constraints and user view requirements for a company that stores information about sales reps, customers, parts, and orders.

User View 1 Requirements: For a sales rep, store the sales rep's number, name, address, total commission, and commission rate.

User View 2 Requirements: For a customer, store the customer's number, name, address, balance, and credit limit. In addition, store the number and name of the sales rep who represents this customer. A sales rep can represent many customers, but a customer must have exactly one sales rep. (A customer *must have* a sales rep and cannot have more than *one* sales rep.)

User View 3 Requirements: For a part, store the part's number, description, units on hand, item class, number of the warehouse in which the part is located, and price. All units of a particular part are stored in the same warehouse.

User View 4 Requirements: For an order, store the order number; order date; number, name, and address of the customer that placed the order; and number of the sales rep who represents that customer. In addition, for each line item within the order, store the part number and description, number of the part that was ordered, and quoted price. The user also has supplied the following constraints:

a. Each order must be placed by a customer that is already in the Customer table.

b. There is only one customer per order.

c. On a given order, there is, at most, one line item for a given part. For example, part DR93 cannot appear in several lines within the same order.

d. The quoted price might not match the current price in the Part table, allowing the company to sell the same parts to different customers at different prices. The user wants to be able to change the price for a part without affecting orders that are currently on file.

What are the user views in the preceding example? In particular, how should the design proceed if you are given requirements that are not specifically stated in the form of user views? Sometimes you might encounter a series of well thought-out user views in a form that you can easily merge into the design. Other times you might be given only a set of requirements, such as those described in Example 1. In another situation, you might be given a list of reports and updates that a system must support. In addition to the requirements, when you are able to interview users and document their needs before beginning the design process, you can make sure that you understand the specifics of their user views *prior* to starting the design process. On the other hand, you may have to take information as you get it and in whatever format it is provided.

When the user views are not clearly defined, you should consider each stated requirement as a separate user view. Thus, you can think of each report or update transaction that the system must support, as well as any other requirement stated in the user views, as an individual user view. In fact, even when the requirements are presented as user views, you may want to split a complex user view into smaller pieces and consider each piece as a separate user view for the design process.

To transform each user view into DBDL, examine the requirements and create the necessary entities, keys, and relationships.

User View 1: For a sales rep, store the sales rep's number, name, address, total commission, and commission rate. You'll need to create only one table to support this view:

```
Rep (RepNum, LastName, FirstName, Street, City, State,
    Zip, Commission, Rate)
```

This table is in third normal form. Because there are no foreign, alternate, or secondary keys, the DBDL representation of the table is the same as the relational model representation.

Notice that you have assumed the sales rep's number (RepNum) is the Rep table's primary key—this is a reasonable assumption. Because the user didn't provide this information, however, you would need to verify its accuracy with the user. In each of the following requirements, you can assume the obvious column (customer number, part number, and order number) is the primary key. Because you are working on the first user view, the "merge" step of the design method will produce a cumulative design consisting of only the Rep table, which is shown in Figure 6-4. This design is simple, so you don't need to represent it with an E-R diagram.

```
Rep (RepNum, LastName, FirstName, Street, City, State, Zip,
     Commission, Rate)
```

FIGURE 6-4   Cumulative design after first user view

User View 2: Because the first user view was simple, you were able to create the necessary table without having to complete each step mentioned in the information-level design method section. The second user view is more complicated, however, so you'll use all the steps to determine the tables. (If you've already determined what the tables should be, you have a natural feel for the process. If so, please be patient and work through the process.)

For a customer, store the customer's number, name, address, balance, and credit limit. In addition, store the number and name of the sales rep who represents this customer. You'll take two different approaches to this requirement, allowing you to see how they both can lead to the same result. The only difference between the two approaches is the entities that you initially identify. In the first approach, suppose you identify two required entities for sales reps and customers. You would begin by listing the following two tables:

```
Rep (
Customer (
```

After determining the unique identifiers, you add the primary keys, which would give the following:

```
Rep (RepNum,
Customer (CustomerNum,
```

Adding columns for the properties of each of these entities would yield this:

```
Rep (RepNum, LastName, FirstName
Customer (CustomerNum, CustomerName, Street, City, State,
    Zip, Balance, CreditLimit
```

Finally, you deal with the relationship: *one* sales rep is related to *many* customers. To implement this one-to-many relationship, include the key of the "one" table as a foreign key in the "many" table. In this case, you would include the RepNum column in the Customer table. Thus, you would have the following:

```
Rep (RepNum, LastName, FirstName)
Customer (CustomerNum, CustomerName, Street, City, State,
    Zip, Balance, CreditLimit, RepNum)
```

Both tables are in third normal form, so you can move on to representing the keys. Before doing that, however, consider another approach that you could have used to determine the tables.

Suppose you didn't realize that there were really two entities, and you created only a single table for customers. You would begin by listing the table as follows:

```
Customer (
```

Adding the unique identifier as the primary key would give this table:

```
Customer (CustomerNum,
```

Finally, adding the other properties as additional columns would yield the following:

```
Customer (CustomerNum, CustomerName, Street, City, State,
    Zip, Balance, CreditLimit, RepNum, LastName, FirstName)
```

A problem occurs, however, when you examine the functional dependencies that exist in the Customer entity. The CustomerNum column determines all the other columns, as it should. However, the RepNum column determines the LastName and FirstName columns; yet RepNum is not an alternate key. This table is in second normal form because no column depends on a portion of the primary key, but it is not in third normal form. Converting the table to third normal form produces the following two tables:

```
Customer (CustomerNum, CustomerName, Street, City, State,
    Zip, Balance, CreditLimit, RepNum)
Rep (RepNum, LastName, FirstName)
```

Notice that these are the same tables you determined with the first approach—it just took a little longer to get there.

Besides the obvious primary keys, CustomerNum for Customer and RepNum for Rep, the Customer table now contains a foreign key, RepNum. There are no alternate keys, nor did the requirements state anything that would require a secondary key. If there were a requirement to retrieve the customer based on the customer's name, for example, you would probably choose to make CustomerName a secondary key.

The next step is to merge these two tables into the cumulative design. You could now represent the Rep table in DBDL in preparation for merging these two tables into the existing cumulative design. Looking ahead, however, you see that because this table has the same primary key as the Rep table from the first user view, you can merge the two tables to form a single table that has the common column RepNum as its primary key and that contains all the other columns from both tables without duplication. For this second user view, the only columns in the Rep table besides the primary key are LastName and FirstName. These columns were already in the Rep table from the first user view that you added to the cumulative design. Thus, you don't need to add anything to the Rep table that already appears in the cumulative design. The cumulative design now contains the Rep and Customer tables shown in Figure 6-5.

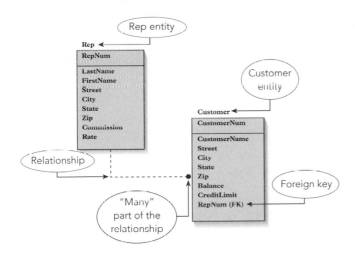

FIGURE 6-5   Cumulative deisgn after second user view

User View 3: Like the first user view, this one poses no special problems. For a part, store the part's number, description, units on hand, item class, number of the warehouse in which the part is located, and price. Only one table is required to support this user view:

```
Part (PartNum, Description, OnHand, Class, Warehouse, Price)
```

This table is in third normal form. The DBDL representation is identical to the relational model representation.

Because PartNum is not the primary key of any table you have already encountered, merging this table into the cumulative design produces the design shown in Figure 6-6, which contains the tables Rep, Customer, and Part.

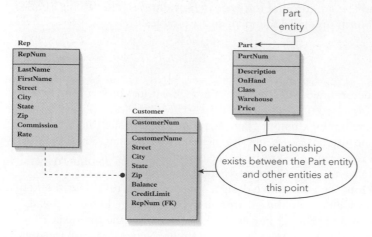

FIGURE 6-6 Cumulative design after third user view

User View 4: This user view is more complicated, and you can approach it in several ways. For an order, store the order number; order date; number, name, and address of the customer that placed the order; and number of the sales rep who represents that customer. In addition, for each line item within the order, store the part number and description, number of the part that was ordered, and quoted price. Suppose on first glance you determine that you need to create only a single entity for orders. You might create the following table:

```
Orders (
```

Because order numbers uniquely identify orders, you would add the OrderNum column as the primary key, giving this table:

```
Orders (OrderNum,
```

Examining the various properties of an order, such as the date, customer number, and so on, as listed in the requirements, you would add the appropriate columns, giving the following:

```
Orders (OrderNum, OrderDate, CustomerNum, CustomerName,
    Street, City, State, Zip, RepNum,
```

What about the fact that you are supposed to store the part number, description, number ordered, and quoted price for each order line in this order? One way of doing this would be to include all these columns within the Orders table as a repeating group (because an order can contain many order lines). This would yield the following:

```
Orders (OrderNum, OrderDate, CustomerNum, CustomerName,
    Street, City, State, Zip, RepNum, (PartNum, Description,
    NumOrdered, QuotedPrice) )
```

At this point, you have a table that contains all the necessary columns. Now you must convert this table to an equivalent collection of tables that are in third normal form. Because this table is not in first normal form, you would remove the repeating group and expand the primary key to produce the following:

```
Orders (OrderNum, OrderDate, CustomerNum, CustomerName, Street,
    City, State, Zip, RepNum, PartNum, Description, NumOrdered,
    QuotedPrice)
```

In the new Orders table, you have the following functional dependencies:

```
OrderNum → OrderDate, CustomerNum, CustomerName, Street,
    City, State, Zip, RepNum
CustomerNum → CustomerName, Street, City, State, Zip, RepNum
PartNum → Description
OrderNum, PartNum → NumOrdered, QuotedPrice
```

From the discussion of the quoted price in the requirement, you should note that a quoted price depends on *both* the order number and the part number, not on the part number alone. Because some columns depend on only a portion of the primary key, the Orders table is not in second normal form. Converting to second normal form would yield the following:

```
Orders (OrderNum, OrderDate, CustomerNum, CustomerName,
    Street, City, State, Zip, RepNum)
Part (PartNum, Description)
OrderLine (OrderNum, PartNum, NumOrdered, QuotedPrice)
```

The Part and OrderLine tables are in third normal form. The Orders table is not in third normal form because CustomerNum determines CustomerName, Street, City, State, Zip, and RepNum; but CustomerNum is not an alternate key. Converting the Orders table to third normal form and leaving the other tables as written would produce the following design for this requirement:

```
Orders (OrderNum, OrderDate, CustomerNum)
Customer (CustomerNum, CustomerName, Street, City, State,
    Zip, RepNum)
Part (PartNum, Description)
OrderLine (OrderNum, PartNum, NumOrdered, QuotedPrice)
```

You can represent this collection of tables in DBDL and then merge them into the cumulative design. Again, however, you can look ahead and see that you can merge this Customer table with the existing Customer table and this Part table with the existing Part table. In both cases, you won't need to add anything to the Customer and Part tables already in the cumulative design; so the Customer and Part tables for this user view will not affect the overall design. The DBDL representation for the Orders and OrderLine tables appears in Figure 6-7.

```
Orders (OrderNum, OrderDate, CustomerNum)
    FK    CustomerNum → Customer

OrderLine (OrderNum, PartNum, NumOrdered, QuotedPrice)
    FK    OrderNum → Orders
    FK    PartNum → Part
```

FIGURE 6-7   DBDL for the Orders and OrderLine tables

At this point, you have completed the process for each user view. Now it's time to review the design to make sure it will fulfill all the stated requirements. If the design contains problems or new information arises, you must modify the design to meet the new user views. Based on the assumption that you do not have to modify the design further, the final information-level design appears in Figure 6-8.

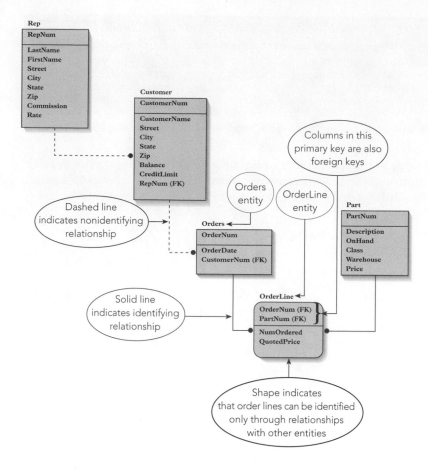

FIGURE 6-8 Final information-level design

There are some differences between the E-R diagram shown in Figure 6-8 and the ones you have seen so far. The OrderLine entity appears as a rectangle with curved corners. Further, the relationships from Orders to OrderLine and from Part to OrderLine are represented with solid lines instead of dashed lines.

Both of these differences are due to the fact that the primary key of the OrderLine entity contains foreign keys. In the OrderLine entity, both columns that compose the primary key (OrderNum and PartNum) are foreign keys. Thus, to identify an OrderLine, you need to know the order and the part to which the order corresponds.

This situation is different from one in which the primary key does not contain one or more foreign keys. Consider the Customer table, for example, in which the primary key is CustomerNum, which is not a foreign key. The Customer table does contain a foreign key, RepNum, which identifies the Rep table. To identify a customer, however, all you need is the customer number; you do not need to know the rep number. In other words, you do not need to know the sales rep to which the customer corresponds.

An entity that does not require a relationship to another entity for identification is called an **independent entity**, and one that does require such a relationship is called a **dependent entity**. Thus, the Customer table is independent, whereas the OrderLine table is dependent. Independent entities have square corners in the diagram, and dependent entities have rounded corners.

A relationship that is necessary for identification is called an **identifying relationship**, whereas one that is not necessary is called a **nonidentifying relationship**. Thus, the relationship between the Rep and Customer tables is nonidentifying, and the relationship between the Orders and OrderLine tables is identifying. In an E-R diagram, a solid line represents an identifying relationship and a dashed line represents a nonidentifying relationship.

EXAMPLE 2

Ray Henry, the owner of a bookstore chain named Henry Books, gathers and organizes information about branches, publishers, authors, and books. Each branch has a number that uniquely identifies the branch. In addition, Ray tracks the branch's name, location, and number of employees. Each publisher has a code that uniquely identifies the publisher. In addition, Ray tracks the publisher's name and city. The only user of the Book database is Ray, but you don't want to treat the entire project as a single user view. Ray has provided you with all the reports the system must produce, and you will treat each report as a user view. Ray has given you the following requirements:

User View 1 Requirements: For each publisher, list the publisher code, publisher name, and city in which the publisher is located.

User View 2 Requirements: For each branch, list the number, name, location, and number of employees.

User View 3 Requirements: For each book, list its code, title, publisher code and name, and price and whether it is a paperback.

User View 4 Requirements: For each book, list its code, title, price, and type. In addition, list the book's author(s) and the name(s) of the author(s). If a book has more than one author, all names must appear in the order in which they are listed on the book's cover. The author order is not always alphabetical.

User View 5 Requirements: For each branch, list its number and name. In addition, list the code and title of each book currently in the branch as well as the number of copies the branch has available.

User View 6 Requirements: For each book, list its code and title. In addition, for each branch that currently has the book in stock, list the number and name of the branch along with the number of copies available.

To transform each user view into DBDL, examine the requirements and create the necessary entities, keys, and relationships.

User View 1: For each publisher, list the publisher code, publisher name, and city in which the publisher is located. The only entity in this user view is Publisher.

```
Publisher (PublisherCode, PublisherName, City)
```

This table is in third normal form; the primary key is PublisherCode. There are no alternate or foreign keys. Assume Ray wants to be able to access a publisher rapidly on the basis of its name. You'll need to specify the PublisherName column as a secondary key.

Because this is the first user view, there is no previous cumulative design. Thus, at this point, the new cumulative design will consist only of the design for this user view, as shown in Figure 6-9. There is no need for an E-R diagram at this point.

```
Publisher (PublisherCode, PublisherName, City)
    SK    PublisherName
```

FIGURE 6-9 DBDL for Book database after first user view

User View 2: For each branch, list the number, name, location, and number of employees. The only entity in this user view is Branch.

```
Branch (BranchNum, BranchName, BranchLocation, NumEmployees)
```

This table is in third normal form. The primary key is BranchNum, and there are no alternate or foreign keys. Ray wants to be able to access a branch rapidly on the basis of its name, so you'll make the BranchName column a secondary key.

Because there is no table in the cumulative design with the BranchNum column as its primary key, you can add the Branch table to the cumulative design during the merge step, as shown in Figure 6-10. Again, there is no need for an E-R diagram with this simple design.

```
Publisher (PublisherCode, PublisherName, City)
    SK    PublisherName

Branch (BranchNum, BranchName, BranchLocation, NumEmployees)
    SK    BranchName
```

FIGURE 6-10 DBDL for Book database after second user view

User View 3: For each book, list its code, title, publisher code and name, and price and whether it is paperback. To satisfy this user requirement, you'll need to create entities for publishers and books and establish a one-to-many relationship between them. This leads to the following:

```
Publisher (PublisherCode, PublisherName)
Book (BookCode, Title, PublisherCode, Price, Paperback)
```

The PublisherCode column in the Book table is a foreign key identifying the publisher. Merging these tables with the ones you already created does not add any new columns to the Publisher table, but it does add columns to the Book table. The result of merging the Book table with the cumulative design is shown in Figure 6-11. Assuming Ray will need to access books based on their titles, you'll designate the Title column as a secondary key.

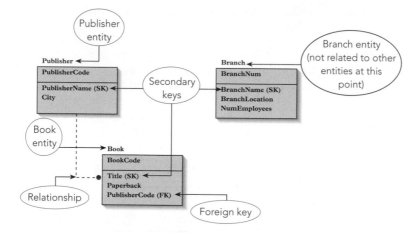

FIGURE 6-11 Cumulative design after third user view

User View 4: For each book, list its code, title, price, and type. In addition, list the book's author(s) and the name(s) of the author(s). If a book has more than one author, all names must appear in the order in which they are listed on the book's cover. There are two entities in the user view for books and authors. The relationship between them is many-to-many (one author can write many books and one book can have many authors). Creating tables for each entity and the relationship between them gives the following:

```
Author (AuthorNum, AuthorLast, AuthorFirst)
Book (BookCode, Title, Price, Type)
Wrote (BookCode, AuthorNum)
```

The third table is named Wrote because it represents the fact that an author *wrote* a particular book. In this user view, you need to be able to list the authors for a book in the appropriate order. To accomplish this goal, you will add a sequence number column to the Wrote table. This completes the tables for this user view, which are as follows:

```
Author (AuthorNum, AuthorLast, AuthorFirst)
Book (BookCode, Title, Price)
Wrote (BookCode, AuthorNum, Sequence)
```

The Author and Wrote tables are new; merging the Book table adds nothing new. Because it may be important to find an author based on the author's last name, the AuthorLast column is a secondary key. The result of the merge step is shown in Figure 6-12.

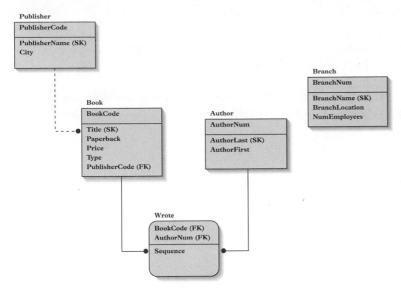

FIGURE 6-12   Cumulative design after fourth user view

User View 5: For each branch, list its number and name. In addition, list the code and title of each book currently in the branch as well as the number of copies the branch has available. Suppose you decide that the only entity mentioned in this requirement contains information about branches. You would create the following table:

```
Branch (
```

You would then add the BranchNum column as the primary key, producing the following:

```
Branch (BranchNum,
```

The other columns include the branch name as well as the book code, book title, and number of units on hand. Because a branch will have several books, the last three columns will form a repeating group. Thus, you have the following:

```
Branch (BranchNum, BranchName, (BookCode, Title, OnHand) )
```

You convert this table to first normal form by removing the repeating group and expanding the primary key. This gives the following:

```
Branch (BranchNum, BranchName, BookCode, Title, OnHand)
```

In this table, you have the following functional dependencies:

```
BranchNum → BranchName
BookCode → Title
BranchNum, BookCode → OnHand
```

The table is not in second normal form because some columns depend on just a portion of the primary key. Converting to second normal form gives the following:

```
Branch (BranchNum, BranchName)
Book (BookCode, Title)
Inventory (BranchNum, BookCode, OnHand)
```

You can name the new table Inventory because it represents each branch's inventory. In the Inventory table, the BranchNum column is a foreign key that identifies the Branch table and the BookCode column is a foreign key that identifies the Book table. In other words, for a row to exist in the Inventory table, *both* the branch number *and* the book code must already be in the database.

You can merge this Branch table with the existing Branch table without adding any new columns or relationships to the database. After adding the Inventory table to the existing cumulative design, you have the design shown in Figure 6-13.

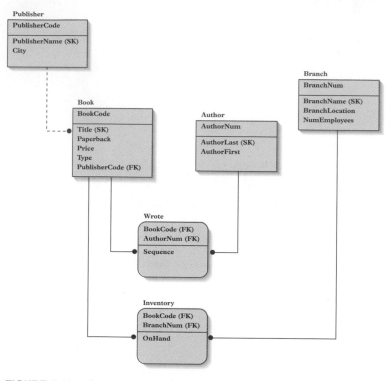

FIGURE 6-13   Cumulative design after fifth user view

NOTE

When you are using a software tool to produce these diagrams, the software may reverse the order of the columns that make up the primary key from the order you intended. For example, the diagram in Figure 6-13 indicates that the primary key for the Inventory table is BookCode and BranchNum, even though you intended it to be BranchNum and BookCode. This change in order is not a problem. Indicating the fields that make up the primary key is what is significant, not the order in which they appear.

Q & A

Question: How would the design for this user view turn out if you began with two entities, Branch and Book, instead of just the single entity Branch?

Answer: In the first step, you would create the following tables:

```
Branch (
Book (
```

Adding the primary keys would give this table:

```
Branch (BranchNum,
Book (BookCode,
```

Adding the other columns would give this table:

```
Branch (BranchNum, BranchName)
Book (BookCode, Title)
```

Finally, you have to implement the relationship between the Branch and Book tables. Because a branch can have many books and a book can be in stock at many branches, the relationship is many-to-many. To implement a many-to-many relationship, you add a new table whose primary key is the combination of the primary keys of the other tables. Doing this, you produce the following:

```
Branch (BranchNum, BranchName)
Book (BookCode, Title)
Inventory (BranchNum, BookCode)
```

Finally, you add any column that depends on both the BranchNum and PublisherCode columns to the Inventory table, giving the following:

```
Branch (BranchNum, BranchName)
Book (BookCode, Title)
Inventory (BranchNum, BookCode, OnHand)
```

Thus, you end up with exactly the same collection of tables, which illustrates a point made earlier: there's more than one way of arriving at the correct result.

User View 6: For each book, list its code and title. In addition, for each branch currently having the book in stock, list the number and name of each branch along with the number of copies available. This user view leads to precisely the same set of tables that were created for User View 5.

You have satisfied all the requirements, and the design shown in Figure 6-13 represents the complete information-level design.

PHYSICAL-LEVEL DESIGN

After the information-level design is complete, you are ready to begin the physical-level design process by implementing the design for the specific DBMS selected by the organization.

Because most DBMSs are relational and the final information-level design already exists in a relational format, producing the design for the chosen DBMS is usually an easy task—you simply use the same tables and columns. At this point, you also need to supply format details, such as specifying that the CustomerNum field will store characters and that its length is three.

Most DBMSs support primary, alternate, secondary, and foreign keys. If you are using a system that supports these keys, you simply use these features to implement the various types of keys that are listed in the final DBDL version of the information-level design. When working in DBMSs that don't support these keys, you need to devise a scheme for handling them to ensure the uniqueness of primary and alternate keys. In addition, you must ensure that values in foreign keys are legitimate; they must match the value of the primary key in some row in another table. For secondary keys, you must ensure that it is possible to retrieve data rapidly on the basis of a value of the secondary key.

For instance, suppose you are implementing the Employee table shown in Figure 6-1 and it has the following DBDL:

```
Employee (EmployeeNum, LastName, FirstName, Street, City,
    State, Zip, WageRate, SocSecNum, DepartmentNum)
    AK    SocSecNum
    SK    LastName
    FK    DepartmentNum → Department
```

The Employee table uses the EmployeeNum column as its primary key, the SocSecNum column as its alternate key, the LastName column as its secondary key, and the DepartmentNum column as a foreign key that matches the DepartmentNum column in the Department table. You must find a way for the DBMS to ensure that the following conditions hold true:

- Employee numbers are unique.
- Social Security numbers are unique.
- Access to an employee on the basis of his or her last name is rapid. (This restriction differs in that it merely states that a certain type of activity must be efficient, but it is an important restriction nonetheless.)
- Department numbers must match the number of a department currently in the database.

When the DBMS can't enforce these restrictions, who should enforce them? Two choices are possible: the users of the system or the programmers. If users must enforce these restrictions, they must be careful not to enter two employees with the same EmployeeNum, an employee with an invalid DepartmentNum, and so on. Clearly, this type of enforcement would put a tremendous burden on users.

When the DBMS can't enforce these restrictions, the appropriate place for the enforcement to take place is in the programs written to access the data in the database. Thus, the responsibility for this enforcement should fall on the programmers who write these programs. Incidentally, users *must* update the data through these programs and *not* through the built-in features of the DBMS in such circumstances; otherwise, the users would be able to bypass all the controls you are attempting to program into the system.

To enforce restrictions, programmers must include logic in their programs. With respect to the DBDL for the Employee table, this means the following:

1. Before an employee is added, the program should determine and process three restrictions:
   a. Determine whether an employee with the same EmployeeNum is already in the database; if so, the program should reject the update.
   b. Determine whether an employee with the same Social Security number is already in the database; if so, the program should reject the update.
   c. Determine that the inputted department number matches a department number that is already in the database; if it doesn't, the program should reject the update.
2. When a user changes the department number of an existing employee, the program should check to make sure the new number matches a department number that is already in the database. If it doesn't, the program should reject the update.
3. When a user deletes a department number, the program should verify that no employees work in the department. If the employees do work in the department and the program allows the deletion of the department, these employees will have invalid department numbers. In this case, the program should reject the update.

Programs must perform these verifications efficiently; in most systems, this means the database administrator will create indexes for each column (or combination of columns) that is a primary key, an alternate key, a secondary key, or a foreign key.

TOP-DOWN VERSUS BOTTOM-UP

Another way to design a database is to use a **bottom-up design method** in which specific user requirements are synthesized into a design. The opposite of a bottom-up design method is a **top-down design method**, which begins with a general database design that models the overall enterprise and repeatedly refines the model to achieve a design that supports all necessary applications. The original design and refinements are often represented with E-R diagrams.

Both strategies have their advantages. The top-down approach lends a more global feel to the project; you at least have some idea where you are headed, which is not so with a strictly bottom-up approach. On the other hand, a bottom-up approach provides a rigorous way of tackling each separate requirement and ensuring that it will be met. In particular, tables are created to satisfy each user view or requirement precisely. When these tables are correctly merged into the cumulative design, you can be sure that you have satisfied the requirements for each user view.

The ideal strategy combines the best of both approaches. Assuming the design problem is sufficiently complicated to warrant the benefits of the top-down approach, you could begin the design process for Premiere Products using a top-down approach by completing the following steps:

1. After gathering data on all user views, review them without attempting to create any tables. In other words, try to get a general feel for the task at hand.
2. From this information, determine the basic entities of interest to the organization (sales reps, customers, orders, and parts). Do not be overly concerned that you might miss an entity. If you do miss one, it will show up in later steps of the design method.
3. For each entity, start a table. For example, if the entities are sales reps, customers, orders, and parts, you will have the following:

```
Rep (
Customer (
Orders (
Part (
```

4. Determine and list a primary key for each table. In this example, you might have the following:

```
Rep (RepNum,
Customer (CustomerNum,
Orders (OrderNum,
Part (PartNum,
```

5. For each one-to-many relationship you can identify among these entities, optionally create and document an appropriate foreign key. For example, if there is a one-to-many relationship between the Rep and Customer tables, add the foreign key RepNum to the Customer table. If you omit this step or fail to list any foreign keys, you'll usually find them when you examine the individual users views later.

After completing the steps for a top-down approach, you can then apply the bottom-up method for examining individual user views. As you design each user view, keep in mind the tables you have created in the initial top-down approach and their keys. When you need to determine the primary key for a table, look for a primary key in your cumulative design. When it is time to determine a foreign key, check the entity's primary key to see if a match exists in the cumulative design. In either case, if the primary key already exists, use the existing name as a foreign key to ensure that you can merge the tables properly. At the end of the design process, you can consider removing any tables that do not contain columns and that have no foreign keys matching them.

Adding these steps to the process brings the benefits of the top-down approach to the approach you have been using. As you proceed through the design process for the individual user views, you will have a general idea of the overall picture.

SURVEY FORM

When designing a database, you might find it helpful to design a survey form to obtain the required information from users. You can ask users to complete the form, or you may want to complete the form yourself during an interview with the user. Before beginning the interview, you can identify all existing data by viewing various reports, documents, and so on. In any case, it is imperative that the completed survey form contain all the information necessary for the design process.

To be truly valuable to the design process, the survey form must contain the following information:

- *Entity information.* For each entity (reps, customers, parts, and so on), record a name and description and identify any synonyms for the entity. For example, at Premiere Products, your survey might reveal that what one user calls "parts" another user calls "products." In addition, record any general information about the entity, such as its use within the organization.
- *Attribute (column) information.* For each attribute of an entity, list its name, description, synonyms, and physical characteristics (such as being 20 characters long and alphanumeric or a number with five digits), along with general information concerning its use. In addition, list any restrictions on values and the place from which the values for the item originate. (For example, the values might originate from time cards or from orders placed by customers or be computed from values from other attributes, such as when subtracting the balance from the credit limit to obtain available credit). Finally, list any security restrictions that apply to the attribute.

- **Relationships.** For any relationship, the survey form should include the entities involved, the type of relationship (one-to-one, one-to-many, or many-to-many), the significance of the relationship (that is, what determines when two objects are related), and any restrictions on the relationship.

- **Functional dependencies.** The survey form should include information concerning the functional dependencies that exist among the columns. To obtain this information, you might ask the user a question such as this: If you know a particular employee number, can you establish other information, such as the name? If so, you can determine that the name is functionally dependent on the employee number. Another question you might ask is this: Do you know the number of the department to which the employee is assigned? If so, you can determine that the department number is functionally dependent on the employee number. If a given employee can be assigned to more than one department, you would not know the department number and the department number would not be dependent on the employee number. Users probably won't understand that term *functional dependency*; therefore, it is important to ask the right questions so that you can identify any functional dependencies. An accurate list of functional dependencies is absolutely essential to the design process.

- **Processing information.** The survey form should include a description of the manner in which the various types of processing (updates to the database, reports that must be produced, and so on) are to take place. To obtain this information, you would pose questions such as these:
  - How exactly is the report to be produced?
  - From where do the entries on the report come?
  - How are the report entries calculated?
  - When a user enters a new order, from where does the data come?
  - Which entities and columns must be updated and how?

  In addition, you need to obtain estimates on processing volumes by asking questions such as these:

  - How often is the report produced?
  - On average, how many pages or screens is the report?
  - What is the maximum length of the report?
  - What is the maximum number of orders the system receives per day?
  - What is the average number of orders the system receives per day?
  - What is the maximum number of invoices the system prints per day?
  - What is the average number of invoices the system prints per day?

OBTAINING INFORMATION FROM EXISTING DOCUMENTS

Existing documents can often furnish helpful information concerning the database design. You need to take an existing document, like the invoice for the company named Holt Distributors shown in Figure 6-14, and determine the tables and columns that would be required to produce the document.

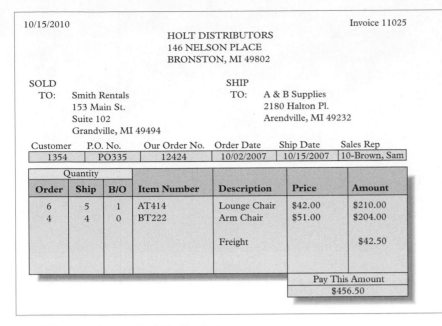

FIGURE 6-14   Invoice for Holt Distributors

The first step in obtaining information from an existing document is to identify and list all columns and give them appropriate names. Figure 6-15 lists the columns you can determine from the invoice shown in Figure 6-14.

```
Invoice Number
Invoice Date
CustomerNumber
CustomerSoldToName
CustomerSoldToAddressLine1
CustomerSoldToAddressLine2
CustomerSoldToCity
CustomerSoldToState
CustomerSoldToZip
CustomerShipToName
CustomerShipToAddress
CustomerShipToCity
CustomerShipToState
CustomerShipToZip
CustomerPONumber
OrderNumber
OrderDate
ShipDate
CustomerRepNumber
CustomerRepLastName
CustomerRepFirstName
ItemNumber
ItemDescription
ItemQuantityOrdered
ItemQuantityShipped
ItemQuantityBackordered
ItemPrice
ItemAmount
Freight
InvoiceTotal
```

FIGURE 6-15   List of possible attributes for the Holt Distributors invoice

The names the user chose for many of these columns might differ from the names you select, but this difference isn't important at this stage. After interviewing the user, you might learn that a required column was not apparent on the document you reviewed. For example, the shipping address for the customer shown in Figure 6-14 didn't require a second line, so you simply listed CustomerShipToAddress rather than CustomerShipToAddressLine1 and CustomerShipToAddressLine2 in your preliminary list of columns (see Figure 6-15). Some columns that you identify may not be required. For example, when the ship date is the same as the invoice date, a separate ShipDate column is unnecessary. Clearly, the user's help is needed to clarify these types of issues.

Next, you need to identify functional dependencies. If you are unfamiliar with the document you are examining, you might not be able to determine its functional dependencies. In this case, you'll need to interview the user to determine the functional dependencies that exist. Sometimes you can make intelligent guesses based on your general knowledge of the type of document you are studying. You may make mistakes, of course, but you can correct them when you interview the user. After initially determining the functional dependencies shown in Figure 6-16, you may find additional information.

```
CustomerNumber →

        CustomerSoldToName
        CustomerSoldToAddressLine1
        CustomerSoldToAddressLine2
        CustomerSoldToCity
        CustomerSoldToState
        CustomerSoldToZip
        CustomerShipToName
        CustomerShipToAddressLine1
        CustomerShipToAddressLine2
        CustomerShipToCity
        CustomerShipToState
        CustomerShipToZip
        CustomerRepNumber
        CustomerRepLastName
        CustomerRepFirstName

ItemNumber →
        ItemDescription
        ItemPrice

InvoiceNumber →
        InvoiceDate
        CustomerNumber
        OrderNumber
        OrderDate
        ShipDate
        Freight
        InvoiceTotal

InvoiceNumber, ItemNumber →
        ItemQuantityOrdered
        ItemQuantityShipped
        ItemQuantityBackordered
        ItemAmount
```

FIGURE 6-16   Tentative list of functional dependencies for the Holt Distributors invoice

Based on your list of functional dependencies, you may learn that the shipping address for a given customer will vary from one invoice to another. In other words, the shipping address depends on the invoice number, not the customer number. A default shipping address may be defined for a given customer in case no shipping address is entered with an order. However, the address that actually appears on the invoice would depend on the invoice number. You may also determine that several columns actually depend on the order that was initially entered. The order date, customer, shipping address, and quantities ordered on each line of the invoice may have been entered as part of the initial order. At the time the invoice was printed, additional information, such as the quantities shipped, the quantities back-ordered, and the freight charges, may have been added. You may also find that the price is not necessarily the one stored with the item and that the price can vary from one order to another. Given all these corrections, a revised list of functional dependencies might look like Figure 6-17.

```
CustomerNumber →
        CustomerSoldToName
        CustomerSoldToAddressLine1
        CustomerSoldToAddressLine2
        CustomerSoldToCity
        CustomerSoldToState
        CustomerSoldToZip
        CustomerRepNumber
        CustomerRepLastName
        CustomerRepFirstName

ItemNumber →
        ItemDescription
        ItemPrice

InvoiceNumber →
        InvoiceDate
        OrderNumber
        ShipDate
        Freight
        InvoiceTotal

OrderNumber →
        OrderDate
        CustomerPONumber
        CustomerShipToName
        CustomerShipToAddressLine1
        CustomerShipToAddressLine2
        CustomerShipToCity
        CustomerShipToState
        CustomerShipToZip

OrderNumber, ItemNumber →
        ItemQuantityOrdered (added when order is entered)
        ItemQuantityShipped (added during invoicing)
        ItemQuantityBackordered (added during invoicing)
        ItemPrice (added when order is entered)
```

FIGURE 6-17   Revised list of functional dependencies for the Holt Distributors invoice

After you have determined the preliminary functional dependencies, you can begin determining the tables and assigning columns. You could create tables with the determinant (the column or columns to the left of the arrow) as the primary key and with the columns to the right of the arrow as the remaining columns. This would lead to the following initial collection of tables:

```
Customer (CustomerNumber, CustomerSoldToName,
    CustomerSoldToAddressLine1, CustomerSoldToAddressLine2,
    CustomerSoldToCity, CustomerSoldToState, CustomerSoldToZip,
    CustomerRepNumber, CustomerRepLastName,
    CustomerRepFirstName)
Part (ItemNumber, ItemDescription, ItemPrice)
Invoice (InvoiceNumber, InvoiceDate, OrderNumber, ShipDate,
    Freight, InvoiceTotal)
Order (OrderNumber, OrderDate, CustomerPONumber,
    CustomerShipToName, CustomerShipToAddressLine1,
    CustomerShipToAddressLine2, CustomerShipToCity,
    CustomerShipToState, CustomerShipToZip)
OrderLine (OrderNumber, ItemNumber, ItemQuantityOrdered,
    ItemQuantityShipped, ItemQuantityBackordered, ItemPrice)
```

These tables would then need to be converted to third normal form and the result merged into the cumulative design.

Some people prefer not to get so specific at this point. Rather, they will examine the various columns and determine a preliminary list of entities, as shown in Figure 6-18.

```
Orders
Customer
Rep
Part
```

FIGURE 6-18 Tentative list of entities

After examining the functional dependencies, they will refine this list, producing a list similar to the one shown in Figure 6-19. At this point, they will create tables for these entities and position each column in the table in which it seems to fit best.

```
Invoice
Customer
Rep
Part
Orders
OrderLine
```

FIGURE 6-19 Expanded list of entities

Whichever approach you take, this kind of effort is certainly worthwhile; it gives you a better feel for the problem when you interact with the user. You can change your work based on your interview with the user. Even if your work proves to be accurate, you still need to ask additional questions of the user. These questions include the following:

- What names do you think are appropriate for the various entities and attributes?
- What synonyms are in use?
- What restrictions exist?
- What are the meanings of the various entities, attributes, and relationships?

If the organization has a computerized system, current file layouts can provide you with additional information about entities and attributes. Current file sizes can provide information on volume. Examining the logic in current programs and their operational instructions can yield processing information. Again, however, this is just a starting point. You still need further information from the user, which you can obtain by asking questions such as these:

- How many invoices do you expect to print?
- Exactly how are the values on the invoice calculated, and where do they come from?
- What updates must be made during the invoicing cycle of processing?
- What fields in the Customer table will be updated?

ONE-TO-ONE RELATIONSHIP CONSIDERATIONS

What, if anything, is wrong with implementing a one-to-one relationship by simply including the primary key of each table as a foreign key in the other table? For example, suppose each Premiere Products customer has a single sales rep and each sales rep represents a single customer. Applying the suggested technique to this one-to-one relationship produces two tables:

```
Rep (RepNum, LastName, FirstName, CustomerNum)
Customer (CustomerNum, CustomerName, RepNum)
```

In practice, these tables would contain any additional sales rep or customer columns of interest in the design problem. For the purposes of illustration, however, assume these are the tables' only columns.

Samples of these tables are shown in Figure 6-20. This design clearly forces a sales rep to be related to a single customer. Because the customer number is a column in the Rep table, there can be only one customer for each sales rep. Likewise, this design forces a sales rep to be related to a single customer.

Rep

RepNum	LastName	FirstName	CustomerNum
20	Kaiser	Valerie	148
35	Hull	Richard	282
65	Perez	Juan	356

Customer

CustomerNum	CustomerName	RepNum
148	Al's Appliance and Sport	20
282	Brookings Direct	35
356	Ferguson's	65

FIGURE 6-20   One-to-one relationship implemented by including the primary key of each table as a foreign key in the other

Q & A

Question: What is the potential problem with this solution?

Answer: There is no guarantee that the information will match. Consider Figure 6-21, for example. The data in the first table indicates that sales rep 20 represents customer 148. The data in the second table, on the other hand, indicates that customer 148 is represented by sales rep 35! This solution may be the simplest way of implementing a one-to-one relationship from a conceptual standpoint, but it clearly suffers from this major deficiency. The programs themselves would have to ensure that the data in the two tables match, a task that the design should be able to accomplish on its own.

Rep

RepNum	LastName	FirstName	CustomerNum
20	Kaiser	Valerie	148
35	Hull	Richard	282
65	Perez	Juan	356

Customer

CustomerNum	CustomerName	RepNum
148	Al's Appliance and Sport	35
282	Brookings Direct	20
356	Ferguson's	65

FIGURE 6-21   Implementation of a one-to-one relationship in which information does not match

Following are possible solutions you might consider for avoiding these problems.

The first possibility is to create a single table such as this:

```
Customer (CustomerNum, CustomerName, RepNum, LastName,FirstName)
```

A sample of this table is shown in Figure 6-22. Which column should be the primary key? If it is the customer number, there is nothing to prevent all three rows from containing the same rep number. On the other hand, if it is the rep number, the same would hold true for the customer number.

Customer

CustomerNum	CustomerName	RepNum	LastName	FirstName
148	Al's Appliance and Sport	20	Kaiser	Valerie
282	Brookings Direct	35	Hull	Richard
356	Ferguson's	65	Perez	Juan

FIGURE 6-22   One-to-one relationship implemented in a single table

The solution is to choose either the customer number or the rep number as the primary key and make the other column the alternate key. In other words, the uniqueness of both customer numbers and rep numbers should be enforced. Because each customer and each sales rep will appear in exactly one row, there is a one-to-one relationship between them.

Although this solution is workable, it has two features that are not particularly attractive. First, it combines columns of two different entities into a single table. It certainly would seem more natural to have one table with customer columns and a separate table with sales rep columns. Second, if it is possible for one entity to exist without the other (for example, if a customer has no sales rep), this structure is going to cause problems.

A better solution is to create two separate tables for customers and sales reps and to include the primary key of one of them as a foreign key in the other. This foreign key would also be designated as an alternate key. Thus, you could choose either

```
Rep (RepNum, LastName, FirstName, CustomerNum)
Customer (CustomerNum, CustomerName)
```

or

```
Rep (RepNum, LastName, FirstName)
Customer (CustomerNum, CustomerName, RepNum)
```

Samples of these two possibilities are shown in Figure 6-23. In either case, you must enforce the uniqueness of the foreign key that you added. In the first solution, for example, if customer numbers need not be unique, all three rows might contain customer 148, violating the one-to-one relationship. You can enforce the uniqueness by designating these foreign keys as alternate keys. They will also be foreign keys because they must match an actual row in the other table.

Solution 1:

Rep

RepNum	LastName	FirstName	CustomerNum
20	Kaiser	Valerie	148
35	Hull	Richard	282
65	Perez	Juan	356

Customer

CustomerNum	CustomerName
148	Al's Appliance and Sport
282	Brookings Direct
356	Ferguson's

Solution 2:

Rep

RepNum	LastName	FirstName
20	Kaiser	Valerie
35	Hull	Richard
65	Perez	Juan

Customer

CustomerNum	CustomerName	RepNum
148	Al's Appliance and Sport	20
282	Brookings Direct	35
356	Ferguson's	65

FIGURE 6-23   One-to-one relationship implemented by including the primary key of one table as the foreign key (and alternate key) in the other table

How do you make a choice between the possibilities? In some cases, it really makes no difference which arrangement you choose. Suppose, however, you anticipate the possibility that this relationship may not always be one-to-one. Suppose there is a likelihood in the future that a sales rep might represent more than one customer but that each customer still will be assigned to exactly one sales rep.

The relationship would then be one-to-many, and it would be implemented with a structure similar to Solution 2. In fact, the structure would differ only in that the rep number in the Customer table would not be an alternate key. Thus, to convert from the second alternative to the appropriate structure would be a simple matter—you would remove the restriction that the rep number in the Customer table is an alternate key. This situation would lead you to favor the second alternative.

MANY-TO-MANY RELATIONSHIP CONSIDERATIONS

Complex issues arise when more than two entities are related in a many-to-many relationship. For example, suppose Premiere Products needs to know which sales reps sold which parts to which customers. In this example, there are no restrictions on which customers a given sales rep may sell to or on the parts that a sales rep may sell. Sample data for this relationship is shown in Figure 6-24.

Sales

RepNum	CustomerNum	PartNum
20	148	AT94
20	282	DR93
35	148	DR93
35	148	DW11
65	282	AT94
65	282	DR93
65	356	AT94

FIGURE 6-24   Sample Sales data

The first row in the table indicates that sales rep 20 sold part AT94 to customer 148. (The number of units sold to the customer is not important in this example.) The second row indicates that sales rep 20 sold part DR93 to customer 282.

Q & A

Question: What is the primary key of the Sales table?
Answer: Clearly, none of the three columns (RepNum, CustomerNum, and PartNum) alone will uniquely identify a record. The combination of RepNum and CustomerNum does not work because there are two rows on which the rep number is 35 and the customer number is 148. The combination of RepNum and PartNum does not work because there are two rows on which the rep number is 65 and the part number is AT94. Finally, the combination of CustomerNum and PartNum does not work because there are two rows on which the customer number is 282 and the part number is DR93. Thus, the primary key for the Sales table must be the combination of all three columns, as follows:

```
Sales (RepNum, CustomerNum, PartNum)
```

Attempting to model this particular situation as two (or three) many-to-many relationships is not legitimate. Consider the following code and the data shown in Figure 6-25, for example, in which the same data is split into three tables.

```
RepCustomer (RepNum, CustomerNum)
CustomerPart (CustomerNum, PartNum)
PartRep (PartNum, RepNum)
```

RepCustomer

RepNum	CustomerNum
20	148
20	282
35	148
65	282
65	356

CustomerPart

CustomerNum	PartNum
148	AT94
148	DR93
148	DW11
282	AT94
282	DR93
356	AT94

PartRep

PartNum	RepNum
AT94	20
AT94	65
DR93	20
DR93	35
DR93	65
DW11	35

FIGURE 6-25   Result obtained by splitting the Sales table into three tables

Figure 6-26 shows the result of joining these three tables. Note that it contains inaccurate information. The second row, for example, indicates that rep 20 sold part DR93 to customer 148. If you look back to Figure 6-24, you will see that is not the case.

Sales

RepNum	CustomerNum	PartNum
20	148	AT94
20	148	DR93 !!!!
20	282	AT94 !!!!
20	282	DR93
35	148	DR93
35	148	DW11
65	282	AT94
65	282	DR93
65	356	AT94

FIGURE 6-26   Result obtained by joining three tables—the second and third rows are in error!

The second row appears in the join because rep 20 is related to customer 148 in the RepCustomer table (rep 20 sold a part to customer 148), customer 148 is related to part DR93 in the CustomerPart table (customer 148 bought part DR93 from a rep), and part DR93 is related to rep 20 in the RepPart table (rep 20 sold part DR93 to a customer). In other words, rep 20 sold a part to customer 148, customer 148 bought part DR93, and rep 20 sold part DR93. Of course, these three facts do not imply that rep 20 sold part DR93 to customer 148. (Customer 148 might have purchased this part from another rep.)

The problem with the preceding relationship is that it involves all three entities—reps, customers, and parts. Splitting the Sales table shown in Figure 6-26 any further is inappropriate. Such a relationship is called a **many-to-many-to-many relationship**.

Remember from the discussion of fourth normal form that there are examples of three-way relationships in which you must split the tables. In particular, if the relationship between sales reps and customers has nothing to do with the relationship between sales reps and parts, this table would violate fourth normal form and would need to be split.

The crucial issue in making the determination between a single many-to-many-to-many relationship and two (or three) many-to-many relationships is the independence. When all three entities are critical in the relationship, the three-way relationship (like Sales) is appropriate. When there is independence among the individual relationships, separate many-to-many relationships are appropriate. Incidentally, if a many-to-many-to-many relationship is created when it is not appropriate to do so, the conversion to fourth normal form will correct the problem.

NULLS AND ENTITY SUBTYPES

Recall that a null is a special value that represents the *absence* of a value in a field. In other words, setting a particular field to null is equivalent to not entering a value in the field. Nulls are used when a value is either unknown or inapplicable. This section focuses on the second possibility—when the value is inapplicable.

Consider, for example, a Student table in which one of the columns, DormNum, is a foreign key that identifies a Dorm (dormitory) table. The DormNum column indicates the number of the dormitory in which a student currently resides. This foreign key is allowed to be null because some students do not live in a dormitory; for these students, DormNum is inapplicable. Thus, for some rows in the Student table, the DormNum column would be null.

When there are many students who do not live in dorms, you can avoid using null values in the DormNum column by removing the DormNum column from the Student table and creating a separate table named StudentDorm that contains the columns StudentNum (the primary key) and DormNum. Students living in a dorm would have a row in this new table. Students not living in a dorm would have a row in the Student table but not in the StudentDorm table.

This change is illustrated in Figure 6-27. Note that StudentNum, the primary key of the StudentDorm table, is also a foreign key that must match a student number in the Student table.

Student

StudentNum	LastName	FirstName	DormNum
1253	Johnson	Ann	3
1662	Anderson	Tom	1
2108	Lewis	Bill	
2546	Davis	Mary	2
2867	Albers	Cathy	2
2992	Matthew	Mark	
3011	Candela	Tim	3
3574	Talen	Sue	

Student

StudentNum	LastName	FirstName
1253	Johnson	Ann
1662	Anderson	Tom
2108	Lewis	Bill
2546	Davis	Mary
2867	Albers	Cathy
2992	Matthew	Mark
3011	Candela	Tim
3574	Talen	Sue

StudentDorm

StudentNum	DormNum
1253	3
1662	1
2546	2
2867	2
3011	3

FIGURE 6-27   Student table split to avoid use of null values

In the process, you have created what is formally called an **entity subtype**. You can say that the StudentDorm table is a subtype of the Student table. In other words, "students living in dorms" is a subtype (or subset) of "students."

Some design methods have specific ways of denoting entity subtypes, but it is not necessary to denote entity subtypes in DBDL. You can recognize entity subtypes by the fact that the primary key is also a foreign key, as shown in Figure 6-28.

```
Student (StudentNum, LastName, FirstName)

StudentDorm (StudentNum, DormNum)
     FK    StudentNum → Student
     FK    DormNum → Dorm
```

FIGURE 6-28   Sample DBDL with entity subtypes

Most approaches to diagramming database designs have ways of representing subtypes. In IDEF1X, for example, a subtype, which is called a **category** in IDEF1X terminology, is represented in the manner shown in Figure 6-29. The circle is the symbol for category. The single line below the circle indicates that the category is an **incomplete category**; that is, there are students who do not fall into the StudentDorm category.

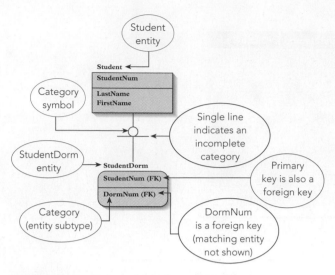

FIGURE 6-29 Entity subtype in an E-R diagram

The issue is more complicated when more than one column can accept null values. Suppose the DormNum, ThesisTitle, and ThesisArea columns in the following Student table can be null.

```
Student (StudentNum, LastName, FirstName, DormNum, ThesisTitle, ThesisArea)
```

In this table, the dorm number is the number of the dorm in which the student resides or is null if the student does not live in a dorm. In addition, students at this college must write a senior thesis. After students attain senior standing, they must select a thesis title in the area in which they will write their thesis. Thus, seniors will have a thesis title and a thesis area, whereas other students will not. You can handle this situation by allowing the fields ThesisTitle and ThesisArea to be null.

The Student table now has three different columns—DormNum, ThesisTitle, and ThesisArea—that can be null. The DormNum column will be null for students who do not live in a dorm. The ThesisTitle and ThesisArea columns will be null for students who have not yet attained senior standing. It wouldn't make much sense to combine all three of these columns into a single table. A better choice would be to create the following table for students living in dorms:

```
StudentDorm (StudentNum, DormNum)
```

For seniors, you could create a second table as follows:

```
SeniorStudent (StudentNum, ThesisTitle, ThesisArea)
```

Samples of these tables are shown in Figure 6-30. Both tables represent entity subtypes. In both tables, the primary key (StudentNum) will also be a foreign key matching the student number in the new Student table.

Student

StudentNum	LastName	FirstName	DormNum	ThesisTitle	ThesisArea
1253	Johnson	Ann	3		
1662	Anderson	Tom	1	P.D.Q. Bach	Music
2108	Lewis	Bill		Cluster sets	Math
2546	Davis	Mary	2		
2867	Albers	Cathy	2	Rad. Treatment	Medicine
2992	Matthew	Mark			
3011	Candela	Tim	3		
3574	Talen	Sue			

Student

StudentNum	LastName	FirstName
1253	Johnson	Ann
1662	Anderson	Tom
2108	Lewis	Bill
2546	Davis	Mary
2867	Albers	Cathy
2992	Matthew	Mark
3011	Candela	Tim
3574	Talen	Sue

StudentDorm

StudentNum	DormNum
1253	3
1662	1
2546	2
2867	2
3011	3

SeniorStudent

StudentNum	ThesisTitle	ThesisArea
1662	P.D.Q. Bach	Music
2108	Cluster sets	Math
2867	Rad. Treatment	Medicine

FIGURE 6-30   Student table split to avoid the use of null values

The DBDL for these tables appears in Figure 6-31. The primary key of the StudentDorm and SeniorStudent tables (StudentNum) is also a foreign key matching the student number in the new Student table.

```
Student (StudentNum, LastName, FirstName)

StudentDorm (StudentNum, DormNum)
     FK    StudentNum → Student
     FK    DormNum →  Dorm

SeniorStudent (StudentNum, ThesisTitle, ThesisArea)
     FK StudentNum → Student
```

FIGURE 6-31   Sample DBDL with entity subtypes

To represent two subtypes (categories) in IDEF1X, you use the same category symbol shown in Figure 6-29. The difference is that there will be two lines coming out of the category symbol—one to each category, as shown in Figure 6-32. Because there are students who do not live in dorms and who are not seniors, these categories are also incomplete; so there is only one line below the category symbol.

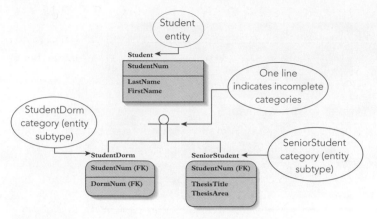

FIGURE 6-32   Two entity subtypes—incomplete categories

By contrast, Figure 6-33 represents a slightly different situation. There are two categories: students who live in dorms (StudentDorm) and students who do not (StudentNonDorm) live in dorms. For students who live in dorms, the attribute of interest is DormNum. For students who do not live in dorms, the attributes of interest are the ones that give the students' local addresses (LocalStreet, LocalCity, LocalState, and LocalZip). The difference between this example and the one shown in Figure 6-32 is that every student *must* be in one of these two categories. These are called **complete categories** and are represented by two lines below the category symbol.

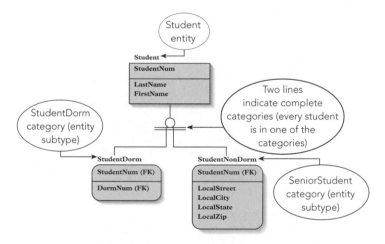

FIGURE 6-33   Two entity subtypes—complete categories

You should group columns that can be null by function. If a given subset of the entity in question can have nulls in a certain collection of columns, you should note this fact. When available, you should strongly consider splitting columns that can have nulls into a separate table (an entity subtype). If you create an entity subtype, you should give the entity subtype a name that suggests the related entity type, such as SeniorStudent for students who are seniors. In addition, you should carefully document the meaning of the entity subtype, especially the conditions that will cause an occurrence of the entity type also to be an occurrence of the entity subtype. If you do not create such an entity subtype, you must at least document precisely when the columns might take on null as a value.

AVOIDING PROBLEMS WITH THIRD NORMAL FORM WHEN MERGING TABLES

When you combine third normal form tables, the result might not be in third normal form. For example, both of the following tables are in third normal form:

Customer (<u>CustomerNum</u>, CustomerName, RepNum)
Customer (<u>CustomerNum</u>, CustomerName, LastName, FirstName)

When you combine them, however, you get the following table:

Customer (<u>CustomerNum</u>, CustomerName, RepNum, LastName, FirstName)

This table is not in third normal form. You would have to convert it to third normal form before proceeding to the next user view.

You can attempt to avoid the problem of creating a table that is not in third normal form by being cautious when representing user views. This problem occurs when a column A in one user view functionally determines a column B in a second user view. Thus, column A is a *determinant* for column B, yet column A is not a column in the second user view.

In the preceding example, the RepNum column in the first table determines the columns LastName and FirstName in the second table, yet the RepNum column is not one of the columns in the second table. If you always attempt to determine whether determinants exist and include them in the tables, you often will avoid this problem. For example, when the second user indicates that the name of a rep is part of that user's view of data, you should ask whether any special way has been provided for sales reps to be uniquely identified within the organization. Even though this user evidently does not need the rep number, he or she might very well be aware of the existence of such a number. If so, you would include this number in the table. Having done this, you would have the following table in this user view:

Customer (<u>CustomerNum</u>, CustomerName, RepNum, LastName, FirstName)

Now the normalization process for this user view would produce the following two tables:

Customer (<u>CustomerNum</u>, CustomerName, RepNum)
Rep (<u>RepNum</u>, LastName, FirstName)

When you merge these two tables into the cumulative design, you do not produce any tables that are not in third normal form. Notice that the determinant RepNum has replaced the columns that it determines, LastName and FirstName, in the Customer table.

THE ENTITY-RELATIONSHIP MODEL

In this chapter, you examined the use of E-R diagrams to visually illustrate the relations and keys represented in DBDL. The **entity-relationship (E-R) model** is an approach to representing data in a database. This model uses E-R diagrams exclusively as the tool for representing entities, attributes, and relationships. The E-R model is widely used and forms the basis of some computerized tools, so it is important that you understand how to use it.

In 1976, Peter Chen of the MIT Sloan School of Management proposed the E-R model; and since then it has been widely accepted as a graphical approach to database representation and database design. The basic constructions in the E-R model are the familiar entities, attributes, and relationships, all of which are represented in E-R diagrams. This section focuses on the standard versions of these diagrams. The versions you examined earlier in this chapter represent one of the common alternative forms of creating these diagrams that is particularly convenient for use with DBDL.

In the standard E-R diagrams, entities are drawn as rectangles and relationships are drawn as diamonds, with lines connecting the entities involved in relationships. Both entities and relationships are named in the E-R model. The lines are labeled to indicate the type of relationship. For example, in Figure 6-34, the one-to-many relationship between sales reps and customers is represented as "1" to "n."

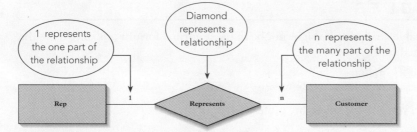

FIGURE 6-34 One-to-many relationship

In Figure 6-35, the many-to-many relationship between orders and parts is represented as "m" to "n."

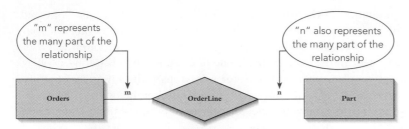

FIGURE 6-35 Many-to-many relationship

Finally, the many-to-many-to-many relationship between sales reps, customers, and parts is represented as "m" to "n" to "p," as shown in Figure 6-36.

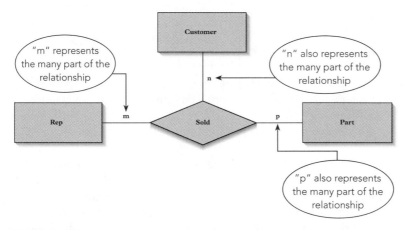

FIGURE 6-36 Many-to-many-to-many relationship

If desired, you can also indicate attributes in the E-R model by placing them in ovals and attaching them to the corresponding rectangles (entities), as shown in Figure 6-37. As in the relational model representation, primary keys are underlined.

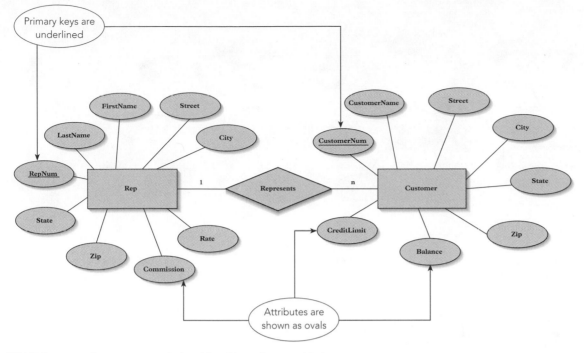

FIGURE 6-37   One-to-many relationship with attributes added

In the original version of the E-R model, attributes and relationships can have attributes. Figure 6-38 shows the many-to-many relationship between the Orders and Part entities, in which the relationship, OrderLine, has two attributes, NumOrdered and QuotedPrice.

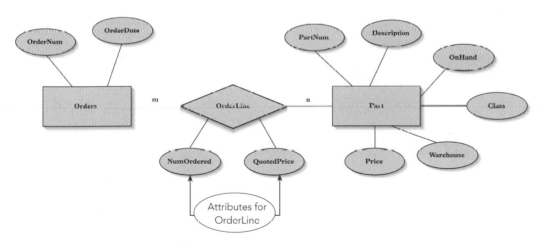

FIGURE 6-38   Many-to-many relationship with attributes

In Figure 6-38, it isn't clear whether OrderLine is an entity or a relationship. To address this confusion, the E-R model was changed slightly so that entities can have attributes but relationships cannot. With this change, OrderLine in Figure 6-38 would be an entity. However, making OrderLine an entity does not effectively communicate the fact that the OrderLine entity is implementing only a many-to-many relationship between orders and parts. To address this problem, an entity that exists to implement a many-to-many relationship is called a **composite entity**. A composite entity is essentially both an entity and a relationship and is represented in an E-R diagram by a diamond within a rectangle. Figure 6-39 shows this new approach.

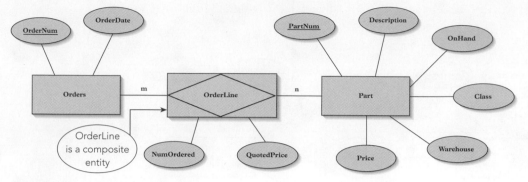

FIGURE 6-39   Composite entity

A complete E-R diagram for the Premiere Products database appears in Figure 6-40. Notice that OrderLine is represented as a composite entity.

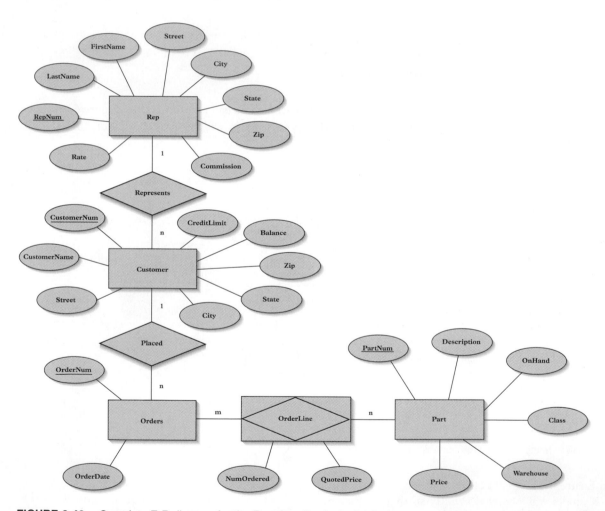

FIGURE 6-40   Complete E-R diagram for the Premiere Products database

When the existence of one entity depends on the existence of another related entity, there is an **existence dependency**. For example, because an order cannot exist without a customer, the relationship between customers and orders is an existence dependency. You indicate an existence dependency by placing an *E* in the relationship diamond, as shown in Figure 6-41. An entity that depends on another entity for its own existence is called a **weak entity**. A double rectangle encloses a weak entity, as shown in Figure 6-41. A weak entity corresponds to the term *dependent entity*, which was previously defined in this chapter.

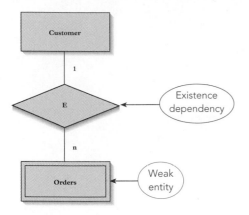

FIGURE 6-41   E-R diagram with an existence dependency and a weak entity

There is another popular way to indicate a one-to-many relationship. In this alternative, you do not label the "one" end of the relationship; instead, you place a crow's foot at the "many" end of the relationship. Figure 6-42 illustrates this style.

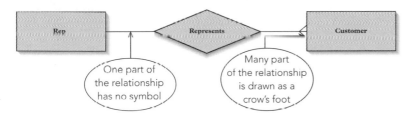

FIGURE 6-42   E-R diagram with a crow's foot

Some people represent **cardinality**, or the number of items that must be included in a relationship, in an E-R diagram. Figure 6-43 shows an E-R diagram that represents cardinality in this way. The two symbols to the right of the Rep rectangle are both the number 1. The 1 closest to the rectangle indicates that the maximum cardinality is one; that is, a customer can have at most one sales rep. The 1 closest to the relationship is the minimum cardinality; that is, a customer must have at least one sales rep. Together the two symbols indicates that a customer must have exactly one sales rep. (If the minimum cardinality were zero, for example, a customer would not be required to have a sales rep.)

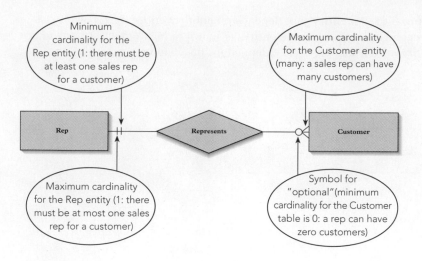

FIGURE 6-43 E-R diagram that represents cardinality

The crow's foot to left of the Customer rectangle indicates that the maximum cardinality is "many." The circle to the left of the crow's foot indicates that the minimum cardinality is zero; that is, a sales rep could be associated with zero customers. An entity in a relationship with minimum cardinality of zero plays an **optional role** in the relationship. An entity with a minimum cardinality of one plays a **mandatory role** in the relationship.

Summary

- Database design is a two-part process of determining an appropriate database structure to satisfy a given set of requirements. In the information-level design, a clean DBMS design that is not dependent on a particular DBMS is created to satisfy the requirements. In the physical-level design, the final information-level design is converted into an appropriate design for the particular DBMS that will be used.

- A user view is the set of necessary requirements to support a particular user's operations. To simplify the design process, the overall set of requirements is split into user views.

- The information-level design method involves applying the following steps to each user view: represent the user view as a collection of tables, normalize these tables (convert the collection into an equivalent collection that is in third normal form), represent all keys (primary, alternate, secondary, and foreign), and merge the results into the cumulative design.

- A database design is represented in a language called Database Design Language (DBDL).

- Designs can be represented visually using entity-relationship (E-R) diagrams. Such diagrams have the following characteristics: there is a rectangle for each entity; the name of the entity appears above the rectangle; the primary key appears above the line in the rectangle; the remaining columns appear below the line; alternate keys, secondary keys, and foreign keys are identified with the letters AK, SK, and FK, respectively; and for each foreign key, there is a dashed line from the rectangle that corresponds to the table being identified to the rectangle that corresponds to the table containing the foreign key. A dot at the end of the line indicates the "many" part of a one-to-many relationship.

- When a relational DBMS is going to be used, the physical-level design process consists of creating a table for each entity in the DBDL design. Any constraints (primary key, alternate key, or foreign key) that the DBMS cannot enforce must be enforced by the programs in the system; this fact must be documented for the programmers.

- The design method presented in this chapter is a bottom-up method. By listing potential relations before beginning the method, you have the advantages of both the top-down and bottom-up approaches.

- A survey form is useful for documenting the information gathered for the database design process.

- To obtain information from existing documents, list all attributes present in the documents, identify potential functional dependencies, make a tentative list of tables, and use the functional dependencies to refine the list.

- To implement a one-to-one relationship, include the primary key of one of the two tables in the other table as a foreign key and then indicate the foreign key as an alternate key.

- If a table's primary key consists of three (or more) columns, you must determine whether there are independent relationships between pairs of these columns. If there are independent relationships, the table is not in fourth normal form and you must split it. If there are no independent relationships, you can't split the table because doing so produces incorrect information.

- If a table contains columns that can be null and the nulls represent the fact that the column is inapplicable for some rows, you can split the table, placing the null column(s) in separate tables. These new tables represent entity subtypes.

- It is possible that the result of merging third normal form tables may not be in third normal form. To avoid this problem, include determinants for columns in the individual tables before merging them.

- The entity-relationship (E-R) model is a method of representing the structure of a database using an E-R diagram. In an E-R diagram, a rectangle represents an entity, a diamond represents a relationship, and an oval represents an attribute.

Key Terms

artificial key	cumulative design
bottom-up design method	Database Design Language (DBDL)
cardinality	dependent entity
category	entity-relationship (E-R) model
complete category	entity subtype
composite entity	existence dependency

IDEF1X
identifying relationship
incomplete category
independent entity
information-level design
intelligent key
logical key

mandatory role
many-to-many relationship
many-to-many-to-many relationship
natural key

nonidentifying relationship
one-to-one relationship
optional role
physical-level design
secondary key
surrogate key
synthetic key
top-down design method
user view
weak entity

Review Questions

1. Define the term *user view* as it applies to database design.

2. What is the purpose of breaking down the overall design problem into a consideration of individual user views?

3. Under what circumstances would you not need to break down an overall design into a consideration of individual user views?

4. The information-level design method presented in this chapter contains steps that must be repeated for each user view. List the steps and briefly describe the kinds of activities that must take place at each step.

5. Describe the function of each of the following types of keys: primary, alternate, secondary, and foreign.

6. Suppose a given user view contains information about employees and projects. Suppose further each employee has a unique EmployeeNum and each project has a unique ProjectNum. Explain how you would implement the relationship between employees and projects in each of the following scenarios:

   a. Many employees can work on a given project, but each employee can work on only a *single* project.

   b. An employee can work on many projects, but each project has a *unique* employee assigned to it.

   c. An employee can work on many projects, and a project can be worked on by many employees.

7. A database at a college is required to support the following requirements. Complete the information-level design for this set of requirements. Use your own experience to determine any constraints you need that are not stated in the problem. Represent the answer in DBDL.

   a. For a department, store its number and name.

   b. For an advisor, store his or her number and name and the number of the department to which he or she is assigned.

   c. For a course, store its code and description (for example, MTH110 or Algebra).

   d. For a student, store his or her number and name. For each course the student has taken, store the course code, course description, and grade received. In addition, store the number and name of the student's advisor. Assume that an advisor may advise any number of students but that each student has just one advisor.

8. List the changes you would need to make in your answer to Question 7 if a student could have more than one advisor.

9. Suppose in addition to the requirements specified in Question 7, you must store the number of the department in which the student is majoring. Indicate the changes this would cause in the design in the following two situations:

   a. The student must be assigned an advisor who is in the department in which the student is majoring.

   b. The student's advisor does not necessarily have to be in the department in which the student is majoring.

10. Illustrate the physical-level design process by means of the design shown in Question 7. List the tables, identify the keys, and list the special restrictions that programs must enforce.

11. Is the database design method top-down or bottom-up? How can you modify this method to gain the advantages to both types of design methods?

12. Design a survey form of your own. Fill it out as it might have been filled out during the database design for Premiere Products. For any questions you have too little information to answer, make a reasonable guess.

13. Using a document at your own school (for example, a report card), determine the attributes present in the document. Using your knowledge of the policies at your school, determine the functional dependencies present in the document. Use these dependencies to create a set of tables and columns that you could use to produce the document.

14. Describe the different ways of implementing one-to-one relationships. Assume you are maintaining information on offices (office numbers, buildings, and phone numbers) and faculty (numbers and names). No office houses more than one faculty member; no faculty member is assigned more than one office. Illustrate the ways of implementing one-to-one relationships using offices and faculty. Which option would be best in each of the following situations?

   a. A faculty member must have an office, and each office must be occupied by a faculty member.

   b. A faculty member must have an office, but some offices are not currently occupied. You must maintain information about the unoccupied offices in an Office relation.

   c. Some faculty members do not have an office, but all offices are occupied.

   d. Some faculty members do not have an office, but some offices are not occupied.

15. For each of the following collections of relations, give the assumptions concerning the relationship between students, courses, and faculty members that are implied by the collection. In each relation, only the primary keys are shown.

   a. Student (StudentNum, CourseNum, FacultyNum)

   b. Student (StudentNum, CourseNum)

       Faculty (CourseNum, FacultyNum)

   c. Student (StudentNum, CourseNum)

       Faculty (CourseNum, FacultyNum)

       StudentFaculty (StudentNum, FacultyNum)

   d. Student (StudentNum, CourseNum, FacultyNum)

   e. Student (StudentNum, CourseNum)

       Faculty (CourseNum, FacultyNum)

       StudentFaculty (StudentNum, FacultyNum)

16. Describe the relationship between columns that can be null and entity subtypes. Under what circumstances would these columns lead to more than one entity subtype?

17. How is it possible to merge a collection of relations that is in third normal form into a cumulative design that is in third normal form but not obtain a collection of relations that is in third normal form? Give an example other than the one in the text.

18. Describe the entity-relationship model. How are entities, relationships, and attributes represented in this model? What is a composite entity? Describe the approach to diagrams that uses a crow's foot. Describe how you would represent cardinality in an E-R diagram.

Premiere Products Exercises

The following exercises are based on the Premiere Products database as designed in Example 1. In each exercise, represent your answer in DBDL and with a diagram. You may use any of the styles presented in this chapter for the diagram.

1. Indicate the changes you need to make to the design of the Premiere Products database to support the following situation: A customer is not necessarily represented by a single sales rep but can be represented by several sales reps. When a customer places an order, the sales rep who gets the commission on the order must be one of the collection of sales reps who represents the customer.

2. Indicate the changes you need to make to the design of the Premiere Products database to support the following situation: There is no relationship between customers and sales reps. When a customer places an order, it may be through any sales rep. On the order, identify both the customer placing the order and the sales rep responsible for the order.

3. Indicate the changes you need to make to the design of the Premiere Products database in the event User View 3 requirements are changed as follows: For a part, store the part's number, description, item class, and price. In addition, for each warehouse in which the part is located, store the number of the warehouse, description of the warehouse, and number of units of the part stored in the warehouse.

4. Indicate the changes you need to make to the Premiere Products database design to support the following situation: The region where customers are located is divided into territories. For each territory, store the territory number (a unique identifier) and territory name. Each sales rep is assigned to a single territory. Each customer is also assigned to a single territory, but the territory must be the same as the territory to which the customer's sales rep is assigned.

5. Indicate the changes you need to make to the the Premiere Products database design to support the following situation: The region where customers are located is divided into territories. For each territory, store the territory number (a unique identifier) and territory name. Each sales rep is assigned to a single territory. Each customer is also assigned to a single territory, which may not be the same as the territory to which the customer's sales rep is assigned.

Henry Books Case

Ray Henry is considering expanding the inventory at his book stores to include movies. He has some special ideas for how he wants to implement this change, and he needs you to help with database design activities. In each exercise, represent your answer in DBDL and with a diagram. You may use any of the styles presented in this chapter for the diagram.

1. Design a database for Ray. He is interested in movies and wants to keep information on movies, actors, and directors in a database. The only user is Ray, and he needs to produce the following reports:

   a. For each director, list his or her number and name and the year he or she was born. If the director is deceased, list the year of death.

   b. For each movie, list its number, its title, the year the movie was made, and its type (for example, Comedy, Drama, or Science Fiction).

   c. For each movie, list its number, its title, the number and name of its director, the critics' rating, the MPAA rating (G, PG, PG-13, or R), the number of awards for which the movie was nominated, and the number of awards the movie won. (The critics rate the movie with a number of "stars." Four stars is the top rating possible. Zero stars is the worst possible rating.)

   d. For each lead actor starring in each movie, list his or her number, name, and birthplace and the year he or she was born. If the actor is deceased, list the year of death.

   e. For each movie, list its number and title, along with the number and name of the actors who appeared in it.

   f. For each lead actor starring in each movie, list his or her number and name, along with the number and name of the other movies in which the actor starred.

2. Expand the database design you created in Exercise 1 so that it will also support the following situation: Ray wants to start a DVD rental program at his stores that he plans to call Henry's DVD Club. He refers to each of his customers as "members." Each member in the club is assigned a number. He also stores the members' names and addresses. In addition, he stores the number of rentals a member has made and the date the member joined the club. He periodically has promotions during which members can earn bonus units that they can later apply to the cost of renting DVDs. He needs to store the number of bonus units a member has earned.

3. Expand the database design you created in Exercise 1 and modified in Exercise 2 so that it will also support the following situation: Ray wants to store information about the DVDs the club owns. When the club purchases a DVD, Ray assigns it a number. Along with the number, he stores the number of the movie on the DVD, the date the DVD was purchased, the number of times it has been rented, and the number of the member who is currently renting it. (If the DVD is not currently being rented, the member number will be null.) Ray also needs to store the number of the branch to which the DVD is assigned. Finally, Ray would like to store the history of the rental of each particular DVD. In particular, he needs to store the DVD number, date of the rental, date it was returned, and number of the member who rented the DVD. Assume a DVD could potentially be rented more than once on the same day. (*Hint:* Review the discussion of the categories of primary keys to determine what type of primary key would be appropriate for this relation.)

Alexamara Marina Group Case

Complete the following tasks. In each exercise, represent your answer in both DBDL and with a diagram. You may use any of the styles presented in this chapter for the diagram.

1. Design a database to produce the following reports. Do not use any surrogate keys in your design.

    a. For each marina, list the marina number, name, address, city, state, and zip code.

    b. For each boat owner, list the owner number, last name, first name, address, city, state, and zip code.

    c. For each marina, list all the slips in the marina. For each slip, list the length of the slip; annual rental fee; name and type of the boat occupying the slip; and boat owner's number, last name, and first name.

    d. For each service category, list the category number and description. In addition, for each service request in each category, list the marina number, slip number of the boat receiving the service, estimated hours for the service, hours already spent on the service, and next scheduled service date.

    e. For each service request, list the marina number, slip number, category description, description of the particular service, and a description of the current status of the service.

2. Expand the database design you created in Exercise 1 so that it will also support the following situation: A specific technician handles each service request. Along with all the details concerning a service request listed in Exercise 1, list the number, last name, and first name of the technician assigned to handle the request.

DBMS FUNCTIONS

INTRODUCTION

In Chapter 6, you learned the steps that the database designer completed to design the Premiere Products database. Management next asks the database administrator (DBA) to explain the functions that a DBMS performs in managing a database. The DBA describes nine functions performed by a DBMS; this chapter explains these nine functions. Some of the functions have been introduced in previous chapters; however, they are re-emphasized here because they are key processing components of a DBMS. The nine functions of a DBMS are:

- **Update and retrieve data.** A DBMS must provide users with the ability to update and retrieve data in a database.

- **Provide catalog services.** A DBMS must store data about the data in a database and make this data accessible to users.

- **Support concurrent update.** A DBMS must ensure that the database is updated correctly when multiple users update the database at the same time.

- **Recover data.** A DBMS must provide methods to recover a database in the event the database is damaged in any way.

- **Provide security services.** A DBMS must provide ways to ensure that only authorized users can access the database.

- **Provide data integrity features.** A DBMS must follow rules so that it updates data accurately and consistently.

- **Support data independence.** A DBMS must provide facilities to support the independence of programs from the structure of a database.

- **Support data replication.** A DBMS must manage multiple copies of the same data at multiple locations.

- **Provide utility services.** A DBMS must provide services that assist in the general maintenance of a database.

UPDATE AND RETRIEVE DATA

A DBMS must provide users with the ability to update and retrieve data in a database; this is the fundamental capability of a DBMS. Unless a DBMS provides this capability, further discussion of what a DBMS does is irrelevant. In updating and retrieving data, users do not need to know how data is physically structured on disk or which processes the DBMS uses to manipulate the data. These structures and manipulations are the sole responsibility of the DBMS.

Updating data in a database includes adding new records and changing and deleting existing records. For example, suppose Elena must update the Premiere Products database by adding data for part AE27, which is a new part. As shown in Figure 7-1, Elena enters the data for part AE27 and then requests that the DBMS add the data to the database. To add this data, the DBMS handles all the work to verify that part AE27 doesn't already exist in the database, stores the part AE27 data in the database, and then informs Elena that the task was completed successfully. How the DBMS performs these steps, where the DBMS stores the data in the database, how the DBMS stores the data, and all other processing details are invisible to Elena.

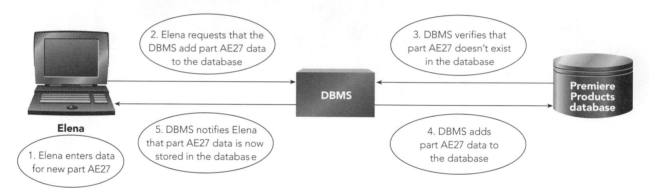

FIGURE 7-1   Adding a new part to the Premiere Products database

Suppose Elena must also update the Premiere Products database by changing the price of part KL62. As shown in Figure 7-2, Elena requests the data for the part and enters the change, but the DBMS performs the tasks of locating and reading the part data, displaying the data for Elena, and changing the price in the database. Once again, Elena does not need to be aware of what tasks the DBMS completes or how the DBMS completes them.

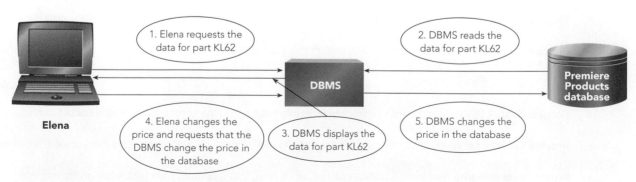

FIGURE 7-2   Changing the price of a part in the Premiere Products database

Figure 7-3 shows Elena retrieving the balance amount for All Season, a customer in the Premiere Products database. The DBMS finds the All Season record using the same strategy it used when it added the customer to the database; Elena doesn't need to know the strategy the DBMS uses to find and read the data. After finding and reading the All Season record in the database, the DBMS displays the customer's balance amount for Elena.

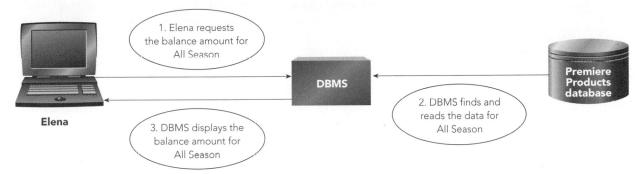

FIGURE 7-3   Retrieving a balance amount from the Premiere Products database

PROVIDE CATALOG SERVICES

A DBMS must store data about the data in a database and make this data accessible to users. Data about the data in a database, or **metadata**, includes table descriptions and field definitions. As described in Chapter 4, the catalog, which is maintained automatically by the DBMS, contains table and field metadata. In addition, the catalog contains metadata about table relationships, views, indexes, users, privileges, and replicated data; the last three items are discussed later in this chapter.

The catalogs for many DBMSs consist of a set of special tables that are included in the database. The DBMS hides these special tables from users of the database. However, the DBMS lets the DBA access and update the tables because the DBA must know the contents of the database and must create and define tables, fields, views, indexes, and other metadata. The DBA can authorize other users access to some catalog tables as necessary.

When the DBA uses the DBMS to access the catalog in the database, the DBA asks questions such as the following:

- What tables and fields are included in the database? What are their names?
- What are the properties of these fields? For example, is the Street field in the Customer table 15 or 30 characters long? Is the CustomerNum field a numeric field, or is it a character field? How many decimal places are in the Rate field in the Rep table?
- What are the possible values for the various fields? For example, are there any restrictions on the possible values for the CreditLimit field in the Customer table or for the Class field in the Part table?
- What are the meanings of the various fields? For example, what exactly is the Class field in the Part table, and what does a Class field value of HW mean?

- What relationships exist between tables in the database? Which relationships are one-to-many, many-to-many, and one-to-one? Must the relationship always exist? For example, must a customer always have a sales rep?
- Which fields and combinations of fields can you rapidly search for specific values because they are indexed? Which fields that are not indexed are candidates for indexes because they are often used in searches?
- Which users have access to the database? For example, which fields can Elena access for retrieval purposes but not update? Which fields can Elena update?
- Which programs or objects (queries, forms, and reports) access which data within the database? How do they access the data? Do these programs merely retrieve the data, or do they update it too? What kinds of updates do the programs perform? Can a certain program add a new customer, for example, or can it merely make changes regarding information about customers that are already in the database? When a program changes customer data, can it change all the fields or only some fields? Which fields?

Enterprise DBMSs, such as Oracle and DB2, often have a catalog called a **data dictionary**, which contains answers to all these questions and more. The data dictionary serves as a super catalog containing metadata beyond what's been described. For example, these DBMSs let the DBA split the data in a database and store the fragmented data on multiple disks at multiple locations. In these cases, the data dictionary must track the location of the data. PC-based DBMSs do not offer a data dictionary, but they have a catalog that provides answers to most of the preceding questions.

SUPPORT CONCURRENT UPDATE

A DBMS must ensure that the database is updated correctly when multiple users update the database at the same time.

Sometimes a person uses a database stored on a single computer. At other times, several people might update a database, but only one person at a time does so. For example, several people might take turns with one PC to update a database. A DBMS handles these situations easily. However, the use of networks and of DBMSs that are capable of running on these networks and that allow several users to update the same database raises a problem that the DBMS must address: concurrent update.

Concurrent update occurs when multiple users make updates to the same database at the same time. On the surface, you might think that a concurrent update doesn't present any problem. Why couldn't two, three, or fifty users update the database simultaneously without causing a problem?

The Concurrent Update Problem

To illustrate the problem with concurrent update, suppose Ryan and Elena are two users who work at Premiere Products. Ryan is currently updating the Premiere Products database to process orders and, among other actions, to increase customers' balances by the amount of their orders. For example, Ryan needs to increase the balance of customer 282 (Brookings Direct) by $100. Elena, on the other hand, is updating the Premiere Products database to post customer payments and, among other things, to decrease customers' balances by the amounts of their payments. Coincidentally, Elena has a $100 payment from Brookings Direct, so she will decrease that customer's balance by $100. The balance for Brookings Direct is $431.50 before the start of these updates; and because the amount of the increase matches the amount of the decrease, the balance should still be $431.50 after the updates. But will it? That depends on how the database handles the updates.

How does the DBMS make the required update for Ryan? First, as shown in Figure 7-4, the DBMS reads the data for Brookings Direct from the database on disk into Ryan's work area in memory (RAM). Second, Ryan enters the order data for Brookings Direct. At this point, Ryan's order entry takes place in his work area in memory, including the addition of the order total of $100 to the balance of $431.50, bringing the balance to $531.50. This change has *not* yet taken place in the database; it has taken place *only* in Ryan's work area in memory. Finally, after Ryan finishes entering the order data for Brookings Direct, the DBMS updates the database with Ryan's changes.

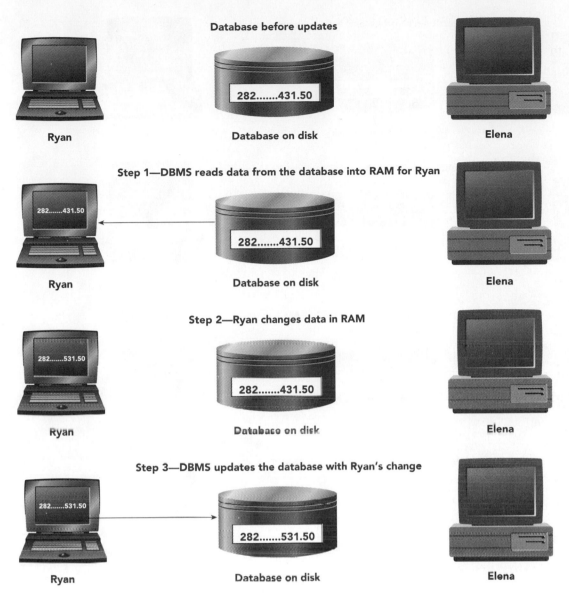

Database before updates

Ryan

Database on disk

Elena

Step 1—DBMS reads data from the database into RAM for Ryan

282.......431.50

Ryan

282.......431.50

Database on disk

Elena

Step 2—Ryan changes data in RAM

282.......531.50

Ryan

282.......431.50

Database on disk

Elena

Step 3—DBMS updates the database with Ryan's change

282.......531.50

Ryan

282.......531.50

Database on disk

Elena

FIGURE 7-4 Ryan updates the database

Suppose Elena begins her update at this point. As shown in Figure 7-5, the DBMS reads the data for Brookings Direct from the database, including the new balance of $531.50. Elena then enters the payment of $100, which decreases the customer balance to $431.50 in her work area in memory. Finally, the DBMS updates the database with Elena's change. The balance for Brookings Direct in the database is now $431.50, which is correct.

Database after Ryan's update and before Elena's update

Ryan   Database on disk   Elena

282.......531.50

Step 1—DBMS reads data from the database into RAM for Elena

Ryan   Database on disk   Elena

282.......531.50   282.......531.50

Step 2—Elena changes data in RAM

Ryan   Database on disk   Elena

282.......531.50   282.......431.50

Step 3—DBMS updates the database with Elena's change

Ryan   Database on disk   Elena

282.......431.50   282.......431.50

FIGURE 7-5   Elena updates the database

In the preceding sequence of updates, everything worked out correctly; but this is not always the case. Do you see how the updates to the database could occur in a way that would lead to an incorrect result?

What if the updates occur in the sequence shown in Figure 7-6 instead? First, the DBMS reads the data from the database into Ryan's work area in memory; and at about the same time, the DBMS reads the data from the database into Elena's separate work area in memory. At this point, both Ryan and Elena have the correct data for Brookings Direct, including a balance of $431.50. Ryan adds $100 to the balance in his work area, and Elena subtracts $100 from the balance in her work area. At this point, in Ryan's work area in memory, the balance is $531.50, while in Elena's work area in memory, the balance is $331.50. The DBMS now updates the database with Ryan's change. At this moment, Brookings Direct has a balance of $531.50 in the database. Finally, the DBMS updates the database with Elena's change. Her update replaces Ryan's. Now the balance in the database for Brookings Direct is $331.50! Had the DBMS updated the database in the reverse order, the final balance would have been $531.50. In either case, you now have incorrect data in the database—one of the updates has been *lost*. The DBMS must prevent these lost updates from occurring to the database.

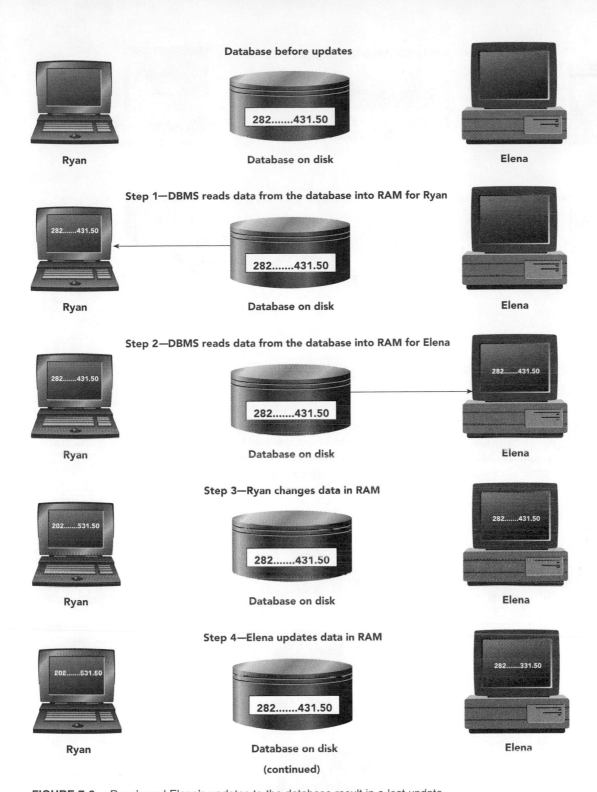

FIGURE 7-6 Ryan's and Elena's updates to the database result in a lost update

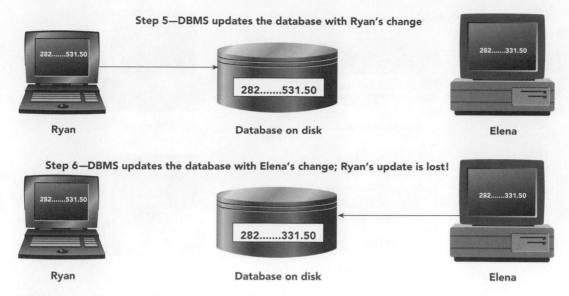

Step 5—DBMS updates the database with Ryan's change

Ryan — Database on disk — Elena

Step 6—DBMS updates the database with Elena's change; Ryan's update is lost!

Ryan — Database on disk — Elena

FIGURE 7-6   Ryan's and Elena's updates to the database result in a lost update (continued)

Avoiding the Lost Update Problem

One way to prevent lost updates is to prohibit concurrent update. This may seem drastic, but it is not so far-fetched. You can let several users access the database at the same time, but for retrieval only; that is, the users can read data from the database, but they can't update any data in the database. When these users need to update the database, such as increasing a customer's balance or changing the price of a part, the database itself is *not* updated. Instead, as shown in Figure 7-7, a special program, which a computer programmer would create for the users to use with the data in their database, adds a record to a separate file.

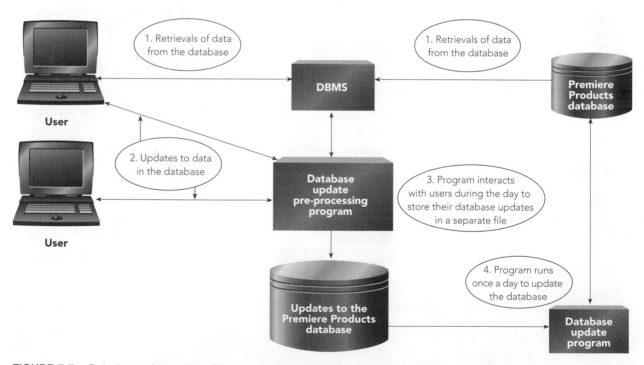

FIGURE 7-7   Delaying updates to the Premiere Products database to avoid the lost update problem

A record in this separate file might indicate, for example, that Premiere Products received a $100 payment from customer 282 on a certain date. Periodically, usually once a day, a single update program reads the *batch* of records in this file one at a time and performs the appropriate updates to the database; this processing technique is called **batch processing**. Because this program is the only way to update the database, you eliminate the problems associated with concurrent update.

Although this alternative approach avoids the lost update problem, it creates another problem. From the time users start updating (adding records to the special batch file) until the time the batch-processing program actually updates the database, the data in the database is out of date. If a customer's balance in the database is $4,500, the true balance is $5,500 if a user had entered an order for this customer that increased its balance by $1,000. If the customer has a $5,000 credit limit, the customer is now over its credit limit by $500.

The batch-processing alternative does not work in any situation that requires the data in the database to be current. These situations include credit card processing, banking, inventory control, and airline reservations. Other simple alternative solutions to the concurrent update problem, such as permitting only one user to update the database, also will not work in these situations because many users need to update the database in a timely way.

Two-Phase Locking

In most situations, you can't solve the concurrent update problem by avoiding it; you need the DBMS to have a strategy for dealing with it. One such strategy is for the DBMS to process an update completely before it begins processing the next update. For example, the DBMS can prevent Elena from beginning her update to the Brookings Direct data until the DBMS completes Ryan's update to that data, or vice versa.

To accomplish such a serial processing of updates, many DBMSs use locking. **Locking** denies other users access to data while the DBMS processes one user's updates to the database. An example of locking using Ryan's and Elena's updates appears in Figure 7-8. After the DBMS reads the data in the database for Ryan's update, the DBMS locks the data, denying access to the data by Elena and any other user. The DBMS retains the locks until Ryan completes his change; then the DBMS updates the database. For the duration of the locks, the DBMS rejects all attempts by Elena to access the data and it notifies Elena that the data is locked. If she chooses to do so, she can keep attempting to access the data until the DBMS releases the locks, at which time the DBMS can process her update. In this simple case at least, the locking technique appears to solve the lost update problem.

Database before updates

Step 1—DBMS reads data from the database into RAM for Ryan and locks the record

Step 2—Elena requests the same record from the database and her request fails

Step 3—Ryan changes data in RAM; Elena's request for the same record again fails

Step 4—DBMS updates the database with Ryan's change; Elena's request for the same record again fails

(continued)

FIGURE 7-8   The DBMS uses a locking scheme to apply Ryan's and Elena's updates to the database

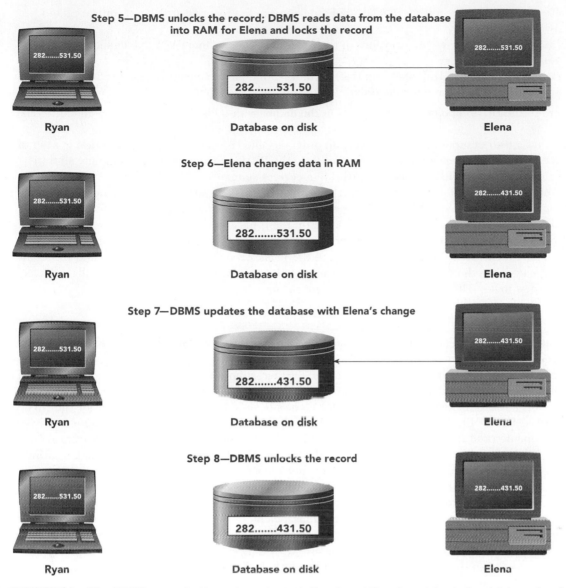

Step 5—DBMS unlocks the record; DBMS reads data from the database into RAM for Elena and locks the record

Ryan   Database on disk   Elena

Step 6—Elena changes data in RAM

Ryan   Database on disk   Elena

Step 7—DBMS updates the database with Elena's change

Ryan   Database on disk   Elena

Step 8—DBMS unlocks the record

Ryan   Database on disk   Elena

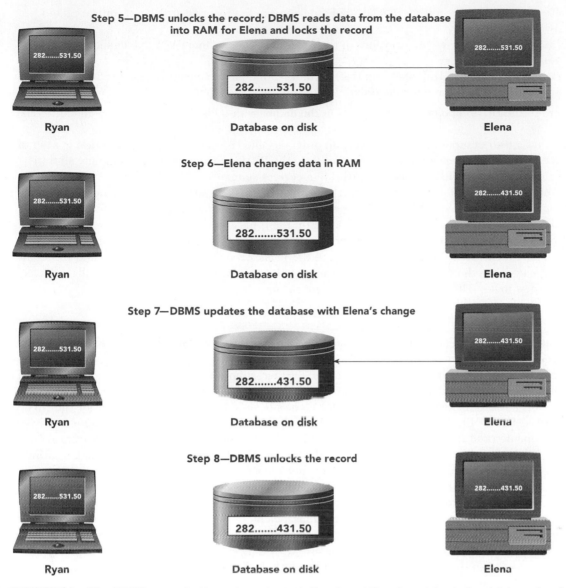

FIGURE 7-8 The DBMS uses a locking scheme to apply Ryan's and Elena's updates to the database (continued)

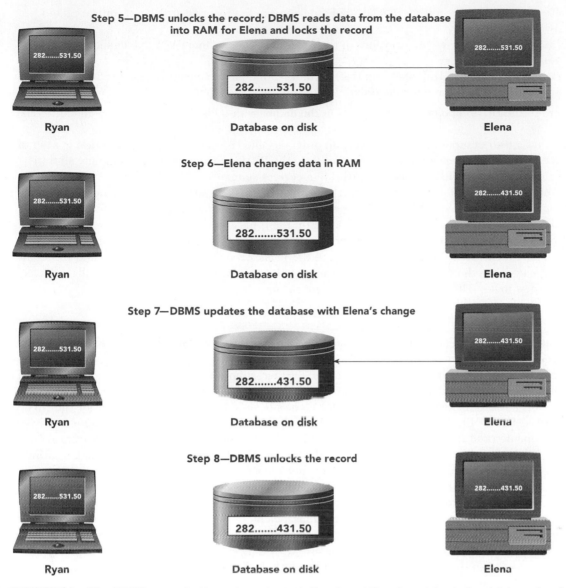

How long should the DBMS hold a lock? If the update involves changing field values in a single row in a single table, such as changing a customer's name and address, the lock no longer is necessary after this row is updated. However, sometimes an update is more involved.

Consider the task of filling an order for Premiere Products. Ryan might think that filling an order involves a single action. He simply indicates that an order currently in the database needs to be filled. Alternatively, Ryan might need to enter data about the order. In either case, Ryan still believes the process is a single action. Behind the scenes, though, filling an order requires that the DBMS update several records in the database. For example, suppose Ryan fills a new order for Brookings Direct that includes the sale of three washers and two dryers; Richard Hull is the rep for Brookings Direct. To fill this order, the DBMS must update the records in the database as follows:

- Add one record to the Orders table for the new order.
- Add one record to the OrderLine table for the sale of the three washers.
- Add one record to the OrderLine table for the sale of the two dryers.
- Change the washer record in the Part table to decrease the record's number of units on hand by three.

- Change the dryer record in the Part table to decrease the record's number of units on hand by two.
- Change the Brookings Direct record in the Customer table to increase the balance by the total amount of the order.
- Change the Richard Hull record in the Rep table to increase the commission by the commission amount owed to the rep for the order.

For this order, the DBMS updates seven records in the database; it adds three records and changes four records.

Each task that a user completes, such as filling an order, is called a transaction. A **transaction** is a set of steps completed by a DBMS to accomplish a single user task; the DBMS must successfully complete all transaction steps or none at all for the database to remain in a correct state.

For transactions such as filling an order in which a single user task requires several updates in the database, what should the DBMS do about locks? How long does the DBMS hold each lock? For safety's sake, the DBMS should hold locks until it completes all the updates in the transaction. This approach for handling locks is called **two-phase locking**. The first phase is the **growing phase**, in which the DBMS locks more rows and releases none of the locks. After the DBMS acquires all the locks needed for the transaction and has completed all database updates, the second phase is the **shrinking phase**, in which the DBMS releases all the locks and acquires no new locks. This two-phase locking approach solves the lost update problem.

Deadlock

Because each user transaction can require more than one lock, another problem may occur. Suppose Ryan is filling the Brookings Direct order for the sale of three washers and two dryers and Elena is filling another customer order that includes the sale of washers and dryers. Further, suppose for Ryan's transaction, the DBMS holds a lock on the washer record and is attempting to lock the dryer record, as shown in Figure 7-9. However, the DBMS has already locked the dryer record for Elena's transaction, so Ryan must wait for the DBMS to release the lock. Before the DBMS releases the lock on the dryer record for Elena's transaction, however, it needs to update (and thus lock) the washer record, which is currently locked for Ryan's transaction. Ryan is waiting for the DBMS to act for Elena (release the lock on the dryer record), while Elena is waiting for the DBMS to act for Ryan (release the lock on the washer record). Without the aid of some intervention, this dilemma could continue indefinitely. Terms used to describe such situations are **deadlock** and the **deadly embrace**. Obviously, some strategy is necessary to prevent, minimize, or manage deadlock. You can minimize the occurrence of deadlock by making sure all programs lock records in the same order whenever possible. For example, all programs for the Premiere Products database should lock records in the Rep table and then lock records in the Customer table consistently. A consistent locking strategy prevents situations in which a user first locks a record in the Rep table, a second user first locks a record in the Customer table, and both users are deadlocked while they wait for the release of records they need to lock next.

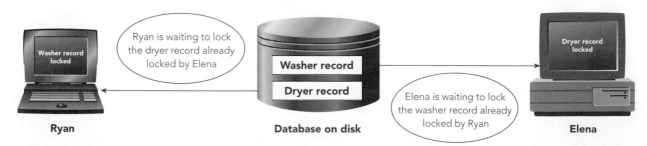

FIGURE 7-9 Two users experiencing deadlock

One strategy for managing deadlock is to let deadlock occur and then have the DBMS detect and break any deadlock. To detect deadlock, the DBMS must keep track of the collection of records it has locked for each transaction, as well as the records it's waiting to lock. If two transactions are waiting for records held by the other, a deadlock has occurred. Actually, more than two users could be involved. Ryan could be waiting for a record held by Elena, while Elena is waiting for a record held by Pat, who, in turn, is waiting for a record held by Ryan.

After the DBMS detects deadlock, the DBMS must break the deadlock. To break the deadlock, the DBMS chooses one deadlocked user to be the **victim**. For the victim's transaction, the DBMS undoes all completed updates, releases all locks, and reschedules the transaction. Using this method of handling deadlocks, the user notices only a delay in the time needed to complete the transaction.

Locking on PC-Based DBMSs

Enterprise DBMSs typically offer sophisticated schemes for locking as well as for detecting and handling deadlocks. PC-based DBMSs provide facilities for the same purposes, but they usually are more limited than the facilities provided by enterprise DBMSs. These limitations, in turn, put an additional burden on the programmers who write the programs that allow concurrent update.

Although the exact features for handling the problems associated with concurrent update vary from one PC-based DBMS to another, the following list is fairly typical of the types of facilities provided:

- Programs can lock an entire table or an individual row within a table, but only one or the other. As long as one program has a row or table locked, no other program can access that row or table.
- Programs can release any or all of the locks they currently hold.
- Programs can inquire whether a given row or table is locked.

This list, although short, makes up the complete set of facilities provided by many PC-based DBMSs. Consequently, the following guidelines have been devised for writing programs for concurrent update:

- If an update transaction must lock more than one row in the same table, you must lock the entire table.
- When a program attempts to read a row that is locked, the program may wait a short period of time and then try to read the row again. This process can continue until the row becomes unlocked. However, it usually is preferable to impose a limit on the number of times a program may attempt to read the row. In this case, reading is done in a loop, which proceeds until the read is successful or the maximum number of times the program can repeat the operation is reached. Programs vary in terms of what action is taken should the loop be terminated without the read being successful. One possibility is to notify the user of the problem and let the user decide whether to try the same update again or move on to something else.
- Because there is no facility to *detect and handle* deadlocks, you must try to *prevent* them. A common approach to this problem is for every program in the system to attempt to lock all the rows and/or tables it needs before beginning an update. Assuming each program is successful in this attempt, it can then perform the required updates. If any row or table that the program needs is already locked, the program should immediately release all the locks that it currently holds, wait some specified period of time, and then try the entire process again. In some cases, it might be better to notify the user of the problem and see whether the user wants to try again. In effect, this means that any program that encounters a problem will immediately get out of the way of all the other programs rather than be involved in a deadlock situation.
- Because locks prevent other users from accessing a portion of the database, it is important that no user keep rows or tables locked any longer than necessary. This is especially significant for update programs. Suppose, for example, that a user is employing an update program to update information about customers. Suppose further that after the user enters the number of the customer to be updated, the customer row is locked and remains locked until the user has entered all the new data and the update has taken place. What if the user is interrupted by a phone call before he or she has finished entering the new data? What if the user goes to lunch? The row might remain locked for an extended period of time. If the update involves several rows, all of which must be locked, the problem becomes that much worse. In fact, in many DBMSs, if more than one row from the same table must be locked, the entire table must be locked, which means that entire tables might be locked for extended periods of time. Clearly, this situation must not be permitted to occur. A variation on the timestamping technique used by some enterprise DBMSs is a programming strategy you can use to overcome this problem.

Timestamping

An alternative to two-phase locking is timestamping. With **timestamping**, the DBMS assigns to each database update the unique time when the update started; this time is called a **timestamp**. In addition, every database row includes the timestamp associated with the last update to the row. The DBMS processes updates to the database in timestamp order. If two users try to change the same row at the same time, the DBMS will process the change that has the earlier timestamp. The other transaction will be restarted and assigned a new timestamp value.

Timestamping avoids the need to lock rows in the database and eliminates the processing time needed to apply and release locks and to detect and resolve deadlocks. On the other hand, additional disk and memory space are required to store the timestamp values; in addition, the DBMS uses extra processing time to update the timestamp values.

One might naturally ask at this point whether the ability to have concurrent update is worth the complexity that it adds to the DBMS. In some cases, the answer is no. Concurrent update may be far from necessary. In most cases, however, concurrent update is necessary to the productivity of the users of the system. In these cases, implementation of locking, timestamping, or some other strategy is essential to the proper performance of the system.

RECOVER DATA

A DBMS must provide methods to recover a database in the event the database is damaged in any way. A database can be damaged or destroyed in many ways. Users can enter data that is incorrect, transactions that are updating the database can end abnormally during an update, a hardware problem can occur, and so on. After any such event has occurred, the database might contain invalid or inconsistent data. It may even be destroyed.

Obviously, a situation in which data has been damaged or destroyed must not be allowed to go uncorrected. The database must be returned to a correct state. **Recovery** is the process of returning the database to a state that is known to be correct from a state known to be incorrect; in performing such a process, you say that you *recover* the database. In situations where indexes or other physical structures in the database have been damaged but the data has not, many DBMSs provide a feature that you can use to repair the database automatically to recover it.

If the data in a database has been damaged, the simplest approach to recovery involves periodically making a copy of the database (called a **backup** or a **save**). If a problem occurs, the database is recovered by copying this backup copy over the database. In effect, the damage is undone by returning the database to the state it was in when the last backup was made.

Unfortunately, other activity besides that which caused the destruction also is undone. Suppose the database is backed up at 10 p.m. and users begin updating it at 8 a.m. the next day. Then suppose that at 11:30 a.m., something happens that destroys the database. If the previous night's backup is used to recover the database, the entire database is returned to the state it was in at 10 p.m. All updates made in the morning are lost, not just the update or updates that were in progress at the time the problem occurred. Thus, during the final part of the recovery process, users would have to redo all the work they had done since 8 a.m.

Journaling

As you might expect, enterprise DBMSs provide sophisticated features to avoid the costly and time-consuming process of having users redo their work. These features include **journaling**, which involves maintaining a **journal** or **log** of all updates to the database. The log is a separate file from the database; thus, the log is still available if a catastrophe destroys the database.

Several types of information are typically kept in the log for each transaction. This information includes the transaction ID and the date and time of each individual update. The log also includes a record of what the data in the row looked like in the database before the update (called a **before image**) and a record of what the data in the row looked like in the database after the update (called an **after image**). In addition, the log contains an entry to indicate the start of a transaction and the successful completion (**commit**) of a transaction.

To illustrate the use of a log by a DBMS, consider the four sample transactions shown in Figure 7-10. Three transactions—1, 3, and 4—require a single update to the database. The second transaction, which is Ryan's order transaction for Brookings Direct, requires seven updates to the database.

Transaction ID	Transaction Description
1	1. Change the Price value for part number DW11 to $389.99
2	1. Add a record to the Orders table: OrderNum of 21700, OrderDate of 10/24/2010, CustomerNum of 282 2. Add a record to the OrderLine table: OrderNum of 21700, PartNum of DW11, NumOrdered of 3, QuotedPrice of $389.00 3. Add a record to the OrderLine table: OrderNum of 21700, PartNum of KL62, NumOrdered of 2, QuotedPrice of $346.50 4. Change the OnHand value for part number DW11 to 9 5. Change the OnHand value for part number KL62 to 10 6. Change the Balance value for CustomerNum 282 to $2,321.50 7. Change the Commission value for RepNum 35 to $39,346.20
3	1. Add customer 510
4	1. Delete part AT94

FIGURE 7-10 Four sample transactions

Suppose these four transactions are the first transactions in a day, immediately following a backup of the database, and they all complete successfully. In this case, the log might look like the sample log shown in Figure 7-11.

Transaction ID	Time	Action	Record Updated	Before Image	After Image
1	8:00	Start			
2	8:01	Start			
2	8:02	Insert	Orders (21700)		(new values)
3	8:03	Start			
1	8:04	Update	Part (DW11)	(old values)	(new values)
2	8:05	Insert	OrderLine (21700, DW11)		
1	8:06	Commit			
4	8:07	Start			
3	8:08	Insert	Customer (510)		(new values)
2	8:09	Insert	OrderLine (21700, KL62)		(new values)
3	8:10	Commit			
2	8:11	Update	Part (DW11)	(old values)	(new values)
2	8:12	Update	Part (KL62)	(old values)	(new values)
4	8:13	Delete	Part (AT94)	(old values)	
2	8:14	Update	Customer (282)	(old values)	(new values)
4	8:15	Commit			
2	8:16	Update	Rep (35)	(old values)	(new values)
2	8:17	Commit			

FIGURE 7-11 Sample log in which all four transactions commit normally

Before studying how the log is used in the recovery process, examine the log itself. Each record in the log includes the ID of the transaction, as well as the time the particular action occurred. The actual time would be more precise than in the example, the DBMS would process the actions much faster, and the date would also be included in the log. For simplicity, each action occurs one minute after the preceding action. The actions are *Start* to indicate the start of a transaction, *Commit* to indicate the transaction completed successfully, *Insert* to identify the addition of a record to the database, *Update* to identify the change of a record, and *Delete* to identify the deletion of a record. For an Insert action, no before image appears in the log because the data did not exist prior to the action. Similarly, for a Delete action, no after image appears in the log.

The sample log shows, for example, that transaction 2 began at 8:01. The database updates to complete the transaction that occurred at 8:02 (order record 21700 inserted), 8:05 (the first order line record inserted), 8:09 (the second order line record inserted), 8:11 (the first part record updated), 8:12 (the second part record updated), 8:14 (customer record 282 updated), and 8:16 (rep record 35 updated). At 8:17, the transaction was committed. During this same time span, the other three transactions were also committed.

Forward Recovery

How is the log used in the recovery process? Suppose a catastrophe destroys the database just after 8:11. In this case, the recovery of the database begins with the most recent database backup from the previous evening at ten o'clock. As shown in Figure 7-12, the DBA copies the backup over the live database. Because the database is no longer current, the DBA executes a DBMS recovery program that applies the after images of committed transactions from the log to bring the database up to date. This method of recovery is called **forward recovery**.

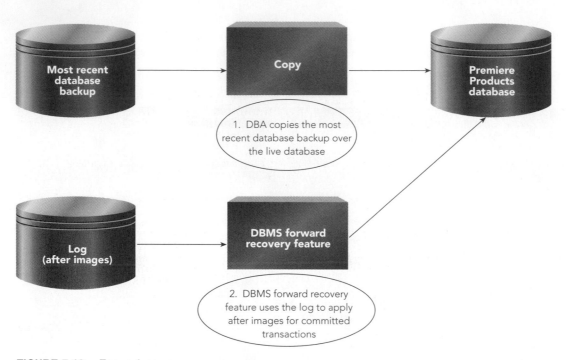

FIGURE 7-12 Forward recovery

In its simplest form, the recovery program copies in chronological order the after image of each record in the log over the actual record in the database. You can improve the recovery process by realizing that if a record was updated 10 times since the last backup, the recovery program would copy the after image records 10 times over the database record. Thus, in reality, the first nine copies are unnecessary. The tenth after image includes all the updates accomplished in the first nine. Thus, you can improve the performance of the recovery program by having it scan the log and then apply the last after image.

Q & A

Question: In the preceding scenario, which transactions in the sample log shown in Figure 7-11 does the recovery program use to update the restored database?
Answer: The catastrophe occurred just after 8:11. Because the recovery program applies transactions committed before the catastrophe, the program applies only transactions 1 and 3. These two transactions committed before 8:11, at which point the DBMS was still processing transactions 2 and 4.

Backward Recovery

If the database has not been destroyed, the problem must involve transactions that were incorrect or, more likely, stopped in midstream. In either case, the database is currently not in a valid state. You can use **backward recovery**, or **rollback**, to recover the database to a valid state by undoing the problem transactions. The DBMS accomplishes the backward recovery by reading the log for the problem transactions and applying the before images to undo their updates, as shown in Figure 7-13.

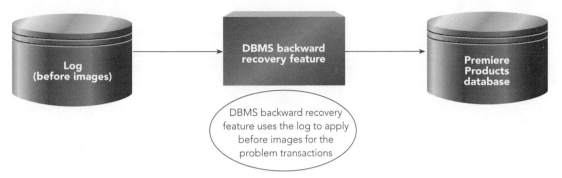

FIGURE 7-13 Backward recovery

Q & A

Question: For the sample log shown in Figure 7-11, what does the DBMS do to roll back transaction 1?
Answer: The DBMS started transaction 1 at 8:00, changed a Part table record at 8:04 for transaction 1, and committed transaction 1 at 8:06. To roll back transaction 1, the DBMS applies the before image of the Part table record.

Q & A

Question: For the sample log shown in Figure 7-11, what does the DBMS do to roll back transaction 3?
Answer: The DBMS started transaction 3 at 8:03, added a Customer table record at 8:08 for transaction 3, and committed transaction 3 at 8:10. Because no before image exists for adding a record, to roll back transaction 3, the DBMS deletes the Customer table record.

Recovery on PC-Based DBMSs

PC-based DBMSs generally don't offer sophisticated recovery features such as journaling. Most of them provide users with a simple way to make backup copies and to recover the database later by copying the backup over the database.

How should you handle recovery in any application system you develop with a PC-based DBMS? You could use the features of the DBMS to periodically make backup copies and use the most recent backup when a recovery is necessary. The more important it is to avoid redoing work, the more often you would make backups. For example, if a backup is made every eight hours, you might have to redo up to eight hours of work. If, on the other hand, a backup is made every two hours, you might have to redo up to two hours of work.

In many situations, this approach, although not particularly desirable, is acceptable. However, for systems with a large number of updates made to the database between backups, this approach is not acceptable. In such cases, the necessary recovery features that are not supplied by the DBMS must be included in the application programs. Each of the programs that updates the database could, for example, also write a record to a separate log file, indicating the update that had taken place. You could write a separate program to read the log file and re-create all the updates indicated by the records in the file. The recovery process would then consist of copying the backup over the actual database and running this special program.

Although this approach does simplify the recovery process for the users of the system, it also causes some problems. First, each of the programs in the system becomes more complicated because of the extra logic involved in adding records to the special log file. Second, you must write a separate program to update the database with the information in this log file. Finally, every time a user completes an update, the system has extra work to do; and this additional processing may slow down the system to an unacceptable level. Thus, in any application, you must determine whether the ease of recovery provided by the approach is worth the price you might have to pay for it. The answer will vary from one system to another.

PROVIDE SECURITY SERVICES

As discussed in Chapter 4, a DBMS must provide ways to ensure that only authorized users can access the database. Security is the prevention of unauthorized access, either intentional or accidental, to a database. The most common security features used by DBMSs are encryption, authentication, authorizations, and views.

Encryption

Encryption converts the data in a database to a format that's indecipherable to a word processor or another program and stores it in an encrypted format. When unauthorized users attempt to bypass the DBMS and get to the data directly, they see only the encrypted version of the data. However, authorized users accessing the data using the DBMS have no problem viewing and working with the data.

When a user updates data in the database, the DBMS encrypts the data before updating the database. Before a legitimate user retrieves the data via the DBMS, the data is decrypted, or decoded, and presented to the user in the normal format. The entire encryption process is transparent to a legitimate user; that is, he or she is not even aware it is happening.

Access lets you encrypt a database with a password; and after you've encrypted the database, you can use Access to decrypt it. **Decrypting** a database reverses the encryption. If your encrypted database takes longer to respond to user requests as it gets larger, you might consider decrypting it to improve its responsiveness.

Using Access to encrypt or decrypt a database is a four-step process:

1. Start Access, but don't open the database you want to encrypt or decrypt.
2. Click More in the Open Recent Database pane on the Getting Started with Microsoft Office Access page, navigate to the drive and folder that contains the database in the Open dialog box, click the database name, click the Open arrow, and then click Open Exclusive.
3. On the Ribbon, click the Database Tools tab; then in the Database Tools group, click the Encrypt with Password button. (To decrypt a database, click the Decrypt Database button.)
4. In the Set Database Password dialog box, type the password for the database in the Password text box, type the same password a second time in the Verify text box, and then click the OK button. (If you are decrypting the database, type the password for the database in the Password text box in the User Database Password dialog box, and then click the OK button.)

Authentication

Authentication refers to techniques for identifying the person who is attempting to access the DBMS. The use of passwords is the most common authentication technique. A **password** is a string of characters assigned by the DBA to a user that the user must enter to access the database. Users also use passwords to access many operating systems, networks, and other computer and Internet resources. Biometric identification techniques and the use of smart cards are increasing in use as an alternative to passwords for authentication. **Biometrics** identify users by physical characteristics such as fingerprints, voiceprints, handwritten signatures, and facial characteristics. **Smart cards** are small plastic cards about the size of a driver's license that have built-in circuits containing processing logic to identify the cardholder.

Unlike individual passwords, a **database password** is a string of characters that the DBA assigns to a database and that users must enter before they can access the database. As long as the database password is known only to authorized database users, unauthorized access to the database is prevented. The DBA should use a database password that is easy for the authorized users to remember but that is not so obvious that others can easily guess the password. If a DBA encrypts an Access database, the DBA must assign a database password, as shown in Figure 7-14. To create the database password, the DBA enters the same password twice to verify that the initial entry is the one that the DBA wants.

FIGURE 7-14   Assigning a database password to the Premiere Products database

After the DBA creates the database password for a database, as shown in Figure 7-15, users must enter it correctly before they can open the database.

FIGURE 7-15   User enters database password to open the Premiere Products database

Authorizations

Using passwords is a security measure that applies to all users of a database; after users enter their passwords successfully, they can retrieve and update all the data in the database. Frequently, the security needs for a database are more individualized. For example, the DBA might need to let some users view and update all data and let other users view only certain data. In this situation, the DBA uses **authorization rules** that specify which users have what type of access to which data in the database.

The DBA grants users specific permissions to tables, queries, and other objects in a database. A user's **permissions** specify what kind of access the user has to objects in the database. The DBA can assign permissions to individual users or to groups of users. The DBA usually creates groups of users, sometimes called **workgroups**; assigns the appropriate permissions to each group; and then assigns each user to the appropriate group based on the permissions the user requires.

Views

Recall from Chapter 4 that a view is a snapshot of certain data in the database at a given moment in time. If a DBMS provides a facility that allows users to have their own views of a database, this facility can be used for security purposes. Tables or fields to which the user does not have access in his or her view effectively do not exist for that user.

Privacy

No discussion of security is complete without at least a brief mention of privacy. Although the terms *security* and *privacy* are often used synonymously, they are different, although related, concepts. **Privacy** refers to the right of individuals to have certain information about them kept confidential. Privacy and security are related because it is only through appropriate *security* measures that *privacy* can be ensured.

Laws and regulations dictate some privacy rules, and companies institute additional privacy rules. Variations in what information is kept confidential occur among organizations. For example, salaries at governmental and many service organizations are public information, but salaries at many private enterprises are kept confidential.

PROVIDE DATA INTEGRITY FEATURES

A DBMS must follow rules so that it updates data accurately and consistently. These rules, called integrity constraints, are categorized as either key integrity constraints or data integrity constraints.

Key integrity constraints consist of primary key constraints and foreign key constraints. Primary key constraints, which are governed by entity integrity (as discussed in Chapter 4), enforce the uniqueness of the primary key. For example, forbidding the addition of a rep whose number matches the number of a rep already in the database is an example of a primary key constraint. Foreign key constraints, which are governed by referential integrity (as discussed in Chapter 4), enforce the fact that a value for a foreign key must match the value of the primary key for some row in a table in the database. Forbidding the addition of a customer whose rep is not already in the database is an example of a foreign key constraint.

Data integrity constraints help to ensure the accuracy and consistency of individual field values. Types of data integrity constraints include the following:

- **Data type.** The value entered for any field should be consistent with the data type for that field. For a numeric field, only numbers should be allowed to be entered. If the field is a date, only a legitimate date should be permitted. For instance, February 30, 2010, is an invalid date that should be rejected.
- **Legal values.** For some fields, not every possible value that is of the assigned data type is legitimate. For example, even though CreditLimit is a numeric field, only the values $5,000.00, $7,500.00, $10,000.00, and $15,000.00 are valid. For the OrderDate field in the Orders table, Premiere Products might insist that only the current date or a future date is an acceptable value when an order is updated. In addition, you should be able to specify which fields can accept null values and which fields can't.
- **Format.** Some fields require a special entry or display format. Although the PartNum field is a character field, for example, only specially formatted strings of characters might be acceptable. Legitimate part numbers might have to consist of two letters followed by two digits; this is an example of an entry format constraint. Users might want the OrderDate field displayed with a four-digit year value instead of a two-digit year value; this is an example of a display format constraint.

Integrity constraints can be handled in one of four ways:

1. The constraint is ignored, in which case no attempt is made to enforce the constraint.
2. The responsibility for constraint enforcement is placed on the users. This means that users must be careful that any updates they make in the database do not violate the constraint.
3. The responsibility for constraint enforcement is placed on programmers. Programmers place into programs the logic to enforce the constraint. Users must update the database only by means of these programs and not through any of the built-in entry facilities provided by the DBMS because these would allow violation of the constraint. Programmers design the programs to reject any attempt by the users to update the database in a way that violates the constraint.
4. The responsibility for constraint enforcement is placed on the DBMS. The DBA specifies the constraint to the DBMS, which then rejects any attempt to update the database in a way that violates the constraint.

Q & A

Question: Which of these four approaches for constraint enforcement is best?
Answer: The first approach, ignoring the constraint, is undesirable because it can lead to invalid data in the database, such as two customers with the same number, part numbers with an invalid format, and invalid credit limits.

The second approach, user constraint enforcement, is a little better because at least an attempt is made to enforce the constraints. However, this approach places the burden of enforcement on users. Besides meaning extra work for users, any mistake on the part of a single user, no matter how innocent, can lead to invalid data in the database.

continued

The third approach removes the burden of enforcement from users and places it on programmers. This solution is better still because it means that users can't violate the constraints. The disadvantage is that all update programs in the system become more complex. This complexity makes programmers less productive and makes programs more difficult to create and modify. This approach also makes changing an integrity constraint more difficult because this may mean changing all the programs that update the database. Furthermore, if the logic in any program used to enforce the constraints is faulty, the program could permit some constraint to be violated; and you might not realize this had happened until a problem occurred at a later date. Finally, you would have to guard against a user bypassing the programs in the system to enter data directly into the database—for example, by using some built-in facility of the DBMS. If a user is able to bypass the programs and enters incorrect data, all the controls that were so diligently placed into the programs are helpless to prevent a violation of the constraints.

The best approach is the one the DBMS enforces. You specify the constraints to the DBMS, and the DBMS ensures that they are never violated.

Nearly all DBMSs include most of the necessary capabilities to enforce the various types of integrity constraints. Consequently, you let the DBMS enforce all the constraints that it is capable of enforcing, then let application programs enforce any other constraints. You also might create a special program whose sole purpose is to examine the data in the database to determine whether any constraints have been violated. You'd run this program periodically and take corrective action to remedy any violations that the program discovers.

NOTE

Access supports key constraints. Access lets you specify a primary key, and then it builds a unique index automatically for the primary key. Access also lets you specify foreign keys, and then it enforces referential integrity automatically. You can use Access to specify data integrity constraints. As shown in Figure 7-16, you can specify the data type for each field and you can specify data format and legal-values integrity constraints.

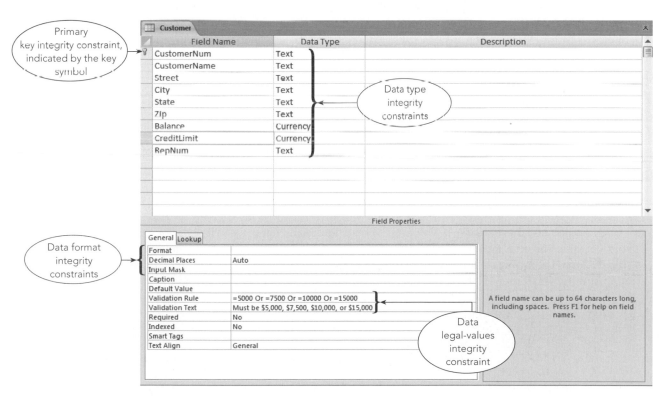

FIGURE 7-16   Example of integrity constraints in Access

SUPPORT DATA INDEPENDENCE

A DBMS must provide facilities to support the independence of programs from the structure of a database. One of the advantages of working with a DBMS is **data independence**, which is a property that lets you change the database structure without requiring you to change the programs that access the database. What types of changes could you or a DBA make to the database structure? A few of these changes are adding a field, changing a field property (such as length), creating an index, and adding or changing a relationship. The following sections describe the data independence considerations for each type of change.

Adding a Field

If you add a new field to a database, you don't need to change any program except, of course, those programs using the new field. However, when a program uses an SQL SELECT * FROM command to select all the fields from a given table, you are presented with an extra field. To prevent this from happening, you need to change the program to restrict the output to only the desired fields. To avoid the imposition of this extra work, you should list all the required fields in an SQL SELECT command instead of using the *.

Changing the Length of a Field

In general, you don't need to change programs because you've changed the length of a field; the DBMS handles all the details concerning this change in length. However, if a program sets aside a certain portion of the screen or a report for the field and the length of the field has increased to the point where the previously allocated space is inadequate, you'll need to change the program.

Creating an Index

To create an index, you enter a simple SQL command or select a few options. Most DBMSs use the new index automatically for all updates and queries. For some DBMSs, you might need to make minor changes in already existing programs to use the new index.

Adding or Changing a Relationship

In terms of data independence considerations, adding or changing a relationship is trickiest of all and is best illustrated with an example. Suppose Premiere Products now has the following requirements:

- Customers are assigned to territories.
- Each territory is assigned to a single rep.
- A rep can have more than one territory.
- A customer is represented by the rep who covers the territory to which the customer is assigned.

To implement these changes, you need to restructure the database. The previous one-to-many relationship between the Rep and Customer tables is no longer valid. Instead, there's now a one-to-many relationship between the Rep table and the new Territory table and a one-to-many relationship between the Territory table and the Customer table, as follows:

```
Rep (RepNum, LastName, FirstName, Street, City, State,
     Zip, Commission, Rate)
Territory (TerritoryNum, TerritoryDescription, RepNum)
Customer (CustomerNum, CustomerName, Street, City, State,
     Zip, Balance, CreditLimit, TerritoryNum)
```

Further, suppose that a user accesses the database via the following view, which is named RepCust:

```
CREATE VIEW RepCust (RNum, RLast, RFirst, CNum, CName) AS
    SELECT Rep.RepNum, LastName, FirstName, Customer.CustomerNum, CustomerName
    FROM Rep, Customer
    WHERE Rep.RepNum=Customer.RepNum
;
```

The defining query is now invalid because there is no RepNum field in the Customer table. A relationship still exists between reps and customers, however. The difference is that you now must go through the Territory table to relate the two tables. If users have been accessing the tables directly to form the relationship, their programs will have to change. If they are using the RepCust view, you will need to change only the definition of the view. The new definition is as follows:

```
CREATE VIEW RepCust (RNum, RLast, RFirst, CNum, CName) AS
    SELECT Rep.RepNum, LastName, FirstName, Customer.CustomerNum, CustomerName
    FROM Rep, Territory, Customer
    WHERE Rep.RepNum=Territory.RepNum
    AND Territory.TerritoryNum=Customer.TerritoryNum
;
```

The defining query is now more complicated than it was before, but this does not affect users of the view. The users continue to access the database the same way they did before, and the DBA won't need to change their programs.

SUPPORT DATA REPLICATION

A DBMS must manage multiple copies of the same data at multiple locations. For performance or other reasons, sometimes data should be duplicated—technically called **replicated**—at more than one physical location. For example, accessing data at a local site is more efficient than accessing data remotely. It's more efficient because using replicated data does not involve data communication and network time delays, users compete for data with fewer other users, and replicated data keeps data available to local users at times when the data might not be available at other sites.

If certain information needs to be accessed frequently from all sites, a company might choose to store the information at all its locations. At other times, users on the road—for example, reps meeting at their customers' sites—might need access to data but would not have this access unless the data was stored on their portable computers.

Replication lets users at different sites use and modify copies of a database and then share their changes with the other users. Replication is a two-step process. First, the DBMS creates copies, called **replicas**, of the database at one or more sites. For example, you could create two replicas, as shown in Figure 7-17, and give the "Replica 1 database" to one user to access at a remote location and give the "Replica 2 database" to a second user to use at a different remote location.

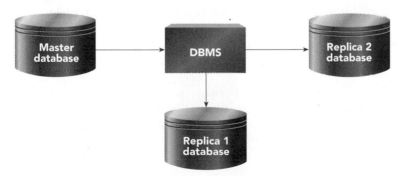

FIGURE 7-17   The DBMS creates replicas from the master database

The master database and all replicas form a replica set. Users then update their individual replicas, just as if they were updating the master database. Periodically, the DBMS exchanges all updated data between the master database and a replica in a process called **synchronization**. For example, after the second user returns from the remote site, the DBA synchronizes the master database and the "Replica 2 database," as shown in Figure 7-18. Later, after the first user returns, the DBA synchronizes the master database and the "Replica 1 database."

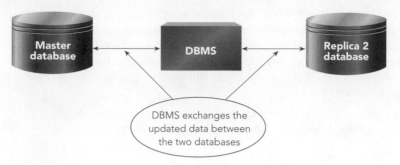

FIGURE 7-18   DBMS sychronizes two databases in a replica set

Ideally, the DBMS should handle all the issues associated with replication for you. The DBMS should do all the work to keep the various copies of data consistent behind the scenes; users should be unaware of the work involved. Access supports replication in the manner described. You'll learn more about replication in Chapter 9.

PROVIDE UTILITY SERVICES

A DBMS must provide services that assist in the general maintenance of a database. In addition to the services already discussed, a DBMS provides a number of **utility services** that assist in the general maintenance of the database. The following is a list of services that might be provided by a PC-based DBMS:

- The DBMS lets you change the database structure—adding new tables and fields, deleting existing tables and fields, changing the name or properties of fields, and so on.
- The DBMS lets you add new indexes and delete indexes that are no longer needed.
- While you are using the database, the DBMS lets you use the services available from your operating system, such as Windows or Linux.
- The DBMS lets you export data to and import data from other software products. For example, you can transfer data easily between the DBMS and a spreadsheet file, a word processing file, a graphics program file, or even another DBMS.
- The DBMS provides support for easy-to-use edit and query capabilities, screen generators, report generators, and so on.
- The DBMS provides support for both procedural and nonprocedural languages. With a **procedural language**, you must tell the computer precisely how a given task is to be accomplished; Basic, C++, and COBOL are examples of procedural languages. With a **nonprocedural language**, you merely describe the task you want the computer to accomplish. The nonprocedural language then determines how the computer will accomplish the task. SQL is an example of a nonprocedural language.
- The DBMS provides an easy-to-use menu-driven or switchboard-driven interface that allows users to tap into the power of the DBMS without having to learn a complicated set of commands.

Summary

- The fundamental capability of a DBMS is to provide users with the ability to update and retrieve data in a database without users needing to know how data is structured on disk or which processes the DBMS uses to manipulate the data.

- A DBMS must store metadata (data about the data) in a database and make this data accessible to users. The metadata is stored in a catalog or data dictionary.

- A DBMS must support concurrent update, allowing multiple users to update the same database at the same time. If concurrent update is not handled correctly, updates might be lost, causing the database to contain invalid data.

- Locking, which denies access by other users to data while the DBMS processes one user's updates, is one approach to concurrent update. Two-phase locking includes a growing phase, in which the DBMS locks more rows and releases none of the locks, followed by a shrinking phase, in which the DBMS releases all locks and acquires no new locks.

- *Deadlock* and *deadly embrace* are terms used to describe the situation in which two or more users are waiting for the other user to release a lock before they can proceed. Enterprise DBMSs have sophisticated facilities for detecting and handling deadlock. Most PC-based DBMSs do not have such facilities, which means that programs that access the database must be written in such a way that deadlocks are avoided.

- An alternative to two-phase locking is timestamping, in which the DBMS processes updates to a database in timestamp order.

- A DBMS must provide methods to recover a database in the event the database is damaged in any way. DBMSs provide facilities for periodically making a backup copy of the database. To recover the database when it is damaged or destroyed, your first step is to copy the backup over the damaged database.

- Enterprise DBMSs maintain a log or journal of all database updates since the last backup. If a database is destroyed, you make the database current from the last backup by using forward recovery to apply the after images of committed transactions. If you need to remove the updates of incorrect or terminated transactions, you use backward recovery or rollback to apply the before images to undo the updates.

- A DBMS must provide security features to prevent unauthorized access, either intentional or accidental, to a database. These security features include encryption (the storing of data in an encoded form), authentication (passwords or biometrics to identify users and database passwords assigned to the database), authorizations (the assigning of authorized users to groups that have permissions for accessing the database), and views (snapshots of certain data in the database that limit a user's access to only the tables and fields included in the view).

- A DBMS must follow rules or integrity constraints so that it updates data accurately and consistently. Key integrity constraints consist of primary key and foreign constraints. Data integrity constraints help to ensure the accuracy and consistency of individual fields and include data type, legal-values, and format integrity constraints.

- A DBMS must provide facilities to support the independence of programs from the structure of a database; *data independence* is the term for this capability.

- A DBMS must provide a facility to handle replication by managing multiple copies of a database at multiple locations.

- A DBMS must provide a set of utility services that assist in the general maintenance of a database.

Key Terms

after image	biometrics
authentication	commit
authorization rule	concurrent update
backup	database password
backward recovery	data dictionary
batch processing	data independence
before image	deadlock

deadly embrace	recovery
decrypting	replica
encryption	replicate
forward recovery	rollback
growing phase	save
journal	shrinking phase
journaling	smart card
locking	synchronization
log	timestamp
metadata	timestamping
nonprocedural language	transaction
password	two-phase locking
permission	utility services
privacy	victim
procedural language	workgroup

Review Questions

1. When users update and retrieve data, what tasks does a DBMS perform that are hidden from the users?

2. What is metadata? Which component of a DBMS maintains metadata?

3. How does a catalog differ from a data dictionary?

4. What is meant by concurrent update?

5. Describe a situation that could cause a lost update.

6. What is locking? What does it accomplish?

7. What is a transaction?

8. Describe two-phase locking.

9. What is deadlock? How does it occur?

10. How do some DBMSs use timestamping to handle concurrent update?

11. What is recovery?

12. What is journaling? What two types of images does a DBMS output to its journal?

13. When does a DBA use forward recovery? What are the forward recovery steps?

14. When does a DBA use backward recovery? What does the DBMS do to perform backward recovery?

15. What is security?

16. What is encryption? How does encryption relate to security?

17. What is authentication? Describe three types of authentication.

18. What are authorization rules?

19. What are permissions? Explain the relationship between permissions and workgroups.

20. How do views relate to security?

21. What is privacy? How is privacy related to security?

22. What are integrity constraints? Describe four different ways to handle integrity constraints. Which approach is most desirable?

23. What is data independence?

24. What is replication? What is synchronization?

25. Describe three utility services that a DBMS should provide.

26. What is a procedural language? What is a nonprocedural language?

Premiere Products Exercises

For the following exercises, you will address problems and answer questions from management at Premiere Products. You do not use the Premiere Products database for any of these exercises.

1. While users were updating the Premiere Products database, one of the transactions was interrupted. You need to explain to management what steps the DBMS will take to correct the database. Using the sample log in Figure 7-11, list and describe the updates that the DBMS will roll back if transaction 2 is interrupted at 8:10.

2. Occasionally, users at Premiere Products obtain incorrect results when they run queries that include built-in (aggregate, summary, or statistical) functions. The DBA told management that unrepeatable reads caused the problems. Use books, articles, and/or the Internet to research the unrepeatable read problem. Write a short report that explains the unrepeatable-read problem to management and use an example with your explanation. (*Note:* Unrepeatable reads are also called inconsistent retrievals, dirty reads, and inconsistent reads.)

3. You've explained replication to management, and some managers ask you for examples of when replication could be useful to them. Describe two situations, other than the ones given in the text, when replication would be useful to an organization.

4. The staff of the marketing department at Premiere Products is scheduled to receive some statistical databases, and they need you to explain these databases to them. (A statistical database is a database that is intended to supply only statistical information to users; a census database is an example of a statistical database.) Using a statistical database, users should not be able to infer information about any individual record in the database. Use books, articles, and/or the Internet to research statistical databases; then write a report that explains them, discusses the problem with using them, and gives the solution to the problem.

5. The DBA at Premiere Products wants you to investigate biometric identification techniques for potential use at the company for computer authentication purposes. Use books, articles, and/or the Internet to research these techniques; then write a report that describes the advantages and disadvantages of each of these techniques. In addition, recommend one technique and provide a justification for your recommendation.

Henry Books Case

Ray Henry plans to upgrade his database and wants you to help him select a different DBMS. To help him, he would like you to complete the following exercises. You do not use the Henry Books database for any of these exercises.

1. Many computer magazines and Web sites present comparisons of several DBMSs. Find one such DBMS comparison and compare the functions in this chapter to the listed features and functions in the comparison. Which functions from this chapter are included in the comparison? Which functions are missing from the comparison? What additional functions are included in the comparison?

2. How well does your school's DBMS fulfill the functions of a DBMS as described in this chapter? Which functions are fully supported? Which are partially supported? Which are not supported at all?

3. Use computer magazines and/or the Internet to investigate one of these DBMSs: DB2, SQL Server, MySQL, Oracle, or Sybase. Then prepare a report that explains how that DBMS handles the following DBMS functions: concurrent update, data recovery, and security. (*Note:* For concurrent update, you might need to review the concurrency control features of the DBMS.)

Alexamara Marina Group Case

For the following exercises, you will address problems and answer questions from the Alexamara Marina Group staff. You do not use the Alexamara database for any of these exercises.

1. The log shown in Figure 7-19 includes four transactions that completed successfully. For each of the four transactions, list the transaction ID and the table(s) modified. Also list whether the modification to the table added, changed, or deleted a record.

Transaction ID	Time	Action	Record Updated	Before Image	After Image
1	11:00	Start			
2	11:01	Start			
1	11:02	Insert	ServiceRequest (16)		(new values)
3	11:03	Start			
2	11:04	Update	Marina (1)	(old values)	(new values)
3	11:05	Update	Owner (EL25)	(old values)	(new values)
1	11:06	Commit			
4	11:07	Start			
3	11:08	Update	MarinaSlip (2)	(old values)	(new values)
3	11:09	Commit			
2	11:10	Update	MarinaSlip (2)	(old values)	(new values)
2	11:11	Commit			
4	11:12	Update	MarinaSlip (2)	(old values)	(new values)
4	11:13	Update	Owner (EL25)	(old values)	(new values)
4	11:14	Commit			

FIGURE 7-19   Sample log in which four transactions commit normally

2. Suppose a catastrophe destroys the database just after 11:10. Which transactions in the sample log shown in Figure 7-19 would the recovery program use to update the restored database? Which transactions would have to be reentered by users?

3. If two of the four transactions shown in Figure 7-19 started at different times, deadlock could have occurred. Adjust the log to create deadlock between these two transactions.

4. Two of the five tables in the Alexamara database are defined as follows:

```
Owner (OwnerNum, LastName, FirstName, Address, City,
        State, Zip)
MarinaSlip (SlipID, MarinaNum, SlipNum, Length,
        RentalFee, BoatName, BoatType, OwnerNum)
```

Suppose a user accesses the database via the following view:

```
CREATE VIEW OwnerBoat AS
        SELECT Owner.OwnerNum, LastName, FirstName, BoatName
        FROM Owner, MarinaSlip
        WHERE Owner.OwnerNum=MarinaSlip.OwnerNum
    ;
```

Suppose further that the database requirements have changed so that a boat can have multiple owners, just as owners can have more than one boat. What's the new database design for the Owner and MarinaSlip tables, as well as any other table(s) needed to satisfy the new requirements? Does the new database design affect the OwnerBoat view? If so, what's the new defining query for the view?

CHAPTER 8

DATABASE ADMINISTRATION

LEARNING OBJECTIVES

Objectives

- Discuss the need for database administration
- Explain the DBA's responsibilities in formulating and enforcing database policies for access privileges, security, disaster planning, and archiving
- Discuss the DBA's administrative responsibilities for DBMS evaluation and selection, DBMS maintenance, data dictionary management, and training
- Discuss the DBA's technical responsibilities for database design, testing, and performance tuning

INTRODUCTION

As you've learned in previous chapters, the database approach has many benefits. At the same time, the use of a DBMS involves potential hazards, especially when a database serves more than one user. For example, concurrent update and security present potential problems. Whom do you allow to access various parts of the database, and in what way? How do you prevent unauthorized accesses?

Note that just managing a database involves fundamental difficulties. So that they can use the database effectively, users must be made aware of the database structure or at least the portion of the database they are allowed to access. Any changes made in the database structure must be communicated to all users, along with information about how the changes will affect them. Backup and recovery must be carefully coordinated, much more so than in a single-user environment; and this coordination presents another complication.

To manage these problems, companies appoint a database administrator (DBA) to manage both the database and the use of the DBMS, that is, to perform database administration tasks. In this chapter, you will learn about the responsibilities of the DBA, which are summarized in Figure 8-1. You'll be focusing on the role of the DBA in a personal computer (PC) environment that is similar to the environment of Premiere Products. You will learn about the DBA's role in formulating and enforcing important policies with respect to the database and its use. Then you will examine the DBA's other administrative responsibilities for DBMS evaluation and selection, DBMS maintenance, data dictionary management, and training. Finally, you will learn about the

DBA's technical responsibilities for database design, testing, and performance tuning.

```
Database Policy Formulation and Enforcement
    Access privileges
    Security
    Disaster planning
    Archiving
Other Database Administrative Functions
    DBMS evaluation and selection
    DBMS maintenance
    Data dictionary management
    Training
Database Technical Functions
    Database design
    Testing
    Performance tuning
```

FIGURE 8-1   DBA responsiblities

DATABASE POLICY FORMULATION AND ENFORCEMENT

The DBA formulates database policies, communicates those policies to users, and enforces them. Among the policies are those covering access privileges, security, disaster planning, and archiving.

Access Privileges

Access to every table and field in a database is not a necessity for every user. Sam, for example, is an employee at Premiere Products; his main responsibility is inventory control. Although he needs access to the entire Part table, does he also need access to the Rep table? It is unlikely. Figure 8-2 illustrates the permitted and denied access privileges for Sam.

Rep

RepNum	LastName	FirstName	Street	City	State	Zip	Commission	Rate
20	Kaiser	Valerie	624 Randall	Grove	FL	33321	$20,542.50	0.05
35	Hull	Richard	532 Jackson	Sheldon	FL	33553	$39,216.00	0.07
65	Perez	Juan	1626 Taylor	Fillmore	FL	33336	$23,487.00	0.05

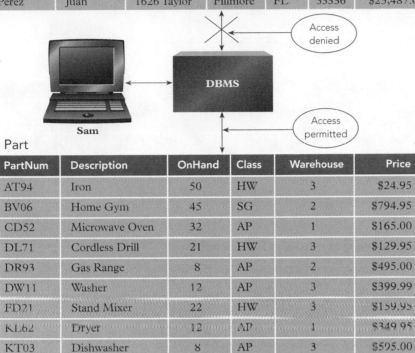

Part

PartNum	Description	OnHand	Class	Warehouse	Price
AT94	Iron	50	HW	3	$24.95
BV06	Home Gym	45	SG	2	$794.95
CD52	Microwave Oven	32	AP	1	$165.00
DL71	Cordless Drill	21	HW	3	$129.95
DR93	Gas Range	8	AP	2	$495.00
DW11	Washer	12	AP	3	$399.99
FD21	Stand Mixer	22	HW	3	$159.95
KL62	Dryer	12	AP	1	$349.95
KT03	Dishwasher	8	AP	3	$595.00
KV29	Treadmill	9	SG	2	$1,390.00

FIGURE 8-2 Permitted and denied access privileges for Sam

Paige, whose responsibility is customer mailings at Premiere Products, clearly requires access to customers' names and addresses, but what about their balances or credit limits? Should she be able to change an address? Should she be able to retrieve customers' balances or credit limits? Figure 8-3 illustrates the permitted and denied access privileges for Paige.

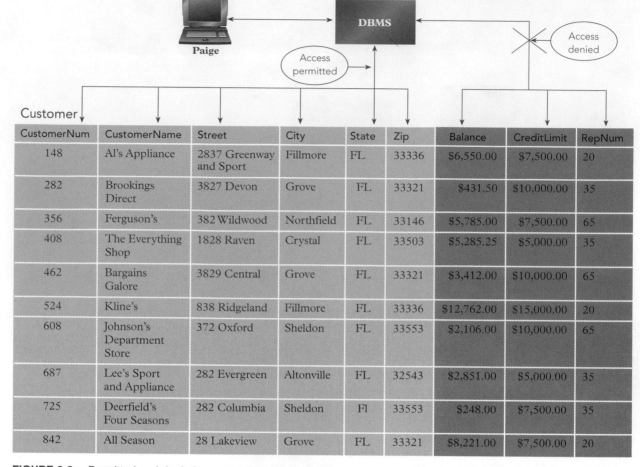

Customer

CustomerNum	CustomerName	Street	City	State	Zip	Balance	CreditLimit	RepNum
148	Al's Appliance and Sport	2837 Greenway	Fillmore	FL	33336	$6,550.00	$7,500.00	20
282	Brookings Direct	3827 Devon	Grove	FL	33321	$431.50	$10,000.00	35
356	Ferguson's	382 Wildwood	Northfield	FL	33146	$5,785.00	$7,500.00	65
408	The Everything Shop	1828 Raven	Crystal	FL	33503	$5,285.25	$5,000.00	35
462	Bargains Galore	3829 Central	Grove	FL	33321	$3,412.00	$10,000.00	65
524	Kline's	838 Ridgeland	Fillmore	FL	33336	$12,762.00	$15,000.00	20
608	Johnson's Department Store	372 Oxford	Sheldon	FL	33553	$2,106.00	$10,000.00	65
687	Lee's Sport and Appliance	282 Evergreen	Altonville	FL	32543	$2,851.00	$5,000.00	35
725	Deerfield's Four Seasons	282 Columbia	Sheldon	Fl	33553	$248.00	$7,500.00	35
842	All Season	28 Lakeview	Grove	FL	33321	$8,221.00	$7,500.00	20

FIGURE 8-3 Permitted and denied access privileges for Paige

Although rep 20 (Valerie Kaiser) should be able to obtain some of the information about her own customers, should she be able to obtain the same information about other customers? Figure 8-4 illustrates the permitted and denied access privileges for Valerie.

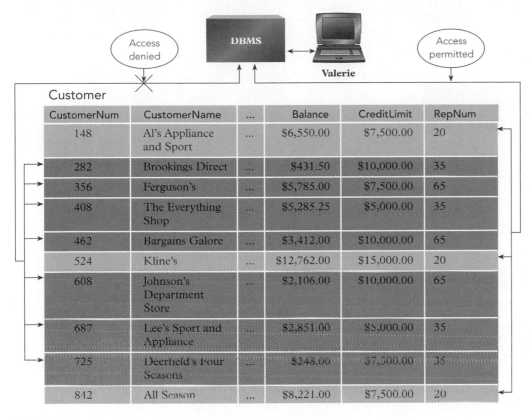

FIGURE 8-4 Permitted and denied access privileges for Valerie

The DBA determines the access privileges for all users and enters the appropriate authorization rules in the DBMS catalog to ensure that users access the database only in ways to which they are entitled. For example, the DBA uses the SQL GRANT statement to define the access privileges users have to the data in the database. The DBA also documents access privilege policy, top-level management approves the policy, and the DBA communicates the policy to management and to all users.

Security

As discussed in previous chapters, security is the prevention of unauthorized access, either intentional or accidental, to a database and the DBA uses views and the SQL GRANT statement as two security mechanisms. Unauthorized access includes access by someone who has no right to access the database at all. For example, as shown in Figure 8-5, the DBMS prevents Brady, who is a programmer at Premiere Products, from accessing the database because the DBA has not authorized Brady as a user.

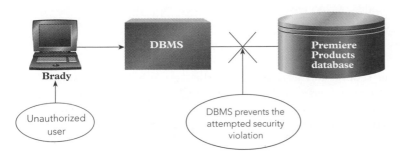

FIGURE 8-5 Attempted security violation by Brady, who's not an authorized user

Unauthorized access also includes users who have legitimate access to some but not all data in a database and who attempt to access data for which they are not authorized. For example, the DBMS prevents Paige from accessing customer balances, as shown in Figure 8-6, because the DBA did not grant her access privileges to that data.

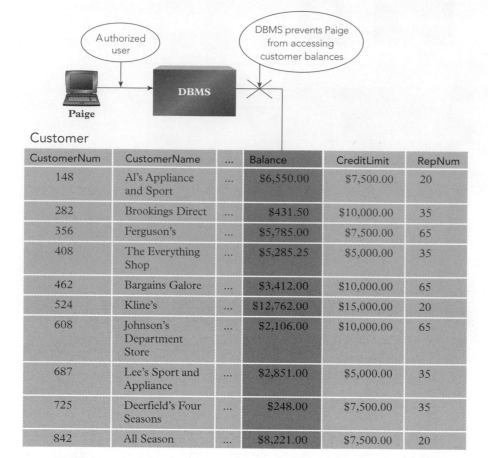

Customer

CustomerNum	CustomerName	...	Balance	CreditLimit	RepNum
148	Al's Appliance and Sport	...	$6,550.00	$7,500.00	20
282	Brookings Direct	...	$431.50	$10,000.00	35
356	Ferguson's	...	$5,785.00	$7,500.00	65
408	The Everything Shop	...	$5,285.25	$5,000.00	35
462	Bargains Galore	...	$3,412.00	$10,000.00	65
524	Kline's	...	$12,762.00	$15,000.00	20
608	Johnson's Department Store	...	$2,106.00	$10,000.00	65
687	Lee's Sport and Appliance	...	$2,851.00	$5,000.00	35
725	Deerfield's Four Seasons	...	$248.00	$7,500.00	35
842	All Season	...	$8,221.00	$7,500.00	20

FIGURE 8-6 Attempted security violation by Paige, who's authorized to access some customer data but is not authorized to access customer balances

The DBA takes the steps necessary to ensure that the database is secure. After the DBA determines the access privileges for each user, the DBA creates security policies and procedures, obtains management approval of the policies and procedures, and then distributes them to authorized users.

To implement and enforce security, the DBA uses the DBMS's security features, such as encryption, authentication, authorizations, and views. If a DBMS lacks essential security features, the DBA might create or purchase special security programs that provide the missing features.

In addition to relying on the security features provided by the DBMS and, if necessary, the special security programs, the DBA monitors database usage to detect potential security violations. If a security violation occurs, the DBA determines who breached security, how the violation occurred, and how to prevent a similar violation in the future.

Disaster Planning

The type of security discussed in the previous section concerns damage to the data in a database caused by authorized and unauthorized users. Damage to a database can also occur through a physical incident such as an abnormally terminated program, a software virus or worm, a disk problem, a power outage, a computer malfunction, a hurricane, a flood, a tornado, or another natural disaster.

To protect an organization's data from physical damage, the DBA creates and implements backup and recovery procedures as part of a disaster recovery plan. A **disaster recovery plan** specifies the ongoing and emergency actions and procedures required to ensure data availability if a disaster occurs.

For example, a disaster recovery plan must include plans for protecting an organization's data against hard drive failures and electrical power loss. To protect against hard drive failures, organizations often use **redundant array of inexpensive/independent drives (RAID)**, in which database updates are replicated to multiple hard drives so that an organization can continue to process database updates after losing one of its hard drives. To protect against electrical power interruptions and outages, organizations use an **uninterruptible power supply (UPS)**, which is a power source such as a battery or fuel cell, for short interruptions and a power generator for longer outages.

For some functions, such as credit card processing, stock exchanges, and airline reservations, data availability must be continuous. In these situations, organizations can switch quickly to duplicate backup systems (usually at a separate backup site) in the event of a malfunction in or a complete destruction of the main system. Other organizations contract with firms using hardware and software similar to their own so that in the event of a catastrophe, they can temporarily use these other facilities as backup sites. Backup sites can be established with different levels of preparedness. A **hot site** is a backup site that an organization can switch to in minutes or hours because the site is completely equipped with duplicate hardware, software, and data. Although hot sites are expensive, businesses such as banks and other financial institutions cannot permit any lengthy service interruptions and must have hot sites. A **warm site** is a backup site that is equipped with duplicate hardware and software but not data, so it takes longer to start processing at a warm site compared to a hot site.

Archiving

Often users need to retain certain data in a database for only a limited time. An order that has been filled, reported on a customer's statement, and paid by the customer is in one sense no longer important. Should you keep the order in the database? If you always keep data in the database as a matter of policy, the database will continually grow. The disk space that is occupied by the database will expand, and programs that access the database might take more time to perform their functions. The increased disk space and longer processing times might be good reasons to remove completed orders and all their associated order lines from the database.

On the other hand, you might need to retain orders and their associated order lines for future reference by users to answer customer inquiries or to check a customer's past history with the company. More critically, you need to retain data legally to satisfy governmental laws and regulations and to meet auditing and financial requirements. Examples of legal reasons for data retention that apply to many organizations are as follows:

- The **Sarbanes-Oxley (SOX) Act** of 2002 is a federal law that specifies data retention and verification requirements for public companies, requires CEOs and CFOs to certify financial statements, and makes it a crime to destroy or tamper with financial records. Congress passed this law in response to major accounting scandals such as Enron, WorldCom, and Tyco.
- The **Patriot Act** of 2001 is a federal law that specifies data retention requirements for the identification of customers opening accounts at financial institutions, allows law enforcement agencies to search companies' and individuals' records and communications, and expands the government's authority to regulate financial transactions. President George W. Bush signed the Patriot Act into law 45 days after the September 11, 2001 terrorist attacks against the United States.
- The Security and Exchange Commission's Rule 17a-4 (**SEC Rule 17a-4**) specifies the retention requirements of all electronic communications and records for financial and investment entities.
- The **Department of Defense (DOD) 5015.2 Standard** of 1997 provides data management requirements for the DOD and for companies supplying or dealing with the DOD.
- The Health Insurance Portability and Accountability Act (**HIPAA**) of 1996 is a federal law that specifies the rules for storing, handling, and protecting health-care transactions.
- The **Presidential Records Act** of 1978 is a federal law that regulates the data retention requirements for all communications, including electronic communications, of U.S. presidents and vice presidents. Congress passed this law after the scandals during the Nixon administration.

Legal compliance to the many data retention laws and regulations is a complicated and expensive process. For example, the length of time organizations must retain data ranges from two to seven years; for some laws, the time period is indefinite. The DBA is responsible for ensuring that data processed by DBMSs is retained in conformance to all laws. Although DBAs need to retain data for legal reasons, they can choose to remove data that's no longer needed for perfomance reasons.

One solution to data retention is to use what is known as a **data archive**, or an **archive**. In ordinary usage, an archive (technically *archives*) is a place where public records and documents are kept. A data archive is similar. It is a place where a record of certain corporate data is kept. In the case of the previously mentioned completed orders and associated order lines, Figure 8-7 shows how you would remove them from the database and place them in the archive, thus storing them for future reference.

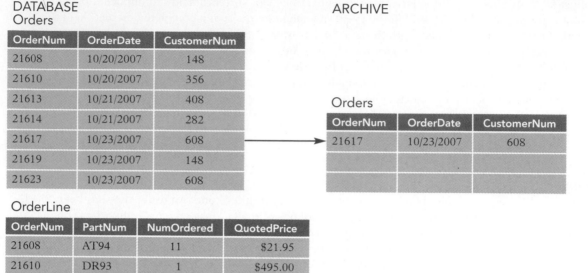

DATABASE
Orders

OrderNum	OrderDate	CustomerNum
21608	10/20/2007	148
21610	10/20/2007	356
21613	10/21/2007	408
21614	10/21/2007	282
21617	10/23/2007	608
21619	10/23/2007	148
21623	10/23/2007	608

ARCHIVE

Orders

OrderNum	OrderDate	CustomerNum
21617	10/23/2007	608

OrderLine

OrderNum	PartNum	NumOrdered	QuotedPrice
21608	AT94	11	$21.95
21610	DR93	1	$495.00
21610	DW11	1	$399.99
21613	KL62	4	$329.95
21614	KT03	2	$595.00
21617	BV06	2	$794.95
21617	CD52	4	$150.00
21619	DR93	1	$495.00
21623	KV29	2	$1,290.00

OrderLine

OrderNum	PartNum	NumOrdered	QuotedPrice
21617	BV06	2	$794.95
21617	CD52	4	$150.00

FIGURE 8-7   Movement of order 21617 from the database to the archive

Typically, the DBA stores the archive on some mass storage device—for example, a disk, tape, CD, or DVD. Whichever medium the DBA uses, the DBA must store copies of both archives and database backups off-site so that recovery can take place even if a company's buildings and contents are destroyed. The off-site location must be a sufficient distance from the main site so that there's no likelihood of a disaster damaging both sites. Once again, it is up to the DBA to establish and implement procedures for the use, maintenance, and storage of the archive.

OTHER DATABASE ADMINISTRATIVE FUNCTIONS

The DBA is also responsible for DBMS evaluation and selection, DBMS maintenance, data dictionary management, and training.

DBMS Evaluation and Selection

When a company decides to purchase a new DBMS, the DBA leads the DBMS evaluation and selection effort. To evaluate the DBMS candidates objectively, the DBA usually prepares a checklist similar to the one shown in Figure 8-8. (This checklist applies specifically to a relational system because most DBMSs are, at least in part, relational. If the DBA had not already selected a data model, such as the relational model, the DBA would have added a "Choice of Data Model" category to the list.) The DBA evaluates each prospective purchase of a DBMS against the categories shown in the figure. An explanation of each category follows Figure 8-8.

```
1. Data Definition
   a. Data types
      (1) Numeric
      (2) Character
      (3) Date
      (4) Logical (T/F)
      (5) Memo
      (6) Currency
      (7) Binary object (pictures, drawings, sounds, and so on)
      (8) Link to an Internet, Web, or other address
      (9) User-defined data types
      (10) Other
   b. Support for nulls
   c. Support for primary keys
   d. Support for foreign keys
   e. Unique indexes
   f. Views
2. Data Restructuring
   a. Possible restructuring
      (1) Add new tables
      (2) Delete existing tables
      (3) Add new columns
      (4) Change the layout of existing columns
      (5) Delete columns
      (6) Add new indexes
      (7) Delete existing indexes
   b. Ease of restructuring
3. Nonprocedural Languages
   a. Nonprocedural languages supported
      (1) SQL
      (2) QBE
      (3) Natural language
      (4) Language unique to the DBMS. Award points on the
          basis of ease of use as well as the types of operations
          (joining, sorting, grouping, calculating various statistics,
          and so on) that are available in the language. You can
          use SQL as a standard against which you can judge the
          language.
   b. Optimization done by one of the following:
      (1) User (in formulating the query)
      (2) DBMS (through built-in optimizer)
      (3) No optimization possible; system does only sequential
          searches
4. Procedural Languages
   a. Procedural languages supported
      (1) Language unique to the DBMS. Award points on the basis
          of the quality of this language both in terms of the types
          of statements and control structures available and the database
          manipulation statements included in the language.
      (2) Java
      (3) C or C++
      (4) GUI language such as Visual Basic
      (5) COBOL
      (6) Other
   b. Can a nonprocedural language be used in conjunction with
      the procedural language (for example, could SQL be
      embedded in a COBOL program)?
5. Data Dictionary
   a. Type of entries
      (1) Tables
      (2) Columns
      (3) Indexes
      (4) Relationships
      (5) Users
      (6) Programs
      (7) Other
   b. Integration of data dictionary with other components of
      the system
```

FIGURE 8-8   DBMS evaluation checklist

6. Concurrent Update
   a. Level of locking
      (1) Field value
      (2) Row
      (3) Page
      (4) Table
      (5) Database
   b. Type of locking
      (1) Shared
      (2) Exclusive
      (3) Both
   c. Responsibility for handling deadlock
      (1) Programs
      (2) DBMS (automatic rollback of transaction causing deadlock)
7. Backup and Recovery
   a. Backup services
   b. Journaling services
   c. Recovery services
      (1) Recover from backup copy only
      (2) Recover using backup copy and journal
   d. Rollback of individual transactions
   e. Incremental backup
8. Security
   a. Encryption
   b. Passwords
   c. Authorization rules
      (1) Access to database only
      (2) Access/update access to any column or combination of columns
   d. Views
   e. Difficulty in bypassing security controls
9. Integrity
   a. Support for entity integrity
   b. Support for referential integrity
   c. Support for data integrity
   d. Support for other types of integrity constraints
10. Replication and Distributed Databases
    a. Partial replicas
    b. Handling of duplicate updates in replicas
    c. Data distribution
    d. Procedure support
       (1) Language used
       (2) Procedures stored in database
       (3) Support for remote stored procedures
       (4) Trigger support
11. Limitations
    a. Number of tables
    b. Number of columns
    c. Length of individual columns
    d. Total length of all columns in a table
    e. Number of rows per table
    f. Number of files that can be open at the same time
    g. Sizes of database, tables, and other objects
    h. Types of hardware supported
    i. Types of LANs supported
    j. Other
12. Documentation and Training
    a. Clearly written manuals
    b. Tutorial
       (1) Online
       (2) Printed
    c. Online help available
       (1) General help
       (2) Context-sensitive help
    d. Training
       (1) Vendor or other company
       (2) Location
       (3) Types (DBA, programmers, users, others)
       (4) Cost
13. Vendor Support
    a. Type of support available
    b. Quality of support available
    c. Cost of support
    d. Reputation of support
14. Performance
    a. External benchmarking done by various organizations
    b. Internal benchmarking
    c. Includes a performance monitor

FIGURE 8-8   DBMS evaluation checklist (continued)

```
15. Portability
    a. Operating systems
       (1) Unix
       (2) Microsoft Windows
       (3) Linux
       (4) Other
    b. Import/export/linking file support
       (1) Other databases
       (2) Other applications (for example, spreadsheets and
           graphics)
    c. Internet and intranet support
16. Cost
    a. Cost of DBMS
    b. Cost of any additional components
    c. Cost of any additional hardware that is required
    d. Cost of network version (if required)
    e. Cost and types of support
17. Future Plans
    a. What does the vendor plan for the future of the system?
    b. What is the history of the vendor in terms of keeping the
       system up to date?
    c. When changes are made in the system, what is involved in
       converting to the new version?
       (1) How easy is the conversion?
       (2) What will it cost?
18. Other Considerations (Fill in your own special requirements.)
    a. ?
    b. ?
    c. ?
    d. ?
```

FIGURE 8-8 DBMS evaluation checklist (continued)

1. **Data definition.** What types of data does the DBMS support? Does it support nulls? What about primary and foreign keys? The DBMS undoubtedly provides support for indexes, but can you specify that an index is unique and then have the system enforce the uniqueness? Does the DBMS support views?

2. **Data restructuring.** What type of database restructuring does the DBMS allow? How easily can the DBA perform the restructuring? Will the system do most of the work, or will the DBA have to create special programs for this purpose?

3. **Nonprocedural languages.** What types of nonprocedural language does the DBMS support? The possibilities are SQL, QBE, natural language, and a DBMS built-in language. If the DBMS supports one of the standard languages, what's the quality of its version? If the DBMS provides its own language, how good is it? How does its functionality compare to that of SQL? How does the DBMS achieve optimization of queries? The DBMS optimizes each query, or the user must do so by the manner in which he or she states the query. If neither happens, no optimization occurs. Most desirable, of course, is the first alternative.

4. **Procedural languages.** What types of procedural languages does the DBMS support? Are they common languages, such as Java, C or C++, and COBOL? Is it a graphical user interface (GUI) language? Does the DBMS provide its own language? In the latter case, how complete is the language? Does it contain all the required types of statements and control structures? What facilities does the language provide for accessing the database? Does the DBMS let you use a nonprocedural language while you are using the procedural language?

5. **Data dictionary.** What kind of data dictionary does the DBMS provide? Is it a simple catalog? Or can it contain more content, such as information about programs and the various data items these programs access? How well is the data dictionary integrated with other components of the system—for example, the nonprocedural language?

6. **Concurrent update.** Does the DBMS support concurrent update? What unit may be locked (field value, row, page, table, or database)? Are exclusive locks the only ones permitted, or are shared locks also allowed? (A **shared lock** permits other users to read the data; with an **exclusive lock**, no other user may access the data in any way.) Does the DBMS resolve deadlock, or must programs resolve it?

7. **Backup and recovery.** What type of backup and recovery services does the DBMS provide? Does the DBMS maintain a journal of changes in the database and use the journal during the recovery process? If a transaction terminates abnormally, does the DBMS roll back its updates? Can the DBMS perform an incremental backup of just the data that has changed?

8. *Security.* What types of security features does the DBMS provide? Does the DBMS support encryption, password support, and authorizations rules? Does the DBMS provide a view mechanism that can be used for security? How difficult is it to bypass the security controls?

9. *Integrity.* What type of integrity constraints does the DBMS support? Does the DBMS support entity integrity (the fact that the primary key cannot be null) and referential integrity (the property that values in foreign keys must match values already in the database)? What types of data integrity does the DBMS support? Does the DBMS support any other types of integrity constraints?

10. *Replication and distributed databases.* Does the DBMS support replication? If so, does the DBMS allow partial replicas (copies of selected rows and fields from tables in a database)? And how does the DBMS handle updates to the same data from two or more replicas? Can the DBMS distribute a database, that is, divide the database into segments and store the segments on different computers? If so, what types of distribution does the DBMS allow and what types of procedure support for distribution does the DBMS provide?

11. *Limitations.* What limitations exist with respect to the number of tables and the number of fields and rows per table? How many files can you open at the same time? (For some databases, each table and each index is in a separate file. Thus, a single table with three indexes, all in use at the same time, would account for *four* files. Problems might arise if the number of files you can open is relatively small and many indexes are in use.) On what types of operating system and hardware is the DBMS supported? What types of local area networks (LANs) can you use with the DBMS? (A **local area network (LAN)** is a configuration of several computers connected together that allows users to share a variety of hardware and software resources. One of these resources is the database. In a LAN, support for concurrent update is very important because many users might be updating the database at the same time. The relevant question here, however, is not how well the DBMS supports concurrent update but which of the LANs you can use with the DBMS.)

12. *Documentation and training.* Does the vendor of the DBMS supply printed or online training manuals? If so, how good are the manuals? Are they easy to use? Is there a good index? Is a tutorial, in either printed or online form, available to assist users in getting started with the system? Is online help available? If so, does the DBMS provide general help and context-sensitive help? (**Context-sensitive help** means that if a user is having trouble and asks for help, the DBMS will provide assistance for the particular feature being used at the time the user asks for the help.) Does the vendor provide training classes? Do other companies offer training? Are the classes on-site or off-site? Are there classes for the DBA and separate classes for programmers and others? What is the cost for each type of training?

13. *Vendor support.* What type of support does the vendor provide for the DBMS, and how good is it? What is the cost? What is the vendor's reputation for support among current users?

14. *Performance.* How well does the DBMS perform, where performance is a measure of how rapidly the DBMS completes its tasks? This is a difficult question to answer because each organization has a different number of users and a different mix of transactions and both factors affect how a DBMS performs. One way to determine relative performance among DBMSs is to look into benchmark tests that various organizations have performed on several DBMSs. Benchmarking typically is done in areas such as sorting, indexing, and reading all rows and then changing data values in all rows. For example, the Transaction Processing Performance Council (www.tpc.org) provides the results of database benchmark tests to its members. Beyond using benchmarks, if an organization has some specialized needs, it may have to set up its own benchmark tests. Does the DBMS provide a performance monitor that measures different types of performance while the DBMS is operational?

15. *Portability.* Which operating systems can you use with the DBMS? What types of files can you import or export? Can the DBMS link to other data sources, such as files and other types of DBMSs? Does the DBMS provide Internet and intranet support? (An **intranet** is an internal company network that uses software tools typically used on the Internet and the World Wide Web.)

16. *Cost.* What is the cost of the DBMS and of any additional components the organization is planning to purchase? Is additional hardware required? If so, what is the associated cost? If the organization requires a special version of the DBMS for a network, what is the additional cost? What is the cost of vendor support, and what types of support plans are available?

17. ***Future plans.*** What plans has the vendor made for the future of the system? This information is often difficult to obtain, but you can get an idea by looking at the performance of the vendor with respect to how it has kept the existing system up to date. How easy has it been for users to convert to new versions of the system?

18. ***Other considerations.*** This is a final catch-all category that contains any special requirements not covered in the other categories. For many organizations, existing financial and other application software and existing hardware limit the DBMS choice.

After the DBA examines each DBMS with respect to all the preceding categories, the DBA and management can compare the results. Unfortunately, this process can be difficult because of the number of categories and their generally subjective nature. To make the process more objective, the DBA can assign a numerical ranking to each DBMS in each category (for example, a number between 0 and 10, where 0 is poor and 10 is excellent). Furthermore, the DBA can assign weights to the categories. Weighting allows an organization to signify which categories are more critical than others. Then you multiply each number used in the numerical ranking by the appropriate weight and add the results, producing a weighted total. Finally, you compare the weighted totals for each DBMS, producing the final evaluation.

How does the DBA arrive at the numbers to assign each DBMS in the various categories? Several methods are used. The DBA can request feedback from other organizations that are currently using the DBMS being considered. The DBA can read journal reviews of the various DBMSs. Sometimes the DBA can obtain a trial version of the DBMS, and members of the staff can give it a hands-on test. In practice, the DBA usually combines all three methods. Whichever method is used, however, the DBA must carefully create the checklist and determine weights before starting the evaluation; otherwise, the findings may be inadvertently slanted in a particular direction.

DBMS Maintenance

After the organization selects and purchases the DBMS, the DBA has primary responsibility for it. The DBA installs the DBMS in a way that is suitable for the organization. If the DBMS configuration needs to be changed, it is the DBA who makes the changes.

When the vendor releases a new version of the DBMS, the DBA reviews it and determines whether the organization should upgrade to it. If the decision is made to convert to the new version or perhaps to a new DBMS, the DBA coordinates the conversion. The DBA also handles any fixes to problems in the DBMS that the vendor releases.

When a problem occurs that affects the database, the DBA coordinates the people required to resolve the problem. Some people, such as programmers and users, are from inside the organization; and others, such as hardware and software vendors, are from outside the organization.

When users have special one-time processing needs or extensive query requirements against the database, the DBA coordinates the users so that their needs are satisfied without unduly affecting other users.

Data Dictionary Management

The DBA also manages the data dictionary. Essentially, the data dictionary is the catalog mentioned in Chapter 7; but it often contains a wider range of information, including information about tables, fields, indexes, programs, and users.

The DBA establishes naming conventions for tables, fields, indexes, and so on. The DBA creates the data definitions for all tables, as well as for any data integrity rules and user views. The DBA also updates the contents of the data dictionary. Finally, the DBA creates and distributes appropriate reports from the data dictionary to users, programmers, and other people in the organization.

Training

The DBA provides training in the use of the DBMS and in how to access the database. The DBA also coordinates the training of users and the technical staff responsible for developing and maintaining database applications. In those cases where the vendor of the DBMS provides training, the DBA handles the scheduling to make sure users receive the training they require. Training is a big expense, but successful organizations make the investment to ensure that their employees are knowledgeable and productive in handling the critical data resource.

TECHNICAL FUNCTIONS

The DBA is also responsible for database design, testing, and performance tuning.

Database Design

The DBA establishes a sound methodology for database design, such as the one discussed in Chapter 6, and ensures that all database designers follow the methodology. The DBA also verifies that the designers obtain all pertinent information from the appropriate users. After the database designers complete the information-level design, the DBA does the physical-level design.

The DBA establishes documentation standards for all the steps in the database design process. The DBA also makes sure that these standards are followed, that the documentation is kept up to date, and that the appropriate personnel have access to the documentation they need.

Requirements don't remain stable over time; they change constantly. The DBA reviews all changes to requirements and determines whether the changes will require that modifications be made to the database. If so, the DBA makes the changes in the design and in the data in the database. The DBA also verifies that programmers modify all programs and documentation affected by the change.

Testing

The hardware, software, and database for the users is called the **production system**, or **live system**. The DBA strictly controls the production system. With just two exceptions, the DBA grants access and update privileges to the production system only to authorized users. The first exception is when problems occur, such as with software. The DBA and others must troubleshoot the problem by accessing the production system. The second exception is when programmers complete new programs or modify existing programs for the production system. For both exceptions, the DBA performs any necessary database modifications or closely controls the activities of others.

Other than for these two exceptions, the DBA does not grant programmers access to the production system. Instead, the DBA and the programmers create a separate system, called the **test system**, or **sandbox**, that programmers use to develop new programs and modify existing programs. After programmers complete the testing of their programs in the test system, a separate quality assurance group performs futher tests, the DBA and the users review and approve the test results, and the DBA reviews and approves the programs and documentation. If the DBA and the users approve, the DBA notifies all affected users when new or corrected features will be available. The DBA then transfers the programs to the production system and makes any required database changes, as shown in Figure 8-9.

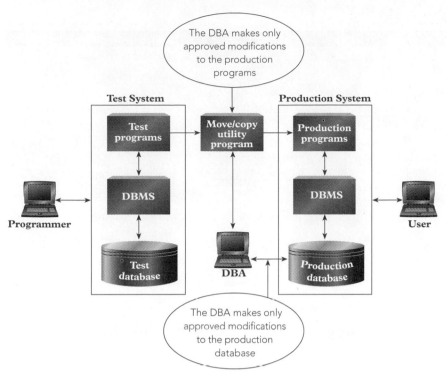

FIGURE 8-9 DBA controls the interaction between the test and production systems

A production system with a DBMS is a complex system. A separate test system reduces the complexity of the production system and provides an extra measure of control.

Performance Tuning

Database performance deals with the ability of the production system to serve users in a timely and responsive manner. Because funding is usually a constraint, the DBA's challenge is to get the best possible performance from the available funds.

Faster computers with faster disks, faster network connections, faster software, and other production system expenditures help improve performance. What can the DBA do if the organization has no additional money for its production system but needs further performance improvements? The DBA can change the database design to improve performance; this process is called **tuning** the design. Some of the performance tuning changes the DBA can make to a database design include creating and deleting indexes, splitting tables, and changing the table design.

By default, Access and some other DBMSs automatically create indexes for primary key and foreign key fields. These indexes make accessing the fields faster than accessing would be without the indexes. Further, indexing common fields improves the speed of joining related tables. If a DBMS doesn't automatically index primary key and foreign key fields, the DBA should create indexes for them. In addition, queries that search indexed fields run faster than comparable queries without indexes for those fields. For example, if users frequently query the Part table to find records based on values for the Class or Warehouse fields, the DBA can improve performance by adding indexes on those fields. On the other hand, a table with many indexes takes longer to update. If users experience delays when they update a table, the DBA can delete some of the table's indexes to improve update performance.

If users access only certain fields in a table, you can improve performance by splitting the table into two or more tables that each have the same primary key as the original and that collectively contain all the fields from the original table. Each resulting table is smaller than the original; the lower amount of data moves faster between disk and memory. For example, suppose dozens of users at Premiere Products access the Customer table shown in Figure 8-10.

Customer

CustomerNum	CustomerName	Street	City	State	Zip	Balance	CreditLimit	RepNum
148	Al's Appliance and Sport	2837 Greenway	Fillmore	FL	33336	$6,550.00	$7,500.00	20
282	Brookings Direct	3827 Devon	Grove	FL	33321	$431.50	$10,000.00	35
356	Ferguson's	382 Wildwood	Northfield	FL	33146	$5,785.00	$7,500.00	65
408	The Everything Shop	1828 Raven	Crystal	FL	33503	$5,285.25	$5,000.00	35
462	Bargains Galore	3829 Central	Grove	FL	33321	$3,412.00	$10,000.00	65
524	Kline's	838 Ridgeland	Fillmore	FL	33336	$12,762.00	$15,000.00	20
608	Johnson's Department Store	372 Oxford	Sheldon	FL	33553	$2,106.00	$10,000.00	65
687	Lee's Sport and Appliance	282 Evergreen	Altonville	FL	32543	$2,851.00	$5,000.00	35
725	Deerfield's Four Seasons	282 Columbia	Sheldon	FL	33553	$248.00	$7,500.00	35
842	All Season	28 Lakeview	Grove	FL	33321	$8,221.00	$7,500.00	20

FIGURE 8-10   Customer table for Premiere Products

If some users access address data from the Customer table and other users access balances and credit limits, the DBA can split the Customer table into two tables, as shown in Figure 8-11, to improve performance. Users needing data from both tables can obtain that data by joining the two split tables on the CustomerNum field.

CustomerAddress

CustomerNum	CustomerName	Street	City	State	Zip
148	Al's Appliance and Sport	2837 Greenway	Fillmore	FL	33336
282	Brookings Direct	3827 Devon	Grove	FL	33321
356	Ferguson's	382 Wildwood	Northfield	FL	33146
408	The Everything Shop	1828 Raven	Crystal	FL	33503
462	Bargains Galore	3829 Central	Grove	FL	33321
524	Kline's	838 Ridgeland	Fillmore	FL	33336
608	Johnson's Department Store	372 Oxford	Sheldon	FL	33553
687	Lee's Sport and Appliance	282 Evergreen	Altonville	FL	32543
725	Deerfield's Four Seasons	282 Columbia	Sheldon	FL	33553
842	All Season	28 Lakeview	Grove	FL	33321

CustomerFinancial

CustomerNum	CustomerName	Balance	CreditLimit	RepNum
148	Al's Appliance and Sport	$6,550.00	$7,500.00	20
282	Brookings Direct	$431.50	$10,000.00	35
356	Ferguson's	$5,785.00	$7,500.00	65
408	The Everything Shop	$5,285.25	$5,000.00	35
462	Bargains Galore	$3,412.00	$10,000.00	65
524	Kline's	$12,762.00	$15,000.00	20
608	Johnson's Department Store	$2,106.00	$10,000.00	65
687	Lee's Sport and Appliance	$2,851.00	$5,000.00	35
725	Deerfield's Four Seasons	$248.00	$7,500.00	35
842	All Season	$8,221.00	$7,500.00	20

FIGURE 8-11   Result of splitting the Customer table into two tables

The DBA can also split tables for security purposes. In Figure 8-11, the CustomerAddress table contains customer address data and the CustomerFinancial table contains customer financial data. Those users granted access only to the CustomerAddress table have no access to customer financial data, thus providing an added measure of security.

Although you design database tables in third normal form to prevent the anomaly problems discussed in Chapter 5, the DBA occasionally denormalizes tables to improve performance. **Denormalizing** converts a table that is in third normal form to a table that is no longer in third normal form. Usually, the conversion produces tables that are in first normal form or second normal form. Denormalizing introduces anomaly problems but can decrease the number of disk accesses that certain types of transactions require, thus improving performance. For example, suppose users who are processing order lines need part descriptions. The DBA might include part descriptions in the OrderLine table, as shown in Figure 8-12.

OrderLine

OrderNum	PartNum	Description	NumOrdered	QuotedPrice
21608	AT94	Iron	11	$21.95
21610	DR93	Gas Range	1	$495.00
21610	DW11	Washer	1	$399.99
21613	KL62	Dryer	4	$329.95
21614	KT03	Dishwasher	2	$595.00
21617	BV06	Home Gym	2	$794.95
21617	CD52	Microwave Oven	4	$150.00
21619	DR93	Gas Range	1	$495.00
21623	KV29	Treadmill	2	$1,290.00

FIGURE 8-12 Including part descriptions in the OrderLine table, which creates a first normal form table

The OrderLine table in Figure 8-12 is in first normal form because there are no repeating groups. Because a part description depends only on the part number, which is just a portion of the primary key for the table, the OrderLine table is not in second normal form and, consequently, is not in third normal form either. As a result, the table has redundancy and anomaly problems that are inherent in tables that are not in third normal form. However, users processing order lines no longer need to join the OrderLine and Part tables to obtain part descriptions, thus improving performance.

Large databases with thousands of users often suffer periodic performance problems as users change their transaction mix. In these cases, the DBA must tune the databases to provide improved performance to all users.

Summary

- The database administrator (DBA) is the person who is responsible for supervising the database and the use of the DBMS.
- The DBA formulates and enforces policies about those users who can access the database, the portions of the database they may access, and in what manner they can access the database.
- The DBA formulates and enforces policies about security, which is the prevention of unauthorized access, either intentional or accidental, to a database. The DBA uses the DBMS's security features and special security programs, if necessary, and monitors database usage to detect potential security violations.
- The DBA creates and implements backup and recovery procedures as part of a disaster recovery plan to protect an organization's data from physical damage.
- The DBA formulates and enforces policies that govern the management of an archive for data that is no longer needed in the database but that must be retained for reference purposes or for compliance with federal laws.
- The DBA leads the effort to evaluate and select a new DBMS. The DBA develops a checklist of desirable features for a DBMS and evaluates each prospective purchase of a DBMS against this checklist.
- The DBA installs and maintains the DBMS after it has been selected and procured.
- The DBA maintains the data dictionary, establishes naming conventions for its contents, and provides information from it to others in the organization.
- The DBA provides database and DBMS training and coordinates and schedules training by outside vendors.
- The DBA verifies all information-level database designs, completes all physical-level database designs, and creates documentation standards. The DBA also evaluates changes in requirements to determine whether he or she needs to change the database design and the data in the database.
- The DBA controls the production system, which is accessible only to authorized users. Other than when authorized by the DBA to access the production system in exceptional situations, programmers access a separate test system. The DBA migrates tested programs to the production system and makes any required database changes.
- The DBA tunes the database design to improve performance. Included among the performance tuning changes the DBA makes are creating and deleting indexes, splitting tables, and denormalizing tables.

Key Terms

archive

context-sensitive help

data archive

denormalizing

Department of Defense (DOD) 5015.2 Standard

disaster recovery plan

exclusive lock

HIPAA

hot site

intranet

live system

local area network (LAN)

Patriot Act

Presidential Records Act

production system

RAID (redundant array of inexpensive/independent drives)

sandbox

Sarbanes-Oxley (SOX) Act

SEC Rule 17a-4

shared lock

test system

tuning

UPS (uninterruptible power supply)

warm site

Review Questions

1. What is a DBA? Why is this position necessary?
2. What are the DBA's responsibilities regarding access privileges?
3. What are the DBA's responsibilities regarding security?

4. What is a disaster recovery plan?

5. What are data archives? What purpose do they serve? What is the relationship between a database and its data archives?

6. Name five categories that you usually find on a DBMS evaluation and selection checklist.

7. What is a shared lock? What is an exclusive lock?

8. What is a LAN?

9. What is context-sensitive help?

10. What is an intranet?

11. After a DBMS has been selected, what is the DBA's role in DBMS maintenance?

12. What are the DBA's responsibilities with regard to the data dictionary?

13. Who trains computer users in an organization? What is the DBA's role in this training?

14. What are the DBA's database design responsibilities?

15. What is the difference between production and test systems?

16. What is meant by "tuning a design"?

17. How can splitting a table improve performance?

18. What is denormalization?

Premiere Products Exercises

For the following exercises, you do not use the Premiere Products database.

1. The DBA asks for your help in planning the data archive for the following Premiere Products database:

```
Rep (RepNum, LastName, FirstName, Street, City, State,
    Zip, Commission, Rate)
Customer (CustomerNum, CustomerName, Street, City,
    State, Zip, Balance, CreditLimit, RepNum)
Orders (OrderNum, OrderDate, CustomerNum)
OrderLine (OrderNum, PartNum, NumOrdered, QuotedPrice)
Part (PartNum, Description, OnHand, Class, Warehouse, Price)
```

Determine which data from the database to archive; that is, for each table, specify whether data needs to be archived. If it does, specify which data, when it should be archived, and whether it should be archived with data from another table.

2. The DBA denormalized some of the data in the Premiere Products database to improve performance, and one of the resulting tables is the following:

```
Customer (CustomerNum, CustomerName, Street, City,
    State, Zip, Balance, CreditLimit, RepNum, RepName)
```

Which field or fields cause the table to no longer be in third normal form? In which normal form is the denormalized table?

3. Does your school have a formal disaster recovery plan? If it does, describe the general steps in the plan. If it does not, describe the informal steps that would be taken if a disaster occurred.

Henry Books Case

Ray Henry wants you to complete the following exercises. You do not use the Henry Books database for any of these exercises.

1. The DBA asks for your help in planning the data archive for the following Henry Books database:

```
Branch (BranchNum, BranchName, BranchLocation, NumEmployees)
Publisher (PublisherCode, PublisherName, City)
Author (AuthorNum, AuthorLast, AuthorFirst)
Book (BookCode, Title, PublisherCode, Type, Price, Paperback)
Wrote (BookCode, AuthorNum, Sequence)
Inventory (BookCode, BranchNum, OnHand)
```

Determine which data from the database to archive; that is, for each table, specify whether data needs to be archived. If it does, specify which data, when it should be archived, and whether it should be archived with data from another table.

2. The DBA denormalized some of the data in the Henry Books database to improve performance, and one of the resulting tables is the following:

    Wrote (BookCode, AuthorNum, Sequence, PublisherCode, PublisherName)

    Which field or fields cause the table to no longer be in third normal form? In which normal form is the denormalized table?

3. Interview the DBA at your school or at a local business to determine the safeguards used to segregate the production system from the test system.

Alexamara Marina Group Case

For the following exercises, you do not use the Alexamara database.

1. The DBA asks for your help in planning the data archive for the following Alexamara database:

    Marina (MarinaNum, MarinaName, Address, City, State, Zip)
    Owner (OwnerNum, LastName, FirstName, Address, City, State, Zip)
    MarinaSlip (SlipID, MarinaNum, SlipNum, Length,
        RentalFee, BoatName, BoatType, OwnerNum)
    ServiceCategory (CategoryNum, CategoryDescription)
    ServiceRequest (ServiceID, SlipID, CategoryNum,
        Description, Status, EstHours, SpentHours,
        NextServiceDate)

    Determine which data from the database to archive; that is, for each table, specify whether data needs to be archived. If it does, specify which data to archive, when it should be archived, and whether it should be archived with data from another table.

2. The DBA denormalized some of the data in the Alexamara database to improve performance, and one of the resulting tables is the following:

    MarinaSlip (SlipID, MarinaNum, MarinaName, SlipNum,
        Length, RentalFee, BoatName, BoatType, OwnerNum,
        LastName, FirstName)

    Which field or fields cause the table to no longer be in third normal form? In which normal form is the denormalized table?

3. Interview the DBA at your school or at a local business to determine the security and access privilege procedures used to safeguard data.

CHAPTER 9

DATABASE MANAGEMENT APPROACHES

LEARNING OBJECTIVES

Objectives

- Describe distributed database management systems (DDBMSs)
- Discuss client/server systems
- Examine the ways databases are accessed on the Web
- Discuss XML and related document specification standards
- Define data warehouses and explain their structure and access
- Discuss the general concepts of object-oriented DBMSs

INTRODUCTION

In previous chapters, you learned about relational DBMSs (RDBMSs), which dominate the database market today. In this chapter, you will examine several database management topics, most of which are applicable to relational systems.

The centralized approach to processing data, in which users access a central computer through personal computers (PCs) and workstations, dominated organizations from the late 1960s through the mid-1980s because there was no alternative approach to compete with it. The introduction of reasonably priced PCs during the 1980s, however, facilitated the placement of computers at various locations within an organization; users could access a database directly at those locations. Networks connected these computers, so users could access not only data located on their local computers but also data located anywhere along the entire network. In the next section, you will study the issues involved in distributed databases where a database is stored on more than one computer.

Organizations often off-load, or shift, data communications functions from central computers to smaller computers to improve processing speed. Similarly, organizations often use client/server systems to off-load database access functions from central computers to other computers; you'll study these client/server systems. You will also learn about accessing databases on the Web and the growing importance of XML and related document standard specifications. Then you'll learn about special database systems, called data warehouses, that allow you to retrieve data rapidly. Finally, you will study object-oriented systems, which treat data as objects and include the actions that operate on the objects.

DISTRIBUTED DATABASES

Premiere Products has multiple locations nationwide. Each location has its own sales reps and customer base, and each location maintains its own inventory. Instead of using a single centralized computer accessed by all the separate locations, Premiere Products is considering installing a computer at each site. If it does so, each site would maintain its own data about its sales reps, customers, parts, and orders. Occasionally, an order at one site might involve parts from another site. In addition, a customer serviced at one site might place orders for its subsidiaries that are located closer to other sites. Consequently, the computer at a particular site would need to communicate with the computers at all the other sites. The computers would have to be connected in a **communications network**, or **network**, as illustrated in Figure 9-1.

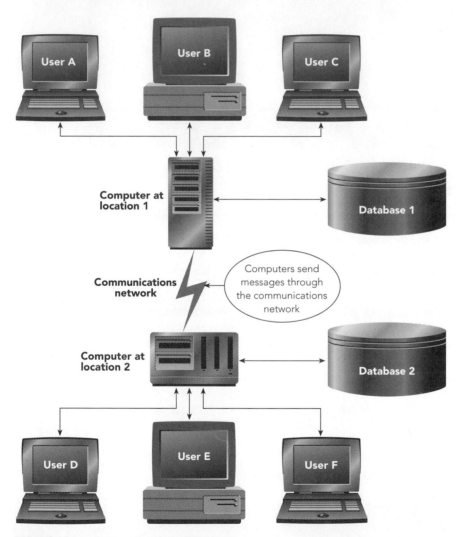

FIGURE 9-1   Communications network

Premiere Products would also divide its existing database and distribute to each site the data needed at that site. In doing so, Premiere Products would be creating a distributed database. A **distributed database** is a single logical database that is physically divided among computers at several sites on a network. To make such a distributed database work properly, Premiere Products needs to purchase a **distributed database management system (DDBMS)**, which is a DBMS capable of supporting and manipulating distributed databases.

Computers in a network communicate through **messages**; that is, one computer sends a message to another. The word *message* is used in a fairly broad way here. A computer might send a message to request data from another computer, or a computer might send a message to indicate a problem. For example, one computer might send a message to another computer to indicate that the requested data is not available. Finally, a computer might send the requested data as a message to another computer.

Accessing data using messages over a network is substantially slower than accessing data on a disk. To access data rapidly in a centralized database, you make design decisions to minimize the number of disk accesses. In general, to access data rapidly in a distributed database, you must attempt to minimize the number of messages. The length of time required to send one message depends on the length of the message and the characteristics of the network. A fixed amount of time, sometimes called the **access delay**, is required for every message. In addition, the time to send a message includes the time it takes to transmit all the characters in the message. The formula for message transmission time is as follows:

```
Communication time = access delay + (data volume / transmission rate)
```

To illustrate the importance of minimizing the number of messages, suppose you have a network with an access delay of 2 seconds and a transmission rate of 750,000 bits per second. Also suppose you send a message that consists of 10,000 records, each of which is 800 bits long. (The 10,000 records is equivalent to approximately 250 pages of single-spaced text.) In this example, you calculate the communication time as follows:

```
Communication time = 2 + ((10,000 * 800) / 750,000)
                   = 2 + (8,000,000 / 750,000)
                   = 2 + 10.67
                   = 12.67 seconds
```

If you send a message that is 100 bits long, your communication time calculation is as follows:

```
Communication time = 2 + (100 / 750,000)
                   = 2 + .0001
                   = 2.0001 seconds or, for practical purposes,
                   = 2 seconds
```

As you can see, in short messages, the access delay becomes the dominant factor. Thus, in general, it is preferable to send a small number of lengthy messages rather than a large number of short messages.

CHARACTERISTICS OF DISTRIBUTED DBMSs

Because a DDBMS effectively contains a local DBMS at each site, an important property of DDBMSs is that they are either homogeneous or heterogeneous. A **homogeneous DDBMS** is one that has the same local DBMS at each site. A **heterogeneous DDBMS** is one that does not; there are at least two sites at which the local DBMSs are different. Heterogeneous DDBMSs are more complex than homogeneous DDBMSs and, consequently, have more problems and are more difficult to manage.

All DDBMSs share several important characteristics. Among these characteristics are location transparency, replication transparency, and fragmentation transparency.

Location Transparency

The definition of a distributed database says nothing about the *ease* with which users access data that is stored at other sites. Systems that support distributed databases should let a user access data at a **remote site**—a site other than the one at which the user is located—just as easily as the user accesses data from the **local site**—the site at which the user is located. Response times for accessing data stored at a remote site might be much slower; but except for this difference, a user should feel as though the entire database is stored at the local site. **Location transparency** is the characteristic of a DDBMS that users do not need to be aware of the location of data in a distributed database.

Replication Transparency

As described in Chapter 7, replication lets users at different sites use and update copies of a database and then share their updates with other users. However, data replication creates update problems that can lead to data inconsistencies. If you update the record of a single part at Premiere Products, the DDBMS must make the update at every location at which data concerning this part is stored. Not only do multiple updates make the process more time-consuming and complicated, but should one of the copies of data for this part be overlooked, the database would contain inconsistent data. Ideally, the DDBMS should correctly handle the updating of replicated data. The steps taken by the DDBMS to update the various copies of data should be done behind the scenes; users should be unaware of the steps. This DDBMS characteristic is called **replication transparency**.

Fragmentation Transparency

A DDBMS supports **data fragmentation** if the DDBMS can divide and manage a logical object, such as the records in a table, among the various locations under its control. The main purpose of data fragmentation is to place data at the location where the data is most often accessed.

Suppose Premiere Products has a local DBMS at each of its three warehouses and wants to fragment its Part table data, which is shown in Figure 9-2, by placing the data for the parts stored in a warehouse in that warehouse's local database.

Part

PartNum	Description	OnHand	Class	Warehouse	Price
AT94	Iron	50	HW	3	$24.95
BV06	Home Gym	45	SG	2	$794.95
CD52	Microwave Oven	32	AP	1	$165.00
DL71	Cordless Drill	21	HW	3	$129.95
DR93	Gas Range	8	AP	2	$495.00
DW11	Washer	12	AP	3	$399.99
FD21	Stand Mixer	22	HW	3	$159.95
KL62	Dryer	12	AP	1	$349.95
KT03	Dishwasher	8	AP	3	$595.00
KV29	Treadmill	9	SG	2	$1,390.00

FIGURE 9-2   Premiere Products Part table data

Using SQL-type statements, you can define the following fragments:

```
DEFINE FRAGMENT Part1 AS
SELECT PartNum, Description, OnHand, Class, Warehouse, Price
FROM Part
WHERE Warehouse='1'

DEFINE FRAGMENT Part2 AS
SELECT PartNum, Description, OnHand, Class, Warehouse, Price
FROM Part
WHERE Warehouse='2'

DEFINE FRAGMENT Part3 AS
SELECT PartNum, Description, OnHand, Class, Warehouse, Price
FROM Part
WHERE Warehouse='3'
```

Each fragment definition indicates which Part table data to select for the fragment. Note that the Part table does not actually exist in any one place. Rather, the Part table exists in three pieces. You assign these pieces, or fragments, to the databases located at the warehouses, as shown in Figure 9-3.

Fragment Part1

PartNum	Description	OnHand	Class	Warehouse	Price
CD52	Microwave Oven	32	AP	1	$165.00
KL62	Dryer	12	AP	1	$349.95

Fragment Part2

PartNum	Description	OnHand	Class	Warehouse	Price
BV06	Home Gym	45	SG	2	$794.95
DR93	Gas Range	8	AP	2	$495.00
KV29	Treadmill	9	SG	2	$1,390.00

Fragment Part3

PartNum	Description	OnHand	Class	Warehouse	Price
AT94	Iron	50	HW	3	$24.95
DL71	Cordless Drill	21	HW	3	$129.95
DW11	Washer	12	AP	3	$399.99
FD21	Stand Mixer	22	HW	3	$159.95
KT03	Dishwasher	8	AP	3	$595.00

FIGURE 9-3   Fragmentation of Part table data by warehouse

You assign Fragment Part1 to the database at warehouse 1, Fragment Part2 to the database at warehouse 2, and Fragment Part3 to the database at warehouse 3. The effect of these assignments is that data about each part is stored in the database at the warehouse where the part is stored. You can access the complete Part table by taking the union of the three fragments.

Users should not be aware of the fragmentation—they should feel as if they are using a single central database. When users are unaware of fragmentation, the DDBMS has **fragmentation transparency**.

ADVANTAGES OF DISTRIBUTED DATABASES

When compared with a single centralized database, distributed databases offer the following advantages:

- *Local control of data.* Because each location retains its own data, a location can exercise greater control over that data. With a single centralized database, on the other hand, the central site that maintains the database is usually unaware of all the local issues at the various sites served by the database.
- *Increasing database capacity.* In a properly designed and installed distributed database, the process of increasing system capacity is often simpler than in a centralized database. If the size of the disk at a single site becomes inadequate for its database, you need to increase the capacity of the disk only at that site. Furthermore, you can increase the capacity of the entire database simply by adding a new site.
- *System availability.* When a centralized database becomes unavailable for any reason, *no* users can continue processing. In contrast, if one of the local databases in a distributed database becomes unavailable, only users who need data in that particular database are affected; other users can continue processing in a normal fashion. In addition, if the data has been replicated (another copy of it exists in other local databases), potentially all users can continue processing. However, processing for users at the site of the unavailable database will be much less efficient because data that was formerly obtained locally must now be obtained through communication with a remote site.
- *Improved performance.* When data is available locally, you eliminate network communication delays and can retrieve data faster than with a remote centralized database.

DISADVANTAGES OF DISTRIBUTED DATABASES

Distributed databases have the following disadvantages:

- **Update of replicated data.** Replicating data can improve processing speed and ensure that the overall system remains available even when the database at one site is unavailable. However, replication can cause update problems, most obviously in terms of the extra time needed to update all the copies. Instead of updating a single copy of the data, the DBMS must update several copies. Because most of these copies are at sites other than the site initiating the update, each update transaction requires extra time to update each copy and extra time to communicate all the update messages over the network.

    Replicated data causes another, slightly more serious problem. Assume an update transaction must update data that is replicated at five sites and that the fifth site is currently unavailable. If all updates must be made or none at all, the update transaction fails. Because the data at a single site is unavailable for update, that data is unavailable for update at *all* sites. This situation certainly contradicts the earlier advantage of increased system availability. On the other hand, if you do not require that all updates be made, the data will be inconsistent.

    Often a DDBMS uses a compromise strategy. The DDBMS designates one copy of the data to be the **primary copy**. As long as the primary copy is updated, the DDBMS considers the update to be complete. The primary site and the DDBMS must ensure that all the other copies are in sync. The primary site sends update transactions to the other sites and notes whether any sites are currently unavailable. If a site is unavailable, the primary site must try to send the update again at some later time and continue trying until it succeeds. This strategy overcomes the basic problem, but it obviously uses more time. Further, if the primary site is unavailable, the problem remains unresolved.

- **More complex query processing.** Processing queries is more complex in a distributed database. The complexity occurs due to the difference in the time it takes to send messages between sites and the time it takes to access a disk. As discussed earlier, minimizing message traffic is extremely important in a distributed database environment. To illustrate the complexity involved with query processing, consider the following query for Premiere Products: List all parts in item class SG with a price that is more than $500.00. For this query, assume (1) the Part table contains 1,000 rows and is stored at a remote site; (2) each record in the Part table is 500 bits long; (3) there is no special structure, such as an index, that would be helpful in processing this query faster; and (4) only 10 of the 1,000 rows in the Part table satisfy the conditions. How would you process this query?

    One query strategy involves retrieving each row from the remote site and examining the item class and price to determine whether the row should be included in the result. For each row, this solution requires two messages. The first is a message from the local site to the remote site requesting a row. It is followed by the second message, which is from the remote site to the local site containing the data or, ultimately, an indication that there is no more data because you have retrieved every row in the table. Thus, in addition to the database accesses, this strategy requires 2,000 messages. Once again, suppose you have a network with an access delay of 2 seconds and a transmission rate of 750,000 bits per second. Based on the calculations for communication time earlier in this chapter, each message requires approximately 2 seconds. You calculate the communication time for this query strategy as follows:

```
Communication time = 2 * 2,000
                   = 4,000 seconds, or 66.7 minutes
```

    A second query strategy involves sending a single message from the local site to the remote site requesting the complete answer to the query. The remote site examines each row in the table and finds the 10 rows that satisfy the query. The remote site then sends a single message back to the local site containing all 10 rows in the answer. You calculate the communication time for this query strategy as follows:

```
Communication time = 2 + (2 + ((10 * 500) / 750,000))
                   = 2 + (2 + (5000 / 750,000)
                   = 2 + (2 + 0.006)
                   = 4.006 seconds
```

Even if the second message is lengthy, especially where many rows satisfied the conditions, this second query strategy is a vast improvement over the first strategy. A small number of lengthy messages is preferable to a large number of short messages.

Systems that are record-at-a-time-oriented can create severe performance problems in distributed systems. If the only choice is to transmit every record from one site to another site as a message and then examine it at the other site, the communication time required can become unacceptably high. DDBMSs that permit a request for a set of records, as opposed to an individual record, outperform record-at-a-time systems.

- **More complex treatment of concurrent update.** Concurrent update in a distributed database is treated basically the same way it is treated in nondistributed databases. A user transaction acquires locks, and the locking is two-phase. (Locks are acquired in a growing phase, during which time no locks are released and the DDBMS applies the updates. All locks are released during the shrinking phase.) The DDBMS detects and breaks deadlocks, and then the DDBMS rolls back interrupted transactions. The primary distinction lies not in the kinds of activities that take place, but in the additional level of complexity created by the very nature of a distributed database.

  If all the records to be updated by a particular transaction occur at one site, the problem is essentially the same as in a nondistributed database. However, the records in a distributed database might be stored at many different sites. Furthermore, if the data is replicated, each occurrence might be stored at several sites, each requiring the same update to be performed. Assuming each record occurrence has replicas at three different sites, an update that would affect five record occurrences in a nondistributed system might affect 20 different record occurrences in a distributed system (each record occurrence together with its three replica occurrences).

  Having more record occurrences to update is only part of the problem. Assuming each site keeps its own locks, the DDBMS must send many messages for each record to be updated: a request for a lock; a message indicating that the record is already locked by another user or that the lock has been granted; a message directing that the update be performed; an acknowledgment of the update; and, finally, a message indicating that the record is to be unlocked. Because all those messages must be sent for each record and its occurrences, the total time for an update can be substantially longer in a distributed database.

  A partial solution to minimize the number of messages involves the use of the primary copy mentioned earlier. Recall that one of the replicas of a given record occurrence is designated as the primary copy. Locking the primary copy, rather than all copies, is sufficient and will reduce the number of messages required to lock and unlock records. The number of messages might still be large, however; and the unavailability of the primary copy can cause an entire transaction to fail. Thus, even this partial solution presents problems.

  Just as in a nondistributed database, deadlock is a possibility in a distributed database. In a distributed database, however, deadlock is more complicated because two types of deadlock, local deadlock and global deadlock, are possible. **Local deadlock** is deadlock that occurs at a single site in a distributed database. If each of two transactions is waiting for a record held by the other at the same site, the local DBMS can detect and resolve the deadlock with a minimum number of messages needed to communicate the situation to the other DBMSs in the distributed system.

  On the other hand, **global deadlock** involves one transaction that requires a record held by a second transaction at one site, while the second transaction requires a record held by the first transaction at a different site. In this case, neither site has information individually to allow this deadlock to be detected; this is a global deadlock, and it can be detected and resolved only by sending a large number of messages between the DBMSs at the two sites.

  The various factors involved in supporting concurrent update greatly add to the complexity and the communications time in a distributed database.

- **More complex recovery measures.** Although the basic recovery process for a distributed database is the same as the one described in Chapter 7, there is an additional potential problem. To make sure that the database remains consistent, each database update should be made permanent or aborted and undone, in which case, *none* of its changes will be made. In a distributed database, with an individual transaction updating several local databases, it is possible—due to problems affecting individual sites—for local DBMSs to commit the updates at some sites and undo the updates at other sites, thereby creating an inconsistent state in the distributed database. The DDBMS *must not* allow this inconsistency to occur.

A DDBMS usually prevents this potential inconsistency through the use of **two-phase commit**. The basic idea of two-phase commit is that one site, often the site initiating the update, acts as **coordinator**. In the first phase, the coordinator sends messages to all other sites requesting that they prepare to update the database; in other words, each site acquires all necessary locks. The sites do not update at this point, however; but they do send messages to the coordinator that they are ready to update. If for any reason any site cannot secure the necessary locks or if any site must abort its updates, the site sends a message to the coordinator that all sites must abort the transaction. The coordinator waits for replies from all sites involved before determining whether to commit the update. If all replies are positive, the coordinator sends a message to each site to commit the update. At this point, each site *must* proceed with the commit process. If any reply is negative, the coordinator sends a message to each site to abort the update; and each site *must* follow this instruction. In this way, the DDBMS guarantees consistency.

While a process similar to two-phase commit is essential to the consistency of the database, two problems are associated with it. For one thing, many messages are sent during the process. For another, during the second phase, each site must follow the instructions from the coordinator; otherwise, the process will not accomplish its intended result. This process means that the sites are not as independent as you would like them to be.

- *More difficult management of the data dictionary.* A distributed database introduces further complexity to the management of the data dictionary or catalog. Where should the data dictionary entries be stored? The three possibilities are as follows: choose one site and store the complete data dictionary at this site and this site alone; store a complete copy of the data dictionary at each site; and distribute, possibly with replication, the data dictionary entries among the various sites.

  Although storing the complete data dictionary at a single site is a relatively simple approach to administer, retrieving information in the data dictionary from any other site is more time-consuming because of the communication involved. Storing a complete copy of the data dictionary at every site solves the retrieval problem because a local DBMS can handle any retrieval locally. Because this second approach involves total replication (every data dictionary occurrence is replicated at every site), updates to the data dictionary are more time-consuming. If the data dictionary is updated with any frequency, the extra time needed to update all copies of the data dictionary might be unacceptable. Thus, you usually implement an intermediate strategy.

  One intermediate strategy is to partition the data by storing data dictionary entries at the site at which the data they describe are located. Interestingly, this approach also suffers from a problem. If a user queries the data dictionary to access an entry not stored at the user's site, the system has no way of knowing the entry's location. Satisfying this user's query might involve sending a message to every other site, which involves a considerable amount of network and DDBMS overhead.

- *More complex database design.* A distributed database adds another level of complexity to database design. Distributing data does not affect the information-level design. During the physical-level design in a nondistributed database, disk activity—both the number of disk accesses and the volumes of data to be transported—is one of the principal concerns. Although disk activity is also a factor in a distributed database, communication activity becomes another concern during the physical-level design. Because transmitting data from one site to another is much slower than transferring data to and from disk, in many situations, communication activity is the most important physical-level design factor. In addition, you must consider possible fragmentation and replication during the physical-level design.

- *More complicated security and backup requirements.* With a single central database, you need to secure the central physical site, the central database, and the network connecting users to the database at the central site. The security requirements for a distributed database are more demanding, requiring you to secure every physical site and every database, in addition to securing the network. Backing up a distributed database is also more complicated and is best initiated and controlled from a single site.

RULES FOR DISTRIBUTED DATABASES

C. J. Date (Date, C. J. "Twelve Rules for a Distributed Database." *ComputerWorld* 21.23, June 8, 1987) formulated 12 rules that distributed databases should follow. The basic goal is that a distributed database should feel like a nondistributed database to users; that is, users should not be aware that the database is distributed. The 12 rules serve as a benchmark against which you can measure DDBMSs. The 12 rules are as follows:

1. *Local autonomy.* No site should depend on another site to perform its database functions.
2. *No reliance on a central site.* The DDBMS should not rely on a single central site to control specific types of operations. These operations include data dictionary management, query processing, update management, database recovery, and concurrent update.
3. *Continuous operation.* Performing functions such as adding sites, changing versions of DBMSs, creating backups, and modifying hardware should not require planned shutdowns of the entire distributed database.
4. *Location transparency.* Users should not be concerned with the location of any specific data in the database. Users should feel as if the entire database is stored at their location.
5. *Fragmentation transparency.* Users should not be aware of any data fragmentation that has occurred in the database. Users should feel as if they are using a single central database.
6. *Replication transparency.* Users should not be aware of any data replication. The DDBMS should perform all the work required to keep the replicas consistent; users should be unaware of the data synchronization work carried out by the DDBMS.
7. *Distributed query processing.* You already learned about the complexities of query processing in a distributed database. The DDBMS must process queries as rapidly as possible.
8. *Distributed transaction management.* You already learned about the complexities of update management in a distributed database and the need for the two-phase commit strategy. The DDBMS must effectively manage transaction updates at multiple sites.
9. *Hardware independence.* Organizations usually have many different types of hardware, and a DDBMS must be able to run on this hardware. Without this capability, users are restricted to accessing data stored only on similar computers, disks, and so on.
10. *Operating system independence.* Even if an organization uses similar hardware, different operating systems might exist in the organization. For the same reason that it is desirable for a DDBMS to support different types of hardware, a DDBMS must be able to run on different operating systems.
11. *Network independence.* Because different sites within an organization might use different communications networks, a DDBMS must run on different types of networks and not be restricted to a single type of network.
12. *DBMS independence.* Another way of stating this requirement is that a DDBMS should be heterogeneous; that is, a DDBMS must support different local DBMSs. Supporting heterogeneous DBMSs is a difficult task. In practice, each local DBMS must "speak" a common language; this common language most likely is SQL.

CLIENT/SERVER SYSTEMS

Networks often include a file server, as shown in Figure 9-4. The **file server** stores the files required by the users on the network. When users need data from a file or a group of files, they send requests to the file server. The file server then sends the requested file or files to the user's computer; that is, the file server sends entire files, not just the data needed by users. Although this approach works to supply data to users, sending entire files generates a high level of communication activity on the network. Adding users to the network and larger files to the file server adds higher levels of communication activity and eventually causes longer delays in supplying data to users.

Computers connected to a network

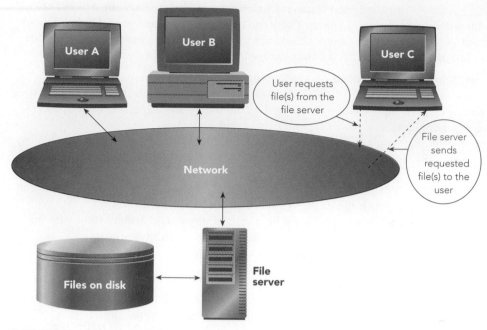

FIGURE 9-4   File server architecture

An alternative architecture, which is called **client/server**, is illustrated in Figure 9-5. In client/server terminology, the **server** is a computer providing data to the **clients**, which are the computers that are connected to a network and that people use to access data stored on the server. A server is also called a **back-end processor** or a **back-end machine**, and a client is also called a **front-end processor** or a **front-end machine**.

Client computers connected to a network

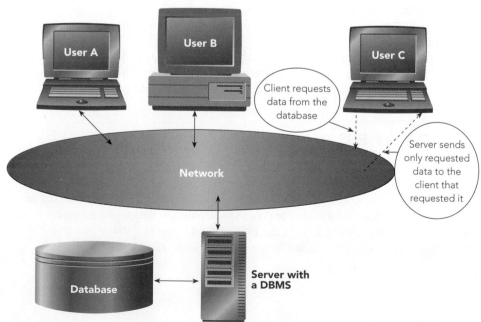

FIGURE 9-5   Two-tier client/server architecture

With this alternative architecture, a DBMS runs on the server. A client sends a request to the server, not for entire *files*, but for specific *data*. The DBMS on the server processes the request, extracts the requested data, and then sends only the requested data back to the client. Compared to a file server architecture, a client/server architecture reduces communication activity on a network, which reduces delays in supplying data to users. Because the clients and the server perform different functions and can run different operating systems, this arrangement of client/server architecture is called a **two-tier architecture**.

In a two-tier architecture, the server performs database functions and the clients perform the presentation functions (or user interface functions), such as determining which form to display on the screen and how to format the form's data. Which of the two tiers, server or clients, performs the business functions, such as the calculations Premiere Products uses to determine commissions, taxes, and order totals? When the clients perform the business functions—each client is called a **fat client** in this arrangement—you have a client maintenance problem. Whenever programmers make changes to the business functions, they must make sure that they place the updated business functions on every client. For organizations with thousands of clients, updating the business functions for all clients is an almost impossible task.

To eliminate the fat client maintenance problem, you can place the business functions on the server. Because clients perform only the presentation functions in this arrangement, each client is called a **thin client**. Although you've now eliminated the fat client maintenance problem by moving the business functions to the server, you've created a scalability problem. **Scalability** is the ability of a computer system to continue to function well as utilization of the system increases. Because the server performs both database and business functions, increasing the number of clients eventually causes a bottleneck on the server and degrades the system's responsiveness to clients. To improve a system's scalability, some organizations use a three-tier client/server architecture, as shown in Figure 9-6. In a **three-tier architecture**, the clients perform the presentation functions, a **database server** performs the database functions, and separate computers (called **application servers**) perform the business functions and serve as an interface between clients and the database server. A three-tier architecture distributes the processing functions so that you eliminate the fat client maintenance problem and maximize the scalability of the system. As the number of users increases, you can upgrade the application and database servers by adding faster processors, disks, and other hardware without changing any client computers. A three-tier architecture is sometimes referred to as an **n-tier architecture** because additional application servers can be added for scalability without impacting the design for the client or the database server.

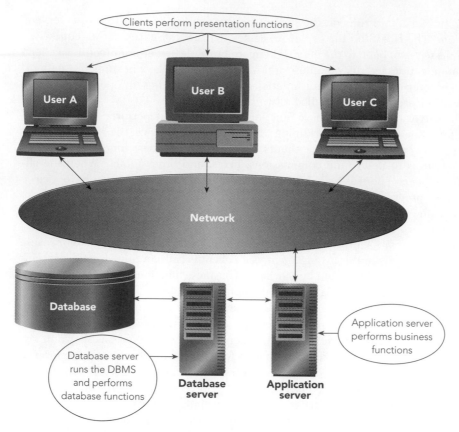

FIGURE 9-6   Three-tier client/server architecture

NOTE

A client/server system stores the database on a single server, and the DBMS resides and processes on that server. Only with a DDBMS is the database itself distributed to multiple computers. However, you can combine a DDBMS with a client/server system to distribute both data and processing functions across multiple computers.

Advantages of Client/Server Systems

Compared to file server systems, a client/server system has the following advantages:

- *Lower network traffic.* A client/server system transmits only the necessary data, rather than entire files, across the network.
- *Improved processing distribution.* A client/server system lets you distribute processing functions among multiple computers.
- *Thinner clients.* Because the application and database servers handle most of the processing in a client/server system, clients do not need to be as powerful or as expensive as they would in a file-server environment.
- *Greater processing transparency.* As far as a user is concerned, all processing occurs on the client just as it does on a stand-alone system. Users do not need to learn any special commands or techniques to work in a client/server environment.
- *Increased network, hardware, and software transparency.* Because client/server systems use SQL as a common language, it is easier for users to access data from a variety of sources. A single operation could access data from different networks, different computers, and different operating systems.
- *Improved security.* Client/server systems can provide a greater level of security than file server systems. In addition to the DBMS security features located on the database server, you can place additional security features on the application servers and on the network.

- *Decreased costs.* Client/server systems have proven to be powerful enough that organizations have replaced, at a considerable cost savings, enterprise applications and mainframe databases with PC applications and databases managed by client/server systems.
- *Increased scalability.* A three-tier client/server system is more scalable than file-server and two-tier architectures. If an application server or database server becomes a bottleneck, you can upgrade the appropriate server or add additional processors to share the processing load.

WEB ACCESS TO DATABASES

The **Internet**, which is a worldwide collection of millions of interconnected computers and computer networks that share resources, is used daily by most people and is an essential portal for all organizations. In particular, people and organizations use the **World Wide Web** (or the **Web**), which is a vast collection of digital documents available on the Internet. Each digital document on the Web is called a **Web page**, each computer on which an individual or organization stores Web pages for access on the Internet is called a **Web server**, and each computer requesting a Web page from a Web server is called a **Web client**. A Web server requires special software to receive and respond to requests for Web pages from Web clients. The dominant Web server software packages are Apache HTTP Server and IIS. **Apache HTTP Server** is a free, open-source package that runs with most operating systems, while **Internet Information Services (IIS)** is a Microsoft package that comes with many versions of its operating systems.

Each Web page is assigned an Internet address called a **Uniform Resource Locator (URL)**; the URL identifies where the Web page is stored—both the location of the Web server and the name and location of the Web page on that server. For example, *http://www.irs.gov/individuals/index.html* is a URL that identifies the Web server (*www.irs.gov*), the location path (*individuals*) on the Web server, and the Web page name (*index.html*). The beginning of the URL (*http*) specifies **Hypertext Transfer Protocol (HTTP)**, which is the data communication method used by Web clients and Web servers to exchange data on the Internet.

You use a computer program called a **Web browser** to retrieve a Web page from a Web client; popular Web browsers include Internet Explorer, Mozilla Firefox, Safari, Netscape, and Opera. As shown in Figure 9-7, a user enters the Web page's URL in a Web browser on a Web client and then sends the request for the Web page over the Internet using HTTP and **Transmission Control Protocol/Internet Protocol (TCP/IP)**, which is the standard protocol for all communication on the Internet. The request for the Web page arrives at the Web server designated in the transmitted URL, and the Web server locates the requested Web page on a disk connected to the Web server and retrieves the Web page. The Web server then responds to the Web client by transmitting the Web page over the Internet using HTTP and TCP/IP, and the Web browser displays the Web page on the user's screen. Note that Web clients on an intranet bypass the Internet and directly access internal company Web pages through the organization's Web server.

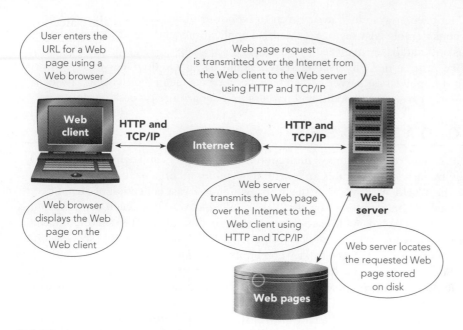

FIGURE 9-7   Retrieving a Web page on the Internet

Each Web page is a text document that contains the necessary codes, called **tags**, that the Web browser interprets to position and format the text in the Web page. A Web page can also contain tags for links to audio files to be played, to graphics and animations to be displayed on the screen, and to other files, which are all sent along with the Web page by the Web server. A Web page can also contain tags for **hyperlinks**, which link one Web page to another or link to another location in the same Web page. Web pages (such as *index.html* in the URL http://www.irs.gov/individuals/index.html) are usually created using a language called **Hypertext Markup Language (HTML)**. You can use a program such as ColdFusion or Adobe Dreamweaver to create the HTML code for Web pages without needing to learn HTML. Many programs, including Microsoft Access, have built-in tools that convert and export objects such as tables and queries to HTML documents.

Web pages that display the same content for all Web clients are called **static Web pages**. At the heart of most Web processing today are activities—such as paying bills, ordering merchandise, buying and selling stocks, and bidding in online auctions—for which the Web pages need to change depending on the Web client's input and responses; these business activities are called **electronic commerce (e-commerce)**. For e-commerce activities, Web servers can't use static Web pages. Instead, Web servers use **dynamic Web pages**, which are pages whose content changes in response to the different inputs and choices made by Web clients. A dynamic Web page includes, or triggers, instructions to tell the Web server how to process the page (**server-side extensions** or **server-side scripts**) and possibly other instructions for the Web browser to process (**client-side extensions** or **client-side scripts**). Client-side extensions can be embedded in HTML documents or contained in separate files that are referenced within the HTML documents, while server-side extensions are usually separately executed programs. Client-side extensions can change the user interface in response to user input actions; JavaScript and VBScript are examples of client-side extension languages. Because of the processing complexities of server-side extensions and the difficulty of creating them, most server-side extensions are created using programming development frameworks, such as ASP.NET and Cold-Fusion, although the PHP scripting language is frequently used with the Apache HTTP Server.

Web servers must have a mechanism for communicating with server-side extensions; Common Gateway Interface (CGI) and Application Program Interface (API) are standard interfaces that provide this capability. In addition, server-side extensions usually include interaction with databases to send Web clients requested data from databases and to update databases with data supplied by Web clients. Several standard software interfaces have been developed to interact with DBMSs; Open Database Connectivity (ODBC), Java Database Connectivity (JDBC), and ADO.NET are examples of these standard interfaces. These standard software interfaces include many DBMS-specific drivers so that a given Web server can work with many different DBMSs.

One common Web-based architecture for dealing with dynamic Web pages, shown in Figure 9-8, uses a three-tier architecture, with the Web clients, a Web server, and a database server as the three tiers. A user on a Web client sends a request for a Web page to the Web server over the Internet using TCP/IP and HTTP. The

Web server receives the request, retrieves the Web page, and then runs server-side extensions associated with the Web page using API. These extensions, among other actions, include instructions for interacting with the database, usually in the form of SQL commands, using API and ODBC in this example. The database server, which contains the DBMS, deals directly with the database and returns the required data back through the ODBC/API interfaces to the Web server. The Web server customizes the HTML document based on the server-side extensions and the data from the database and the Web client; then using TCP/IP and HTTP, the Web server transmits the Web page over the Internet to the Web client. The Web browser displays the Web page on the user's screen, executing any client-side extensions as appropriate. Interaction between the Web client, the Web server, and the database server continues in a similar fashion as the user at the Web client fills in data or chooses options in the delivered Web page and sends follow-up Web page requests to the Web server.

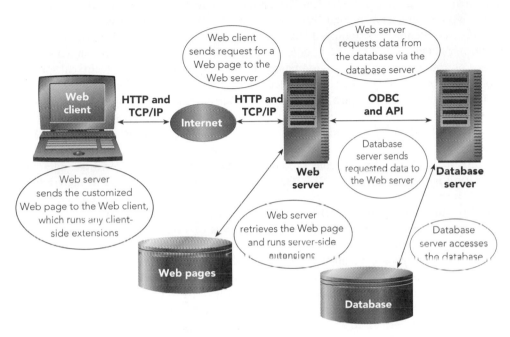

FIGURE 9-8 Three-tier Web-based architecture

A further complication for database processing over the Web is that HTTP is inherently a **stateless** protocol, which means that once the Web server responds to a Web client request for a Web page by delivering the page, the connection between the two is closed and the Web server retains no information about the request or the Web client. The stateless nature of HTTP allows for maximum throughput of Web pages through the Internet. However, the stateless nature of Web pages is at odds with most e-commerce processing. Consider placing an order over the Internet. If you've ever done so, you know that you might view and interact with dozens of Web pages to select the products you want to buy and to place them in a shopping cart. Then you view the shopping cart Web page, making adjustments to the products you are ordering; view another Web page to confirm the order; enter your name and address information in another Web page; enter your credit card information in a different Web page; and finally go through additional Web pages to confirm and place the final order. In this scenario, the vendor's Web server must somehow remember the key data from many different Web pages, even though each delivered Web page is stateless. Organizations use several techniques to remember key data supplied by a Web client. Among the client-side techniques are **cookies** (small files written on a Web client's hard drive by a Web server) and hidden form fields, while server-side solutions usually include storing session information in a database or using other forms of session management, where a **session** is the duration of a Web client's connection to a Web server.

Organizations benefit in many ways from using the Web for database processing. They can transfer data to and from their databases to suppliers, customers, and others outside the company; this provides current information in a timely way to those needing the information. As another example, a company can allow customers to place orders that directly update the organization's database and trigger the processing required to fulfill the orders. Additionally, Web clients can access an organization's Web pages at their convenience 24-7. The tradeoffs for an organization using the Web for database processing include the increased complexities and cost of maintaining an always available Web presence and reliance on the Internet with potential data communication contention difficulties and increased security exposure.

X M L

Many different software languages, software products, computer hardware devices, and standards exist to make e-commerce possible. As e-commerce evolves, these Web components are constantly changing and improving with new components appearing frequently. Since 1994, the international **World Wide Web Consortium (W3C)** has developed Web standards, specifications, guidelines, and recommendations, including HTML standards. HTML is a text-based **markup language**, which means that it contains tags that describe a document's content and appearance. HTML was created and first used in 1990 by Tim Berners-Lee, the founder of the Web, the founder of W3C, and the person who wrote the software for the first Web browser and for the first Web server. As the basis for creating HTML, Berners-Lee used **Standard Generalized Markup Language (SGML)**, which is a **metalanguage** (a language used to define another language) used to create document markup languages; SGML became a standard in 1986. Languages based on the full SGML are used to manage large, complex reports and technical specifications for a variety of computer platforms, printers, and other devices. Berners-Lee borrowed the tagging concepts and some of the tags from SGML for the HTML language, adding a few tags specifically for the processing of Web pages over the Internet.

HTML contains tags that describe the content and appearance of Web pages to Web browsers, but HTML does not describe the structure and meaning of the data it contains. That is, you can't identify in an HTML document which data elements are in the Web page, what each data elements means, and how those data elements are related. This limitation is not a problem for Web pages that are intended to be used in the traditional way, in which a user requests and works with Web pages using a Web browser. However, e-commerce between organizations, called **business to business (B2B)**, is an important part of communication across the Internet. Organizations send data from their databases to the databases of other organizations, and those organizations that send data need to receive data in return. In these situations, the structure and meaning of the transmitted data are of utmost importance because organizations structure common data, such as product data and cost data, in their databases in different ways. Somehow the document containing the data being transmitted between organizations must convey the structure and meaning of the data it contains. To address the inability of HTML to specify the structure and meaning of data and to address the need for the exchange of data between organizations, XML was developed and became a W3C recommendation in 1998.

Extensible Markup Language (XML), a metalanguage derived from a restricted subset of SGML, is designed for the exchange of data on the Web. Using XML, you can create text documents that follow simple, specific rules for their content and you can define new tags that define the data in the document and the structure of the data so that programs running on any platform can interpret and process the document.

Figure 9-9 shows the key portions of a file that was created by exporting the Rep table in the Premiere Products database as an XML document using Access.

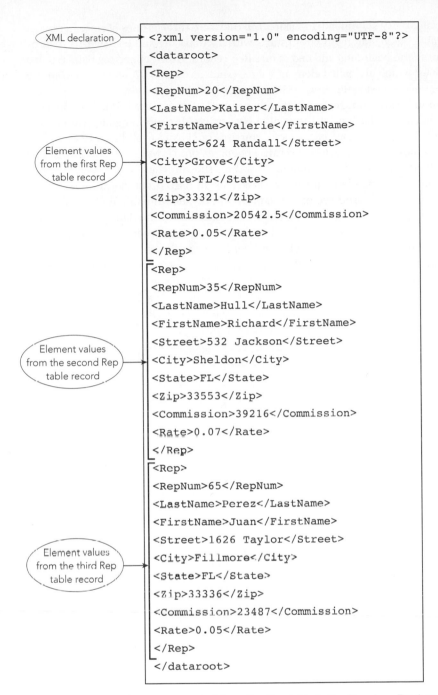

```
<?xml version="1.0" encoding="UTF-8"?>
<dataroot>
<Rep>
<RepNum>20</RepNum>
<LastName>Kaiser</LastName>
<FirstName>Valerie</FirstName>
<Street>624 Randall</Street>
<City>Grove</City>
<State>FL</State>
<Zip>33321</Zip>
<Commission>20542.5</Commission>
<Rate>0.05</Rate>
</Rep>
<Rep>
<RepNum>35</RepNum>
<LastName>Hull</LastName>
<FirstName>Richard</FirstName>
<Street>532 Jackson</Street>
<City>Sheldon</City>
<State>FL</State>
<Zip>33553</Zip>
<Commission>39216</Commission>
<Rate>0.07</Rate>
</Rep>
<Rep>
<RepNum>65</RepNum>
<LastName>Perez</LastName>
<FirstName>Juan</FirstName>
<Street>1626 Taylor</Street>
<City>Fillmore</City>
<State>FL</State>
<Zip>33336</Zip>
<Commission>23487</Commission>
<Rate>0.05</Rate>
</Rep>
</dataroot>
```

XML declaration

Element values from the first Rep table record

Element values from the second Rep table record

Element values from the third Rep table record

FIGURE 9-9 XML document created from the Rep table in the Premiere Products database

An XML document should begin with an **XML declaration** that specifies to an XML processor which version of XML to use. The first line in the XML document shown in Figure 9-9 is the XML declaration:

```
<?xml version="1.0" encoding="UTF-8"?>
```

The XML declaration instructs the XML processor to use version 1.0 of the XML specification. The second, optional clause in the XML declaration (encoding="UTF-8") specifies that the XML document uses Unicode character coding.

Following the XML declaration in Figure 9-9, the <dataroot> tag identifies an element named dataroot, which is a standard element in Office 2007 exported XML documents. The dataroot element serves as a container for all the other elements defined in the XML document, and its matching </dataroot> tag at the end of the document identifies the end of the scope of the dataroot element.

In between the <dataroot> and </dataroot> statements in Figure 9-9, there are three groups of statements, one group for each record from the Rep table. Each statement group starts with a <Rep> tag and ends with a matching closing </Rep> tag; those tags identify the beginning and end of one Rep record. User-defined tag pairs (such as <Rate> and </Rate>) enclose field values, which are called element values (such as 0.05, 0.07, and 0.05) from the Rep records. Each tag must have a matching closing tag in an XML document.

Web pages continue to be written in HTML, but the last W3C recommendation was for HTML 4.01 in 1999. Since then, the W3C has focused on recommendations for **Extensible Hypertext Markup Language (XHTML)**, which is a markup language based on XML and, thus, is a stricter version of HTML. Web browsers continue to support HTML and have been slow to adapt to the XHTML specification. However, as more organizations use XML, more XHTML-based Web pages will be created and used on the Internet.

An XML document contains element tags and element values. How does an XML processor understand the meaning of the tags and the characteristics and structure of the data in an XML document? You use either a Document Type Definition or an XML schema to provide those important facts about the data. A **Document Type Definition (DTD)** specifies the elements (tags), the attributes (characteristics associated with each tag), and the element relationships for an XML document. The DTD can be a separate file with a .dtd extension, or you can include it at the beginning of an XML document. An **XML schema** is a newer form of DTD that more closely matches database features and terminology; you can embed it at the beginning of an XML document or place it in a separate file with an .xsd extension. Figure 9-10 shows the portion of an XML schema specifying the characteristics of the Rate field from the Rep table. Notice how closely the attributes for the Rate element in the XML schema match the properties for the Rate field in the Rep table.

```
<xsd:element name="Rate" minOccurs="0" jetType="double"
        sqlSType="float" type="xsd:double">
<xsd:annotation>
<xsd:appinfo>
<fieldProperty name="ColumnWidth" type="3" value="840"/>
<fieldProperty name="ColumnOrder" type="3" value="0"/>
<fieldProperty name="ColumnHidden" type="1" value="0"/>
<fieldProperty name="DecimalPlaces" type="2" value="255"/>
<fieldProperty name="Required" type="1" value="0"/>
<fieldProperty name="DisplayControl" type="3" value="109"/>
<fieldProperty name="TextAlign" type="2" value="0"/>
<fieldProperty name="AggregateType" type="4" value="-1"/>
<fieldProperty name="GUID" type="9"
        value="CgLbv43o5ECFLODxDEetHA=="/>
</xsd:appinfo>
</xsd:annotation>
</xsd:element>
```

FIGURE 9-10   XML schema for the Rate element from the Rep table

XML documents contain data; and DTDs and XML schemas define the structure, characteristics, and relationships of the data in an XML document. Also, XHTML documents focus on data, not on presentation details. The presentation aspects of an XML or XHTML document can be described by a stylesheet. The **Extensible Stylesheet Language (XSL)** is a standard W3C language for creating stylesheets for XML documents; a **stylesheet** is a document that specifies how to process the data contained in another document and present the data in a Web browser, in a printed report, on a mobile device, in a sound device, or in other presentation media. A related W3C standard language is **XSL Transformations (XSLT)**, which defines the rules to process an XML document and change it into another document; this other document may be another XML document, an XSL document, an HTML or XHTML document, or most any other type of document.

More and more data is being stored, exchanged, and presented using XML; and the W3C has developed recommendations for **XQuery**, which is a language for querying XML, XSL, XHTML, other XML-based documents, and similarly structured data repositories. There is growing interest in XQuery, and several products have been developed based on the XQuery standard.

One example of the inroads made by XML is Microsoft's Office 2007 suite. With this suite, Microsoft has switched from its native file formats to a new file format that it calls Office Open XML for the Excel 2007, PowerPoint 2007, and Word 2007 programs. The **Office Open XML** file format is a compressed version of XML, but you can save each of these files in a more traditional XML-based format.

Figure 9-11 illustrates the interaction between XML and the languages that are closely related to XML. A Web browser can display a Web page by processing an XML document with styles supplied by an XSL document. A Web browser can also display a Web page by processing an HTML or XHTML document with styles supplied by an XSL document; the HTML or XHTML document is created by an XML processor using an XSLT transform on an XML document. Also, a Web client can obtain information from an XML document by using an XQuery processor. Finally, an XML processor can create an XML document from data in a database using a DTD or an XML schema, or the XML processor can update the database using an XML document with a DTD or an XML schema.

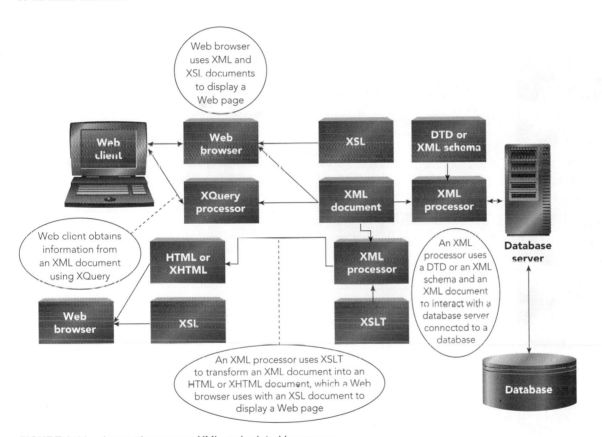

FIGURE 9-11 Interaction among XML and related languages

DATA WAREHOUSES

Among the objectives that organizations have when they use RDBMSs are data integrity, high performance, and ample availability. The leading RDBMSs are able to satisfy these requirements. Typically, when users interact with an RDBMS, they use transactions, such as adding a new order and changing a customer's sales rep. Thus, these types of systems are called **online transaction processing (OLTP)** systems.

For each transaction, OLTP typically deals with a few rows from the tables in a database in a highly structured, repetitive, and predetermined way. If you need to know the status of specific customers, parts, and orders or if you need to update data for specific customers, parts, and orders, an RDBMS and OLTP are the ideal tools to use.

When you need to analyze data from a database, however, an RDBMS and OLTP often suffer from severe performance problems. For example, finding total sales by site and by month requires the joining of all the rows in many tables; such processing takes a considerable number of database accesses and considerable time to accomplish. Consequently, many organizations continue to use RDBMSs and OLTP for their normal day-to-day processing or for *operational purposes*, but the organizations have turned to data warehouses for the *analysis* of their data. The following definition for a data warehouse is credited to W. H. Inmon (Inmon, W. H. *Building the Data Warehouse*. QED, 1990), who originally coined the phrase.

Definition: A **data warehouse** is a subject-oriented, integrated, time-variant, nonvolatile collection of data in support of management's decision-making process.

Subject-oriented means that data is organized by entity rather than by the application that uses the data. For example, Figure 9-12 shows the databases for typical operational applications such as inventory, order entry, production, and accounts payable. When the data from these operational databases is loaded into a data warehouse, it is transformed into subjects such as product, customer, vendor, and financial. Data about products appears once in the warehouse even though it might appear in many files and databases in the operational environment.

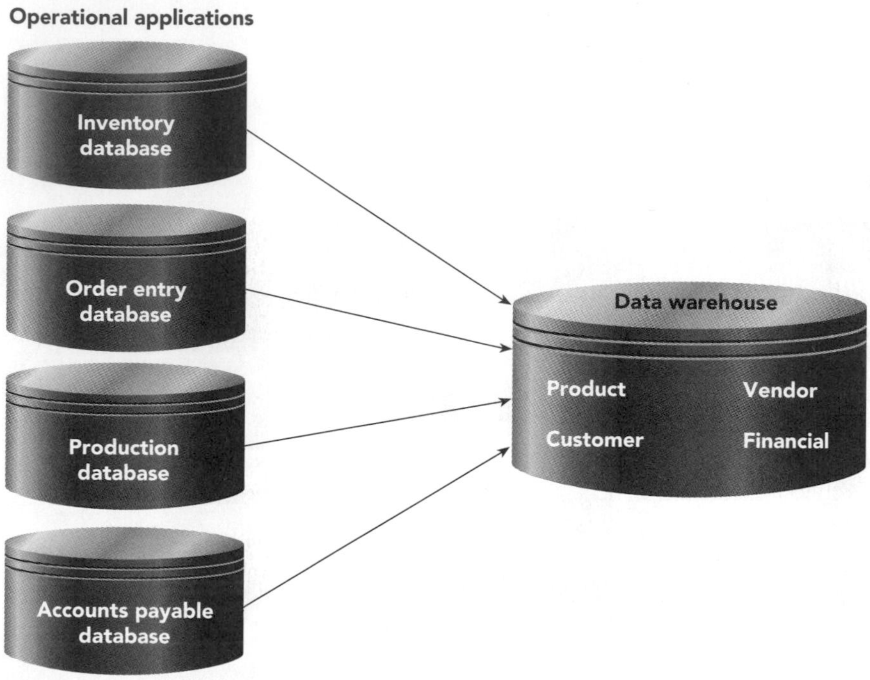

FIGURE 9-12   Data warehouse architecture

NOTE

For the operational applications shown in Figure 9-12, large organizations use a variety of DBMSs and file-processing systems that have been developed over a period of many years.

Integrated means that data is stored in one place in the data warehouse even though the data originates from everywhere in the organization and from a variety of external sources. The data can come from recently developed applications or from legacy systems developed many years ago.

Time-variant means that data in a data warehouse represents snapshots of data at various points in time in the past, such as at the end of each month. This is unlike an operational application, which has data that is accurate as of the moment. Data warehouses also retain historical data for long periods of time; that data is summarized to specific time periods, such as daily, weekly, monthly, and annually.

Nonvolatile means that data is read-only. Data is loaded into a data warehouse periodically, but users cannot update a data warehouse directly.

In summary, a data warehouse contains read-only snapshots of highly consolidated and summarized data from multiple internal and external sources that are refreshed periodically, usually on a daily or weekly basis. Companies use data warehouses in support of their decision-making processing, which typically consists of unstructured and nonrepetitive requests for exactly the type of information contained in a data warehouse.

Data Warehouse Structure and Access

A typical data warehouse structure is shown in Figure 9-13. The central Sales table is called a fact table. A **fact table** consists of rows that contain consolidated and summarized data. The fact table contains a multipart primary key, each part of which is a foreign key to the surrounding dimension tables. Each **dimension table** contains a single-part primary key that serves as an index for the fact table and that contains other fields associated with the primary key value. The overall structure shown in Figure 9-13 is called a **star schema** because of its conceptual shape.

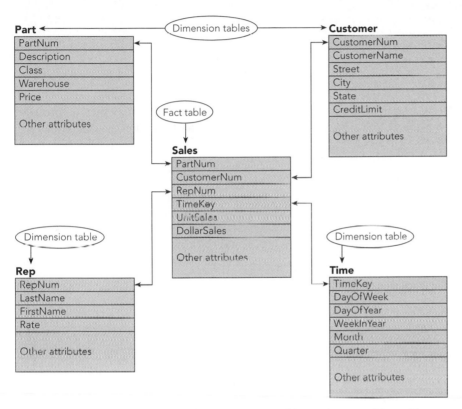

FIGURE 9-13   A star schema with four dimension tables and a central fact table

Access to a data warehouse is accomplished through the use of **online analytical processing (OLAP)** software. OLAP software, whether it's part of the DBMS or a separate product, is optimized to work efficiently with data warehouses.

Users access a data warehouse using OLAP software to answer questions such as the following: How has the average customer balance changed each year over the past five years? What are the total sales by month for this year, and how do they compare to last year's sales?

In posing those types of questions, users perceive the data in a data warehouse as a **multidimensional database**. For example, if users' questions pertain to the Part, Customer, and Time dimensions, which appear in Figure 9-13, they might visualize the data warehouse as a multidimensional database in the shape of a **data cube**, as shown in Figure 9-14. Each axis in the data cube (Part, Customer, and Time) represents data from a dimension table in Figure 9-13, and the cells in the data cube represent unit sales and dollar sales data from the Sales fact table in Figure 9-13.

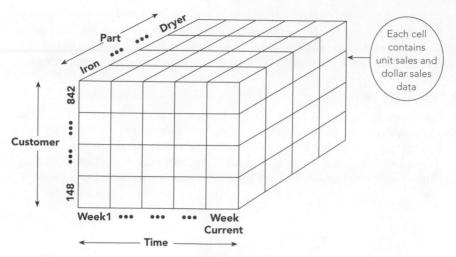

FIGURE 9-14   A data cube representation of the Part, Customer, and Time dimensions

When users access a data warehouse, their queries usually involve aggregate data, such as total sales by month and average sales by customer. As users view the aggregate results from their queries, they often need to perform further analyses of the data they're viewing. OLAP software should let users perform these analyses as easily and quickly as possible.

Users' analyses typically involve actions that include the following:

- *Slice and dice.* Instead of viewing all data in a data cube, users typically view only portions of the data. You **slice and dice** data to select portions of the available data or to reduce the data cube. For example, suppose the Time dimension in the conceptual data cube that appears in Figure 9-14 contains detailed sales data on a weekly basis for Premiere Products. Further, suppose the sales manager queries the data warehouse to view this week's total sales, both in dollars and in units sold, as shown in Figure 9-15.

TotalDollars	TotalUnits
$8,911.14	28

FIGURE 9-15   Total sales query results

Conceptually, the sales manager's query slices the data cube to reduce it to the shaded "Week Current" portion, which is shown in Figure 9-16.

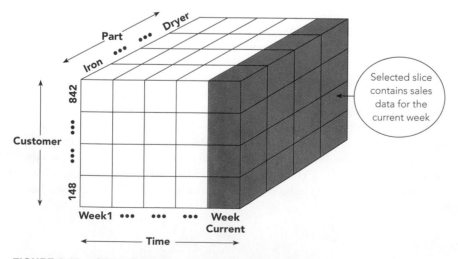

FIGURE 9-16   Slicing the data cube based on the Time dimension

If the sales manager's next query displays this week's total sales for irons, the query dices the sliced data cube, reducing it to the shaded portion shown in Figure 9-17.

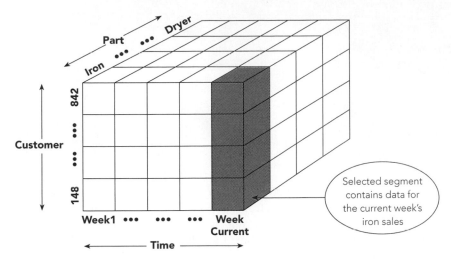

FIGURE 9-17 Dicing the sliced data cube based on the Part dimension

The results for the sales manager's queries for this diced portion of the data cube appear in Figure 9-18.

TotalDollars	TotalUnits
$241.45	11

FIGURE 9-18 Query results for total sales of irons

The sales manager's first query sliced the data cube to focus on the current week's sales, and the second query reduced the slice by dicing only the cells in the data cube that are for irons.

- **Drill down.** When you view specific aggregate data, you **drill down** the data to view and analyze lower levels of aggregation; that is, you go to a more detailed view of the data. For example, suppose again that the sales manager queries the data warehouse to view this week's total sales, as shown in Figure 9-15. To analyze details of these sales, the sales manager might drill down to view total sales by class, as shown in Figure 9-19.

Class	TotalDollars	TotalUnits
AP	$4,499.79	13
HW	$241.45	11
SG	$4,169.90	4

FIGURE 9-19 Query results for total sales by class

Finally, the sales manager might drill down to view total sales by part within class, as shown in Figure 9-20.

Class	Description	TotalDollars	TotalUnits
AP	Dishwasher	$1,190.00	2
AP	Dryer	$1,319.80	4
AP	Gas Range	$990.00	2
AP	Microwave Oven	$600.00	4
AP	Washer	$399.99	1
HW	Iron	$241.45	11
SG	Home Gym	$1,589.90	2
SG	Treadmill	$2,580.00	2

FIGURE 9-20   Query results for total sales by class and part

- *Roll up.* When you view specific aggregate data, you **roll up** the data to view and analyze higher levels of aggregation. Rolling up the data is the exact opposite of drilling down the data. For example, the sales manager might start with the query results for total sales by class and part (see Figure 9-20), click the appropriate button to roll up the data for the query results for the total sales by class (see Figure 9-19), and then click another button to roll up the data for the query results for total sales (see Figure 9-15).

Data mining consists of uncovering new knowledge, patterns, trends, and rules from the data stored in a data warehouse. You use data mining software to answer questions such as the following:

- Which products best attract new customers?
- What factors best predict which customers default in making payments?
- What are the optimal seasonal inventory levels based on predicted economic factors?
- What is the optimal number of customers to assign to each sales rep?

Because data warehouses often contain enormous amounts of data, users can't sift through the data in them to find answers to those questions. Instead, with minimal user interaction, data mining software attempts to answer the questions by using sophisticated analytical, mathematical, and statistical techniques.

Rules for OLAP Systems

E. F. Codd (Codd, E. F., S. B. Codd, and C. T. Salley. "Providing OLAP (On-line Analytical Processing) to User-Analysts: An IT Mandate." Arbor Software, August, 1993) formulated 12 rules that OLAP systems should follow. The 12 rules serve as a benchmark against which you can measure OLAP systems. The 12 rules are as follows:

1. *Multidimensional conceptual view.* Users must be able to view data in a multidimensional way, matching the way data appears naturally in an organization. For example, users can view data about the relationships between data using the dimensions of parts, customer locations, sales reps, and time.
2. *Transparency.* Users should not have to know they are using a multidimensional database nor need to use special software tools to access data. For example, if users usually access data using a spreadsheet, they should still be able to use a spreadsheet to access a multidimensional database.
3. *Accessibility.* Users should perceive data as a single user view even though the data may be physically located in several heterogeneous locations and in different forms, such as relational databases and standard files.
4. *Consistent reporting performance.* Retrieval performance should not degrade as the number of dimensions and the size of the warehouse grow.
5. *Client/server architecture.* The server component of OLAP software must be intelligent enough that a variety of clients can be connected with minimal effort.
6. *Generic dimensionality.* Every dimension table must be equivalent in both its structural and operational capabilities. For example, you should be able to obtain information about parts as easily as you obtain information about sales reps.

7. ***Dynamic sparse matrix handling.*** Missing data should be handled correctly and efficiently and not affect the accuracy or speed of data retrieval.

8. ***Multiuser support.*** OLAP software must provide secure, concurrent retrieval of data. Because you don't update a data warehouse when you're using it, concurrent update is not an issue; so problems of security and access are less difficult than in an OLTP environment.

9. ***Unrestricted, cross-dimensional operations.*** Users must be able to perform the same operations across any number of dimensions. For example, you should be able to ask for statistics based on the dimensions of time, location, and part just as easily as you would ask for statistics based on the single dimension of location.

10. ***Intuitive data manipulation.*** Users should be able to act directly on individual data values without needing to use menus or other interfaces. Of course, these other interfaces can be used, but they should not be the required method of processing.

11. ***Flexible reporting.*** Users should be able to retrieve data results and view them any way they want for analysis.

12. ***Unlimited dimensions and aggregation levels.*** OLAP software should allow at least 15 data dimensions and an unlimited number of aggregation (summary) levels.

OBJECT-ORIENTED DBMSs

Organizations use relational databases to store and access data consisting of text and numbers. Additionally, some organizations store and access graphics, drawings, photographs, video, sound, voice mail, spreadsheets, and other complex objects in their databases. RDBMSs store these complex objects using special data types, generically called **binary large objects (BLOBs)**. Some applications, such as computer-aided design and manufacturing (CAD/CAM) and geographic information systems (GIS), have as their primary focus the storage and management of complex objects. For these systems, many companies use object-oriented DBMSs.

What Is an Object-Oriented DBMS?

The relational model, which has a strong theoretical foundation, is the foundation for RDBMSs. Although object-oriented DBMSs do not have a corresponding theoretical foundation, they all exhibit several common characteristics. Central to all object-oriented systems is the concept of an object. An **object** is a set of related attributes along with the actions that are associated with the set of attributes. A customer object, for example, consists of the attributes associated with customers (number, name, balance, and so on) together with the actions that are associated with customer data (add customer, change credit limit, delete customer, and so on).

In relational systems, you create the actions as part of data manipulation (in the programs that update the database), rather than as part of the data definition. In contrast, in object-oriented systems, you define the actions as part of the data definition and then use the actions whenever they are required. In an object-oriented system, the data and actions are **encapsulated**, which means that you define an object to contain both the data and its associated actions. Thus, an **object-oriented database management system (OODBMS)** is a database management system in which data and the actions that operate on the data are encapsulated into objects.

To become familiar with OODBMSs, you should have a general understanding of the following object-oriented concepts: objects, classes, methods, messages, and inheritance.

Objects and Classes

To understand the distinction between objects and classes, you will examine an object-oriented representation of the following relational model representation of the Premiere Products database.

```
Rep (RepNum, LastName, FirstName, Street, City, State, Zip, Commission, Rate)
Customer (CustomerNum, CustomerName, Street, City, State, Zip, Balance, CreditLimit, RepNum)
Orders (OrderNum, OrderDate, CustomerNum)
OrderLine (OrderNum, PartNum, NumOrdered, QuotedPrice)
Part (PartNum, Description, OnHand, Class, Warehouse, Price, Allocated)
```

This version of the Premiere Products database contains an extra field, Allocated, in the Part table. The Allocated field stores the number of units of a part that are currently on order (allocated). Figure 9-21 shows a representation of this database as a collection of objects.

```
Rep OBJECT
RepNum:                    Sales Rep Numbers
LastName:                  Last Names
FirstName:                 First Names
Street:                    Addresses
City:                      Cities
State:                     States
Zip:                       Zip Codes
Commission:                Commissions
Rate:                      Commission Rates
Customer:                  Customer OBJECT; MV


Customer OBJECT
CustomerNum:               Customer Numbers
CustomerName:              Customer Names
Street:                    Addresses
City:                      Cities
State:                     States
Zip:                       Zip Codes
Balance:                   Balances
CreditLimit:               Credit Limits
Rep:                       Rep OBJECT; SUBSET[RepNum, LastName, FirstName]


Part OBJECT
PartNum:                   Part Numbers
Description:               Part Descriptions
OnHand:                    Units
Class:                     Item Classes
Warehouse:                 Warehouse Numbers
Price:                     Prices
Allocated:                 Units
OrderLine:                 OrderLine OBJECT; MV


Orders OBJECT
OrderNum:                  Order Numbers
OrderDate:                 Dates
Customer:                  Customer OBJECT; SUBSET[CustomerNum, CustomerName, RepNum]
OrderLine:                 OrderLine OBJECT; MV


OrderLine OBJECT
OrderNum:                  Order Numbers
PartNum:                   Part Numbers
NumOrdered:                Units
QuotedPrice:               Prices
```

FIGURE 9-21   Object-oriented representation of the Premiere Products database

NOTE

Figure 9-21 shows just one of many approaches to representing objects. However, all techniques have the same general features.

You'll notice the following differences between the collection of objects in Figure 9-21 and the relational model representation:

- You represent each entity (Rep, Customer, and so on) as an *object* rather than a relation.
- You list the attributes vertically below the object names. In addition, you follow each attribute by the name of the domain associated with the attribute. A **domain** is the set of values permitted for an attribute.

- Objects can contain other objects. For example, the Rep object contains the Customer object as one of its attributes. In the Rep object, the letters *MV* following the Customer object indicate that the Customer object is multivalued. In other words, a single occurrence of the Rep object can contain multiple occurrences of the Customer object. Roughly speaking, this is analogous to a relation containing a repeating group.
- An object can contain a portion of another object. The Customer object, for example, contains the Rep object. The word *SUBSET* indicates, however, that the Customer object contains only a subset of the Rep object. In this case, the Customer object contains three of the Rep object attributes: RepNum, LastName, and FirstName.

Notice that each of two objects can appear to contain the other. The Rep object contains the Customer object, and the Customer object contains the Rep object (or at least a subset of it). The important thing to keep in mind is that users deal with *objects*. If the users of the Customer object require access to the rep's number and name, the rep's number and name are part of the Customer object. If the users of the Rep object require data about all the customers of a sales rep, the Customer object is part of the Rep object. This arrangement is not to imply, of course, that the data is physically stored this way; but this is the way its users perceive the data.

Objects can contain more than one other object. Notice that the Orders object contains the Customer object and the OrderLine object, with the OrderLine object being multivalued. Nevertheless, users of the Orders object perceive it as a single unit.

Technically, the objects in Figure 9-21 are classes. The term **class** refers to the general structure. The term *object* refers to a specific occurrence of a class. Thus, Rep is a class, whereas the data for rep 20 is an object.

Methods and Messages

Methods are the actions defined for a class. Figure 9-22 shows two methods associated with the Orders object. The first method, Add Order, adds an order to the database. In this example, users enter data; then the program places the data temporarily in computer memory in a work area named WOrders. (In this example, the *W* prefix indicates a temporary work area record or field.) The WOrders record consists of a user-entered value for the order number stored in WOrderNum, a user-entered value for the order date stored in WOrderDate, and so on.

```
Add Order (WOrders)
        Add row to Orders table
                OrderNum     = WOrderNum
                OrderDate    = WOrderDate
                CustomerNum = WCustomerNum
        For each order line record in WOrders DO
                Add row to OrderLine table
                        OrderNum    = WOrderNum
                        PartNum     = WPartNum
                        NumOrdered  = WNumOrdered
                        QuotedPrice = WQuotedPrice
                Update Part table (WHERE PartNum = WPartNum)
                        Allocated     = Allocated + WNumOrdered


Delete Order (WOrderNum)
        Delete row from Orders table (WHERE OrderNum = WOrderNum)
        For each OrderLine record (WHERE OrderNum = WOrderNum) DO
                Delete row from OrderLine table
                Update Part table (WHERE Part.PartNum = OrderLine.PartNum)
                        Allocated      = Allocated – NumOrdered
```

FIGURE 9-22   Two methods for the Premiere Products object-oriented database

Q & A

Question: Describe the steps in the Add Order method.
Answer: The steps accomplish the following:

- Add a row to the Orders table for the new order.
- For each order line record associated with the order, add a row to the OrderLine table.
- For each matched order line record, update the Allocated value in the Part table for the corresponding part.

In Figure 9-22, the second method, Delete Order, deletes an order. The only data a user inputs to this method is the order number to be deleted, which is placed temporarily in WOrderNum.

Q & A

Question: Describe the steps in the Delete Order method.
Answer: The steps accomplish the following:

- Delete the order with the user-entered order number (WOrderNum) from the Orders table.
- For each order line record in which the order number matches the value of WOrderNum, delete the record.
- For each matched order line record, subtract the NumOrdered value from the Allocated value for the corresponding part in the Part table. (Because the method deletes the order line record, the parts are no longer allocated.)

NOTE

The two methods in Figure 9-22 are fairly complicated, consisting of several steps with each step involving separate updates to the database. Many methods are much simpler, although some methods are even more complicated.

You define methods during the data definition process. To execute the steps in a method, a user sends a message to the object. A message is a request to execute a method. As part of sending the message to an object, the user sends the required data (for example, full order data for the Add Order method, but only the order number for the Delete Order method). The process is similar to the process of calling a subroutine or invoking a procedure in a standard programming language.

Inheritance

A key feature of object-oriented systems is **inheritance**. For any class, you can define a **subclass**. Every occurrence of the subclass is also considered an occurrence of the class. The subclass *inherits* the structure of the class as well as its methods. In addition, you can define additional attributes and methods for the subclass.

As an example, suppose Premiere Products has a special type of order that has all the characteristics of other orders. In addition, it contains a freight amount and a discount that are calculated in a special way. Rather than create a new class for this type of order, you can define it as a subclass of the Orders class. In that way, the special order type automatically has all the attributes of the Orders class. The new subclass also has all the same methods of the Orders class, including the updating of the Allocated field in the Part table whenever orders are added or deleted. The only thing you would have to add would be those attributes and methods that are specific to this new type of order, thus greatly simplifying the entire process.

Unified Modeling Language (UML)

The **Unified Modeling Language (UML)** is an approach you can use to model all the various aspects of software development for object-oriented systems. UML includes a way to represent database designs.

UML includes several types of diagrams, each with its own special purpose. Figure 9-23 describes the purpose of some of the most commonly used UML diagrams.

Diagram Type	Description
Class	For each class, shows the name, attributes, and methods of the class, as well as the relationships between the classes in the database.
Use Case	Describes how the system is to behave from the standpoint of the system's users.
State	Shows the possible states of an object. (For example, an order could be in the placed, open, filled, or invoiced states.) Also shows the possible transitions between states (for example, placed→open→filled→invoiced).
Sequence	Shows the sequence of possible interactions between objects over time.
Activity	Shows the business and operational step-by-step workflows of components in a system.
Component	Complex software systems are usually subdivided into smaller components. This type of diagram shows these components and their relationships with each other.

FIGURE 9-23 UML diagrams

The type of diagram most relevant to database design is the **class diagram**. Figure 9-24 shows a sample class diagram for the Premiere Products database. A rectangle represents a class. The top portion of a rectangle contains the name of the class, the middle portion contains the attributes, and the bottom portion contains the methods. The lines joining the classes represent the relationships and are called **associations** in UML.

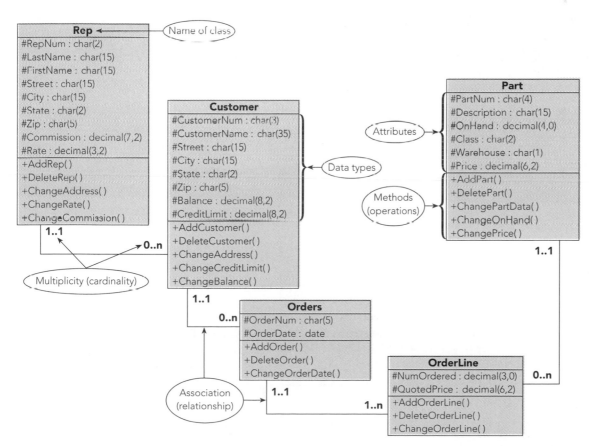

FIGURE 9-24 Class diagram for the Premiere Products database

In a class diagram, a visibility symbol precedes each attribute. The **visibility symbol** indicates whether other classes can view or update the value in the attribute. The possible visibility symbols are public visibility (+), protected visibility (#), and private visibility (-). With **public visibility**, any other class can view or update the value. With **protected visibility**, only the class itself or public or protected subclasses of the class can view or update the value. With **private visibility**, only the class itself can view or update the value. The name of the attribute, a colon, and then the data type for the attribute follow the visibility symbol.

At each end of each association is an expression that represents the multiplicity, or cardinality, of the relationship. **Multiplicity** indicates the number of objects that can be related to an individual object at the other end of the relationship. UML provides various alternatives for representing multiplicity. In the alternative shown in Figure 9-24, two periods separate two symbols. The first symbol represents the minimum number of objects, and the second symbol represents the maximum number of objects. A second number of n indicates that there is no maximum number of objects.

In the association between Customer and Orders, for example, the multiplicity for Customer is 1..1. This multiplicity indicates that an order must correspond to at least one customer and can correspond to, at most, one customer. In other words, an order must correspond to *exactly* one customer. The multiplicity for Orders is 0..n, indicating that a customer can have as few as zero orders (that is, a customer does not have to have any orders currently in the database) and that there is no limit on the number of orders a customer can have. In the association between Orders and OrderLine, the multiplicity for OrderLine is 1..n rather than 0..n. This multiplicity indicates that each order must have *at least one* order line but that the number of order lines is unlimited. If, on the other hand, the multiplicity for OrderLine was 1..5, an order would be required to have anywhere from one to five order lines.

You can also specify constraints, which are restrictions on the data that can be stored in the database. You enter the constraint in the shape shown in Figure 9-25 and then connect the shape to the class to which it applies.

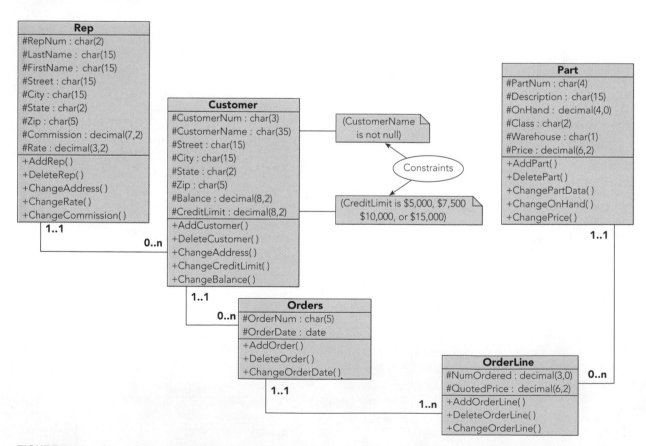

FIGURE 9-25   Class diagram for the Premiere Products database with constraints

You learned about entity subtypes and how to represent them in E-R diagrams. In UML, these entity subtypes are called subclasses. In addition, when one class is a subclass of a second class, you call the second class a **superclass** of the first class. The relationship between a superclass and a subclass is called a **generalization**, which is shown in Figure 9-26. This class diagram represents the relationship between the class of students and the subclass of students who live in dorms.

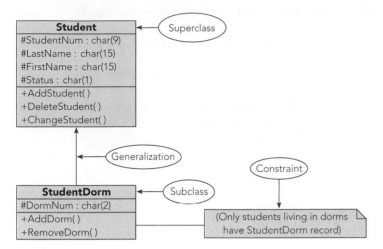

FIGURE 9-26   Class diagram with a generalization and a constraint

Rules for OODBMSs

Just as rules specify desired characteristics for DDBMSs and OLAP, OODBMSs also have a set of rules. These rules serve as a benchmark against which you can measure object-oriented systems. The rules are as follows:

1. *Complex objects.* An OODBMS must support the creation of complex objects from simple objects such as integers and characters.
2. *Object identity.* An OODBMS must provide a way to identify objects; that is, the OODBMS must provide a way to distinguish between one object and another.
3. *Encapsulation.* An OODBMS must encapsulate data and associated methods together in the database.
4. *Information hiding.* An OODBMS must hide from the users of the database the details concerning the way data is stored and the actual implementation of the methods.
5. *Types or classes.* You are already familiar with the idea of a class. Types are very similar to classes and correspond to abstract types in programming languages. The differences between the two are subtle and will not be explored here. It is important to know, however, that an OODBMS supports either abstract types or classes (it doesn't matter which).
6. *Inheritance.* An OODBMS must support inheritance.
7. *Late binding.* In this case, **binding** refers to the association of operations to actual program code. With late binding, this association does not happen until runtime, that is, until some user actually invokes the operation. Late binding lets you use the same name for different operations, which is called **polymorphism** in object-oriented systems. For example, an operation to display an object on the screen requires different program code when the object is a picture than when it is text. With late binding, you can use the same name for both operations. At the time a user invokes this "display" operation, the system determines the object being displayed and then binds the operation to the appropriate program code.
8. *Computational completeness.* You can use functions in the language of the OODBMS to perform various computations.
9. *Extensibility.* Any DBMS, object-oriented or not, comes with a set of predefined data types, such as numeric and character. An OODBMS should be **extensible**, meaning that it is possible to define new data types. Furthermore, the OODBMS should make no distinction between the data types provided by the system and the new data types.

10. *Persistence.* In object-oriented programming, **persistence** refers to the ability to have a program *remember* its data from one execution to the next. Although this is unusual in programming languages, it is common in all database systems. After all, one of the fundamental capabilities of any DBMS is its ability to store data for later use.

11. *Performance.* An OODBMS should have sufficient performance capabilities to manage very large databases effectively.

12. *Concurrent update support.* An OODBMS must support concurrent update. (You learned about concurrent update in Chapter 7.)

13. *Recovery support.* An OODBMS must provide recovery services. (You learned about recovery services in Chapter 7.)

14. *Query facility.* An OODBMS must provide query facilities. (You learned about query facilities such as QBE and SQL in Chapter 2 and Chapter 3, respectively.)

Summary

- A distributed database is a single logical database that is physically divided among computers at several sites on a network. A user at any site can access data at any other site. A DDBMS is a DBMS capable of supporting and manipulating distributed databases.

- Computers in a network communicate through messages. Minimizing the number of messages is important for rapid access to distributed databases.

- A homogenous DDBMS is one that has the same local DBMS at each site, whereas a heterogeneous DDBMS is one that does not.

- Location transparency, replication transparency, and fragmentation transparency are important characteristics of DDBMSs.

- DDBMSs permit local control of data, increased database capacity, improved system availability, and added efficiency.

- DDBMSs are more complicated than non-DDBMSs in the areas of updating replicated data, processing queries, treating concurrent update, providing measures for recovery, managing the data dictionary, designing databases, and managing security and backup requirements.

- The two-phase commit usually uses a coordinator to manage concurrent update.

- C. J. Date presented 12 rules that serve as a benchmark against which you can measure DDBMSs. These rules are local autonomy, no reliance on a central site, continuous operation, location transparency, fragmentation transparency, replication transparency, distributed query processing, distributed transaction management, hardware independence, operating system independence, network independence, and DBMS independence.

- A file server stores the files required by users and sends entire files to the users.

- In a two-tier client/server architecture, a DBMS runs on a file server and the server sends only the requested data to the clients. The server performs database functions, and the clients perform presentation functions. A fat client can perform the business functions, or the server can perform the business functions in a thin client arrangement.

- In a three-tier client/server architecture, the clients perform the presentation functions, database servers perform the database functions, and application servers perform business functions. A three-tier architecture is more scalable than a two-tier architecture.

- The advantages of client/server systems are lower network traffic; improved processing distribution; thinner clients; greater processing transparency; increased network, hardware, and software transparency; improved security; decreased costs; and increased scalability.

- Web servers interact with Web clients using HTTP and TCP/IP to display HTML Web pages on Web clients' screens.

- Dynamic Web pages, not static Web pages, are used in e-commerce; and server-side and client-side extensions provide the dynamic capability, including the capability to interact with databases.

- Cookies and session management techniques are used to counteract the stateless nature of HTTP.

- XML was developed in response to the need for data exchange between organizations and due to the inability of HTML to specify the structure and meaning of its data. XML is a metalanguage designed for the exchange of data on the Web.

- The W3C has developed recommendations for other languages related to XML. These languages include XHTML, a markup language based on XML and a stricter version of HTML; DTD and XML schema, both used to specify the structure and meaning of data in an XML document; XSL, a language for creating stylesheets; XSLT, which transforms an XML document into another document; and XQuery, which is an XML query language.

- OLTP is used with relational database management systems, and OLAP is used with data warehouses.

- A data warehouse is a subject-oriented, integrated, time-variant, nonvolatile collection of data in support of management's decision-making process.

- A typical data warehouse data structure is a star schema consisting of a central fact table surrounded by dimension tables.

- Users perceive the data in a data warehouse as a multidimensional database in the shape of a data cube. OLAP software lets users slice and dice data, drill down data, and roll up data.

- Data mining consists of uncovering new knowledge, patterns, trends, and rules from the data stored in a data warehouse.

- E. F. Codd presented 12 rules that serve as a benchmark against which you can measure OLAP systems. These rules are multidimensional conceptual view; transparency; accessibility; consistent reporting performance; client/server architecture; generic dimensionality; dynamic sparse matrix handling; multiuser support; unrestricted, cross-dimensional operations; intuitive data manipulation; flexible reporting; and unlimited dimensions and aggregation levels.

- Object-oriented DBMSs deal with data as objects. An object is a set of related attributes along with the actions that are associated with the set of attributes. An OODBMS is a database management system in which data and the actions that operate on the data are encapsulated into objects. A domain is the set of values that are permitted for an attribute. The term *class* refers to the general structure, and the term *object* refers to a specific occurrence of a class. Methods are the actions defined for a class, and a message is a request to execute a method. A subclass inherits the structure and methods of its superclass.

- The UML is an approach to model all the various aspects of software development for object-oriented systems. The class diagram represents the design of an object-oriented database. Relationships are called associations, and visibility symbols indicate whether other classes can view or change the value in an attribute. Multiplicity indicates the number of objects that can be related to an individual object at the other end of the relationship. Generalization is the relationship between a superclass and a subclass.

- Rules that serve as a benchmark against which you can measure object-oriented systems are complex objects, object identity, encapsulation, information hiding, types or classes, inheritance, late binding, computational completeness, extensibility, persistence, performance, concurrent update support, recovery support, and query facility.

Key Terms

access delay	distributed database
Apache HTTP Server	distributed database management system (DDBMS)
application server	Document Type Definition (DTD)
association	domain
back-end machine	drill down
back-end processor	dynamic Web page
binary large objects (BLOBs)	electronic commerce (e-commerce)
binding	encapsulated
business to business (B2B)	extensible
class	Extensible Hypertext Markup Language (XHTML)
class diagram	Extensible Markup Language (XML)
client	Extensible Stylesheet Language (XSL)
client/server	fact table
client-side extension	fat client
client-side script	file server
communications network	fragmentation transparency
cookies	front-end machine
coordinator	front-end processor
database server	generalization
data cube	global deadlock
data fragmentation	heterogeneous DDBMS
data mining	homogeneous DDBMS
data warehouse	hyperlink
dimension table	Hypertext Markup Language (HTML)

Hypertext Transfer Protocol (HTTP)

inheritance

Internet

Internet Information Services (IIS)

local deadlock

local site

location transparency

markup language

message

metalanguage

method

multidimensional database

multiplicity

network

n-tier architecture

object

object-oriented database management system (OODBMS)

Office Open XML

online analytical processing (OLAP)

online transaction processing (OLTP)

persistence

polymorphism

primary copy

private visibility

protected visibility

public visibility

remote site

replication transparency

roll up

scalability

server

server-side extension

server-side script

session

slice and dice

Standard Generalized Markup Language (SGML)

star schema

stateless

static Web page

stylesheet

subclass

superclass

tag

thin client

three-tier architecture

Transmission Control Protocol/Internet Protocol (TCP/IP)

two-phase commit

two-tier architecture

Unified Modeling Language (UML)

Uniform Resource Locator (URL)

visibility symbol

Web browser

Web client

Web page

Web server

World Wide Web (Web)

World Wide Web Consortium (W3C)

XML declaration

XML schema

XQuery

XSL Transformations (XSLT)

Review Questions

1. What is a distributed database? What is a DDBMS?

2. What different design decisions do you make to access data rapidly in a centralized database compared to a distributed database?

3. How does a homogeneous DDBMS differ from a heterogeneous DDBMS? Which is more complex?

4. What is meant by a local site? by a remote site?

5. What is location transparency?

6. What is replication? Why is it used? What benefit is derived from using it? What are the biggest potential problems?

7. What is replication transparency?

8. What is data fragmentation? What purpose does data fragmentation serve?

9. What is fragmentation transparency?

10. Why is local control of data an advantage in a distributed database?

11. Why is the ability to increase system capacity an advantage in a distributed database?

12. Why is system availability an advantage in a distributed database?

13. Why is increased efficiency an advantage in a distributed database?

14. What are two disadvantages of updating replicated data in a distributed database?

15. What causes query processing to be more complex in a distributed database?

16. What is meant by local deadlock? by global deadlock?

17. Describe the two-phase commit process. How does it work? Why is it necessary?

18. Describe three possible approaches to storing data dictionary entries in a distributed system.

19. What additional factors must you consider during the information-level design of a distributed database?

20. What additional factors must you consider during the physical-level design of a distributed database?

21. List and briefly describe the 12 rules against which you can measure DDBMSs.

22. What is the difference between a file server and a client/server system?

23. In a two-tier client/server architecture, what problems occur when you place the business functions on the clients? on the server?

24. What is a fat client? What is a thin client?

25. What is scalability?

26. What is a three-tier architecture?

27. List the advantages of a client/server architecture as compared to a file server.

28. What are HTTP and TCP/IP?

29. What are dynamic Web pages? How can you augment HTML to provide the dynamic capability?

30. Explain why HTTP is a stateless protocol and what types of techniques are used in e-commerce to deal with this complication.

31. What is XML? Why was it developed?

32. What is the purpose of DTDs and XML schemas?

33. What does XSLT accomplish?

34. What are the characteristics of OLTP systems?

35. What is a data warehouse?

36. What does it mean when a data warehouse is nonvolatile?

37. What is a fact table in a data warehouse?

38. When do you use OLAP?

39. What three types of actions do users typically perform when they use OLAP software?

40. What is data mining?

41. What are the 12 rules against which you can measure OLAP systems?

42. What is an OODBMS?

43. What is a domain?

44. How do classes relate to objects?

45. What is a method? What is a message? How do messages relate to methods?

46. What is inheritance? What are the benefits to inheritance?

47. What is the UML?

48. What are relationships called in UML?

49. What is a visibility symbol in UML?

50. What is multiplicity?

51. What is generalization?

52. What are the 14 rules against which you can measure object-oriented systems?

Premiere Products Exercises

For the following exercises, you will answer problems and questions from management at Premiere Products. You do not use the Premiere Products database for any of these exercises.

1. Fragment the Customer table so that customers of rep 20 form a fragment named CustomerRep20, customers of rep 35 form a fragment named CustomerRep35, and customers of rep 65 form a fragment named CustomerRep65. (Include all fields from the Customer table in each fragment.) In addition, you need to fragment the Orders table so that orders are distributed and stored with the customers that placed the orders. For example, fragment OrdersRep20 consists of those orders placed by customers of rep 20. Write the SQL-type statements to create these fragments.

2. Create a class diagram for the Premiere Products database, as shown in Chapter 1, Figure 1-5. If you need to make any assumptions in preparing the diagram, document those assumptions.

3. A user queries the Part table in the Premiere Products database over the company intranet. Assume the Part table contains 5,000 rows, each row is 1,000 bits long, the access delay is 2.5 seconds, the transmission rate is 50,000 bits per second, and only 20 of the 5,000 rows in the Part table satisfy the query conditions. Calculate the total communication time required for this query based on retrieving all table rows one row at a time; then calculate the total communication time required based on retrieving the 20 rows that satisfy the query conditions in a single message.

Henry Books Case

Ray Henry asks you to research improvements he might make to his database processing. To help him, he would like you to complete the following exercises. You do not use the Henry Books database for any of these exercises.

1. Use computer magazines or the Internet to investigate one of these DBMSs: DB2, SQL Server, MySQL, Oracle, or Sybase. Then prepare a report that explains how that DBMS handles two of the following distributed database functions: deadlock, fragmentation, replication, the data dictionary or log, and distributed queries.

2. Create a class diagram for the Henry Books database, as shown in Chapter 1, Figures 1-15 through 1-18. If you need to make any assumptions in preparing the diagram, document those assumptions.

Alexamara Marina Group Case

For the following exercises, you will answer problems and questions from the Alexamara Marina Group staff. You do not use the Alexamara database for any of these exercises.

1. Use computer magazines, books, or the Internet to investigate one of the following Web services: Application Programming Interface (API); Common Gateway Interface (CGI); Simple Object Access Protocol (SOAP); Universal Description, Discovery, and Integration (UDDI); or Web Services Description Language (WSDI). Then prepare a report that defines the Web service and explains its purpose and features.

2. Create a class diagram for the Alexamara database, as shown in Chapter 1, Figures 1-20 through 1-24. If you need to make any assumptions in preparing the diagram, document those assumptions.

311

COMPREHENSIVE DESIGN EXAMPLE: MARVEL COLLEGE

INTRODUCTION

Marvel College has decided to computerize its operations. In this appendix, you will design a database that satisfies many user

requirements by applying the design techniques you learned in Chapter 6 to a significant set of requirements.

MARVEL COLLEGE REQUIREMENTS

Marvel College has provided you with the following requirements that its new system must satisfy. You will use these requirements to design a new database.

General Description

Marvel College is organized by department (math, physics, English, and so on). Most departments offer more than one major; for example, the math department might offer majors in calculus, applied mathematics, and statistics. Each major, however, is offered by only one department. Each faculty member is assigned to a single department. Students can have more than one major, but most students have only one. Each student is assigned a faculty member as an advisor for his or her major; students who have more than one major are assigned a faculty advisor for each major. The faculty member may or may not be assigned to the department offering the major.

A code that has up to three characters (CS for Computer Science, MTH for Mathematics, PHY for Physics, ENG for English, and so on) identifies each department. Each course is identified by the combination of the department code and a three-digit number (CS 162 for Programming I, MTH 201 for Calculus I, ENG 102 for Creative Writing, and so on). The number of credits offered by a particular course does not vary; that is, all students who pass the same course receive the same amount of credit.

A two-character code identifies the semester in which a course is taught (FA for fall, SP for spring, and SU for summer). The code is combined with two digits that designate the year (for example, FA10 represents the fall semester of 2010). For a given semester, a department assigns each section of each course a four-digit schedule code (schedule code 1295 for section A of MTH 201, code 1297 for section B of MTH 201, code 1302 for section C of MTH 201, and so on). The schedule codes might vary from semester to semester. The schedule codes are listed in the school's time schedule, and students use them to indicate the sections in which they want to enroll. (You'll learn more about the enrollment process later in this section.)

After all students have completed the enrollment process for a given semester, each faculty member receives a class list for each section he or she will be teaching. In addition to listing the students in each section, the class list provides space to record the grade each student earns in the course. At the end of the semester, the faculty member enters the students' grades in this list and sends a copy of the list to the records office, where the grades are entered into the database. (In the future, the college plans to automate this part of the process.)

After an employee of the records office posts the grades (by entering them into the database), the DBMS generates a report card for each student; then the report cards are mailed to the addresses printed on the report card. The grades earned by a student become part of his or her permanent record and will appear on the student's transcript.

Report Requirements

Employees at Marvel College require several reports to manage students, classes, schedules, and faculty members; these reports have the following requirements.

Report card: At the end of each semester, the system must produce a report card for each student. A sample report card is shown in Figure A-1.

MARVEL COLLEGE

Department	Course Number	Course Description	Grade	Credits Taken	Credits Earned	Grade Points
Computer Science	CS 162	Programming I	A	4	4	16.0
Mathematics	MTH 201	Calculus I	B+	3	3	9.9

Current Semester Totals

7	7	3.70	25.9
Credits Taken	Credits Earned	GPA	Total Points

Semester: FA10

Cumulative Totals

44	44	3.39	149.2
Credits Taken	Credits Earned	GPA	Total Points

Student Number: 381124188

Student Name & Address	Local Address (IF DIFFERENT)
Brian Connors 686 Franklin Hart, MI 48282	4672 Westchester Trent, MI 48222

FIGURE A-1 Sample report card for Marvel College

Class list: The system must produce a class list for each section of each course; a sample class list is shown in Figure A-2. Note that space is provided for the grades. At the end of the semester, the instructor enters each student's grade and sends a copy of the class list to the records office.

CLASS LIST

Department: CS   Computer Science   Term: FA10
Course: 162   Programming I   (4 CREDITS)
Section: B
Schedule Code: 2366

Time: 1:00 - 1:50 M, T, W, F
PLACE: 118 SCR

Instructor: 462   Diane Johnson

Student Number	Student Name	Class Standing	Grade
381124188	Brian Connors	2	
.	.	.	
.	.	.	
.	.	.	

FIGURE A-2 Sample class list for Marvel College

Grade verification report: After the records office processes the class list, it returns the class list to the instructor with the grades entered in the report. The instructor uses the report to verify that the records office entered the students' grades correctly.

Time schedule: The time schedule shown in Figure A-3 lists all sections of all courses offered during a given semester. Each section has a unique four-digit schedule code. The time schedule lists the schedule code; the department offering the course; the course's number, section letter, and title; the instructor teaching the course; the time the course meets; the room in which the course meets; the number of credits generated by the course; and the prerequisites for the course. In addition to the information shown in Figure A-3, the time schedule includes the date the semester begins and ends, the date final exams begin and end, and the last withdrawal date (the last date on which students may withdraw from a course for a refund and without academic penalty).

```
                    TIME SCHEDULE      Term: FA10

 Course #   Code #  Sect   Time              Room      Faculty
              .       .      .                .          .
              .       .      .                .          .
              .       .      .                .          .


 CHEMISTRY (CHM) Office: 341 NSB

   111  Chemistry I                          4 CREDITS
             1740    A     10:00-10:50 M, T, W, F   102 WRN    Johnson
             1745    B     12:00-12:50 M, T, W, F   102 WRN    Lawrence
              .       .      .                .          .
              .       .      .                .          .
              .       .      .                .          .

            Prerequisite: MTH 110
   112  Chemistry II                         4 CREDITS
             1790    A     10:00-11:50 M, W         109 WRN    Adams
             1795    B     12:00-1:50 T, R          102 WRN    Nelson
              .       .      .                .          .
              .       .      .                .          .
              .       .      .                .          .

            Prerequisite: CHM 111
   114  ....
```

FIGURE A-3 Sample time schedule for Marvel College

Registration request form: A sample registration request form is shown in Figure A-4. A student uses this form to request classes for the upcoming semester. Students indicate the sections for which they want to register by entering each section's schedule code; for each of these sections, students may also enter a code for an alternative section in case the first requested section is full.

REGISTRATION REQUEST FORM

Student Number: 381124188   Term: SP11
Name: Brian Connors
Permanent Address: 686 Franklin   Local Address: 4672 Westchester
City: Hart   City: Trent
State: MI   State: MI
Zip: 48282   Zip: 48222

SCHEDULE CODES

	PRIMARY	ALTERNATE
1.		
2.		
3.		
4.		
5.		
6.		
7.		
8.		
9.		
10.		

FIGURE A-4   Sample registration request form for Marvel College

Student schedule: After all students have been assigned to sections, the system produces a student schedule form, which is mailed to students so that they know the classes in which they have been enrolled. A sample student schedule form is shown in Figure A-5. This form shows the schedule for an individual student for the indicated semester.

STUDENT SCHEDULE

Student Number: 381124188   Term: SP11
Name: Brian Connors
Permanent Address: 686 Franklin   Local Address: 4672 Westchester
City: Hart   City: Trent
State: MI   State: MI
Zip: 48282   Zip: 48222

Schedule Code	Course Number	Course Description	Section	Credits	Time	Room
2366	CS 253	Programming II	B	4	1:00–1:50 M, T, W, F	118 SCR
.	.	.	.	.		
.	.	.	.	.		
.	.	.	.	.		
		Total Credits:		16		

FIGURE A-5   Sample student schedule for Marvel College

Full student information report: This report lists complete information about a student, including his or her major(s) and all grades received to date. A sample of a full student information report is shown in Figure A-6.

FULL STUDENT INFORMATION

Student Number: 381124188   Term: FA10
    Name: Brian Connors

    City: Hart   City: Trent
    State: MI   State: MI
    Zip: 48282   Zip: 48222

Major 1: Information Sys.   Department: Computer Science   Advisor: Mark Lawerence
Major 2: Accounting   Department: Business   Advisor: Jill Thomas
Major 3:   Department: Advisor:

Term	Course Number			Credits	Grade Earned	
SP10	MTH	123	Trigonometry	4	A	16.0
	HST	201	Western Civilization	3	A-	11.1
	ENG	101	American Literature	3	A	12.0
FA10	CS	162	Programming I	4	A	16.0
	MTH	201	Calculus I	4	B+	9.9
	.	.	.	.	.	.
	.	.	.	.	.	.
	.	.	.	.	.	.
	.	.	.	.	.	.
	.	.	.	.	.	.
	.	.	.	.	.	.

Credits Attempted: 44
Credits Earned: 44
Grade Points: 149.2
Grade Point Avg: 3.39
Class Standing: 2

FIGURE A-6   Sample full student information report for Marvel College

Faculty information report: This report lists all faculty by department and contains each faculty member's ID number, name, address, office location, phone number, current rank (Instructor, Assistant Professor, Associate Professor, or Professor), and starting date of employment. It also lists the number, name, and local and permanent addresses of each faculty member's advisees; the code number and description of the major in which the faculty member is advising each advisee; and the code number and description of the department to which this major is assigned. (Remember that this department need not be the one to which the faculty member is assigned.)

Work version of the time schedule: Although this report is similar to the original time schedule (see Figure A-3), it is designed for the college's internal use. It shows the current enrollments in each section of each course, as well as the maximum enrollment permitted per section. It is more current than the time schedule. (When students register for courses, enrollment figures are updated on the work version of the time schedule. When room or faculty assignments are changed, this information also is updated. A new version of this report that reflects the revised figures is printed after being updated.)

Course report: For each course, this report lists the code and name of the department that is offering the course, the course number, the description of the course, and the number of credits awarded. This report also includes the department and course number for each prerequisite course.

Update (Transaction) Requirements

In addition to being able to add, change, and delete any information in the report requirements, the system must be able to accomplish the following update requirements:

Enrollment: When a student attempts to register for a section of a course, the system must determine whether the student has received credit for all prerequisites to the course. If the student is eligible to enroll in the course and the number of students currently enrolled in the section is less than the maximum enrollment, enroll the student.

Post grades: For each section of each course, the system must post the grades that are indicated on the class list submitted by the instructor and produce a grade verification report. (*Posting the grades* is the formal term for the process of entering the grades permanently in the students' computerized records.)

Purge: Marvel College retains section information, including grades earned by the students in each section, for two semesters following the end of the semester. Then the system removes this information from the system. (Grades assigned to students are retained by course but not by section.)

MARVEL COLLEGE INFORMATION-LEVEL DESIGN

You should give some consideration to the overall requirements before you apply the method to the individual user requirements. For example, by examining the documents shown in Figures A-1 through A-6, you may have identified the following entities: department, major, faculty member, student, course, and semester.

NOTE

Your list might include the section and grade entities. On the other hand, you might not have included the semester entity. In the long run, as long as the list is fairly reasonable, what you include won't make much difference. In fact, you may remember that this step is not even necessary. The better you do your job now, however, the simpler the process will be later on.

After identifying the entities, you assign a primary key to each one. In general, this step will require some type of consultation with users. You may need to ask users directly for the required information, or you may be able to obtain it from some type of survey form. Assume that having had such a consultation, you created a relation for each of these entities and assigned them the following primary keys:

```
Department (DepartmentCode,
Major (MajorNum,
Faculty (FacultyNum,
Student (StudentNum,
Course (DepartmentCode, CourseNum,
Semester (SemesterCode,
```

Note that the primary key for the Course table consists of two attributes, DepartmentCode (such as CS) and CourseNum (such as 153), both of which are required. The database could contain, for example, CS 153 and CS 353. Thus, the department code alone cannot be the primary key. Similarly, the database could contain ART 101 and MUS 101, two courses with the same course number but with different department codes. Thus, the course number alone cannot be the primary key either.

Now you can begin examining the individual user views as stated in the requirements. You can create relations for these user views, represent any keys, and merge the new user views into the cumulative design. Your first task is to determine the individual user views. The term *user view* never appeared in the list of requirements. Instead, Marvel College provided a general description of the system, together with a collection of report requirements and another collection of update requirements. How do these requirements relate to user views?

Certainly, you can think of each report requirement and each update requirement as a user view. But what do you do with the general description? Do you think of each paragraph (or perhaps each sentence) in the report as representing a user view, or do you use each paragraph or sentence to furnish additional information about the report and update requirements? Both approaches are acceptable. Because the second approach is often easier, it is the approach you will follow in this text. Think of the report and update requirements as user views and when needed, use the statements in the general description as additional information about these user views. You will also consider the general description during the review process to ensure that your final design satisfies all the functionality it describes.

First, consider one of the simpler user views, the course report. (Technically, you can examine user views in any order. Sometimes you take them in the order they are listed. In other cases, you may be able to come up with a better order. Often, examining some of the simpler user views first is a reasonable approach.)

Before you proceed with the design, consider the following method. First, with some of the user views, you will attempt to determine the relations involved by carefully determining the entities and relationships between them and using this information when creating the relations. This process means that from the outset, the collection of tables created will be in or close to third normal form. With other user views, you will create a single relation that may contain some number of repeating groups. In these cases, as you will see, the normalization process still produces a correct design, but it also involves more work. In practice, the more experience a designer has, the more likely he or she is to create third normal form relations immediately.

Second, the name of an entity or attribute may vary from one user view to another, and this difference requires resolution. You will attempt to use names that are the same.

User View 1—Course report: For each course, list the code and name of the department that is offering the course, the course number, the course title, and the number of credits awarded. This report also includes the department and course number for each prerequisite course. Forgetting for the moment the requirement to list prerequisite courses, the basic relation necessary to support this report is as follows:

Course (<u>DepartmentCode</u>, DepartmentName, <u>CourseNum</u>, CourseTitle, NumCredits)

The combination of DepartmentCode and CourseNum uniquely determines all the other attributes. In this relation, DepartmentCode determines DepartmentName; thus, the table is not in second normal form. (An attribute depends on only a portion of the key.) To correct this situation, the table is split into the following two tables:

Course (<u>DepartmentCode</u>, <u>CourseNum</u>, CourseTitle, NumCredits)
Department (<u>DepartmentCode</u>, DepartmentName)

The DepartmentCode attribute in the first relation is a foreign key identifying the second relation.
To maintain prerequisite information, you need to create the relation Prereq:

Prereq (<u>DepartmentCode</u>, <u>CourseNum</u>, <u>DepartmentCode/1</u>, <u>CourseNum/1</u>)

In this table, the attributes DepartmentCode and CourseNum refer to the course and the attributes DepartmentCode/1 and CourseNum/1 refer to the prerequisite course. If CS 362 has a prerequisite of MTH 345, for example, there will be a row in the Prereq table in which the DepartmentCode is CS, the CourseNum is 362, the DepartmentCode/1 is MTH, and the CourseNum/1 is 345.

> **NOTE**
>
> Because there are two attributes named DepartmentCode and two attributes named CourseNum, you must be able to distinguish between them. The software used to produce these diagrams makes the distinction by appending the characters /1 to one of the names, which is why these names appear in the Prereq table. In this example, the DepartmentCode/1 and CourseNum/1 attributes represent the department code and course number of the prerequisite course, respectively. When it is time to implement the design, you typically assign them names that are more descriptive. For instance, you might name them PrereqDepartmentCode and PrereqCourseNum, respectively.

The DBDL version of these tables is shown in Figure A-7.

```
Department (DepartmentCode, DepartmentName)

Course (DepartmentCode, CourseNum, CourseTitle, NumCredits)
    FK DepartmentCode → Department

Prereq (DepartmentCode, CourseNum, DepartmentCode/1,
    CourseNum/1)
    FK DepartmentCode, CourseNum → Course
    FK DepartmentCode/1, CourseNum/1 → Course
```

FIGURE A-7   DBDL for User View 1

The result of merging these relations into the cumulative design appears in the E-R diagram shown in Figure A-8. Notice that the Department and Course tables have been merged with the existing Department and Course tables in the cumulative design. In the process, the attribute DepartmentName was added to the Department table and the attributes CourseTitle and NumCredits were added to the Course table. In addition, the attribute DepartmentCode in the Course table is a foreign key. Because the Prereq table is new, it was added to the cumulative design in its entirety. Notice also that you do not yet have any relationships among the entities Student, Major, Faculty, and Semester.

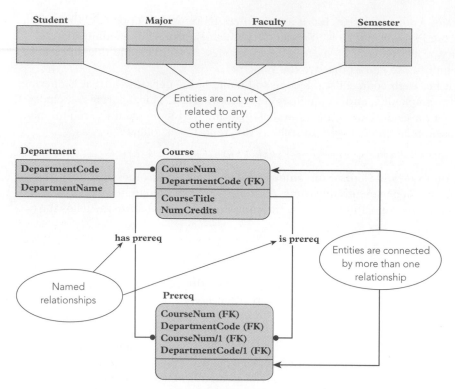

FIGURE A-8 Cumulative design after User View 1

In Figure A-8, there are two relationships between Course and Prereq. To distinguish between them, it is necessary to name the relationships. In the figure, the name for the first relationship is "has prereq" and the name for the second relationship is "is prereq."

NOTE

When using a software tool to produce E-R diagrams, the software might reverse the order of the fields that make up the primary key. For example, the E-R diagram in Figure A-8 indicates that the primary key for the Course table is CourseNum and then DepartmentCode, even though you intended it to be DepartmentCode and then CourseNum. This difference is not a problem. Indicating the fields that make up the primary key is significant, not the order in which they appear.

User View 2—Faculty information report: List all faculty by department and each faculty member's ID number, name, address, office location, phone number, current rank (Instructor, Assistant Professor, Associate Professor, or Full Professor), and starting date of employment. In addition, list the number, name, and local and permanent addresses of each faculty member's advisees; the code number and description of the major in which the faculty member is advising each advisee; and the code number and description of the department to which this major is assigned. This user view involves three entities (departments, faculty, and advisees), so you can create the following three tables:

```
Department (
Faculty (
Advisee (
```

The next step is to assign a primary key to each table. Before doing so, however, you should briefly examine the tables in the cumulative design and use the same names for any existing tables or attributes. In this case, you would use DepartmentCode as the primary key for the Department table and FacultyNum as the primary key for the Faculty table. There is no Advisee table in the cumulative collection, but there is a Student table.

Because advisees and students are the same, rename the Advisee entity to Student and use the StudentNum attribute as the primary key rather than AdvisorNum. Your efforts yield the following tables and primary keys:

```
Department (DepartmentCode,
Faculty (FacultyNum,
Student (StudentNum,
```

Next, add the remaining attributes to the tables:

```
Department (DepartmentCode, DepartmentName)
Faculty (FacultyNum, LastName, FirstName, Street, City, State,
    Zip, OfficeNum, Phone, CurrentRank, StartDate, DepartmentCode)
Student (StudentNum, LastName, FirstName, LocalStreet,
    LocalCity, LocalState, LocalZip, PermStreet, PermCity,
    PermState, PermZip, (MajorNum, Description,
    DepartmentCode, FacultyNum, LastName, FirstName) )
```

The DepartmentCode attribute is included in the Faculty table because there is a one-to-many relationship between departments and faculty members. Because a student can have more than one major, the information about majors (number, description, department, and the number and name of the faculty member who advises this student in this major) is a repeating group.

Because the key to the repeating group in the Student table is MajorNum, removing this repeating group yields the following:

```
Student (StudentNum, LastName, FirstName, LocalStreet,
    LocalCity, LocalState, LocalZip, PermStreet, PermCity,
    PermState, PermZip, MajorNum, Description,
    DepartmentCode, FacultyNum, LastName, FirstName)
```

Converting this relation to second normal form produces the following tables:

```
Student (StudentNum, LastName, FirstName, LocalStreet,
    LocalCity, LocalState, LocalZip, PermStreet, PermCity,
    PermState, PermZip)
Major (MajorNum, Description, DepartmentCode, DepartmentName)
Advises (StudentNum, MajorNum, FacultyNum)
```

In this case, you must remove the following dependencies to create third normal form tables: OfficeNum determines Phone in the Faculty table, and DepartmentCode determines DepartmentName in the Major table. Removing these dependencies produces the following collection of tables:

```
Department (DepartmentCode, DepartmentName)
Faculty (FacultyNum, LastName, FirstName, Street, City, State,
    Zip, OfficeNum, CurrentRank, StartDate, DepartmentCode)
Student (StudentNum, LastName, FirstName, LocalStreet,
    LocalCity, LocalState, LocalZip, PermStreet, PermCity,
    PermState, PermZip)
Advises (StudentNum, MajorNum, FacultyNum)
Office (OfficeNum, Phone)
Major (MajorNum, Description, DepartmentCode)
```

The DBDL representation is shown in Figure A-9.

```
Department (DepartmentCode, DepartmentName)

Student (StudentNum, LastName, FirstName, LocalStreet, LocalCity,
      LocalState, LocalZip, PermStreet, PermCity, PermState, PermZip)

Office (OfficeNum, Phone)

Faculty (FacultyNum, LastName, FirstName, Street, City, State, Zip,
        OfficeNum, CurrentRank, StartDate, DepartmentCode)
     FK OfficeNum → Office
     FK DepartmentCode → Department

Major (MajorNum, Description, DepartmentCode)
     FK DepartmentCode → Department

Advises (StudentNum, MajorNum, FacultyNum)
     FK StudentNum → Student
     FK FacultyNum → Faculty
     FK MajorNum → Major
```

FIGURE A-9   DBDL for User View 2

The result of merging these tables into the cumulative design is shown in Figure A-10. The tables Student, Faculty, Major, and Department are merged with the existing tables with the same primary keys and with the same names. Nothing new is added to the Department table in the process, but the other tables receive additional attributes. In addition, the Faculty table also receives two foreign keys, OfficeNum and DepartmentCode. The Major table receives one foreign key, DepartmentCode. The Advises and Office tables are new and thus are added directly to the cumulative design.

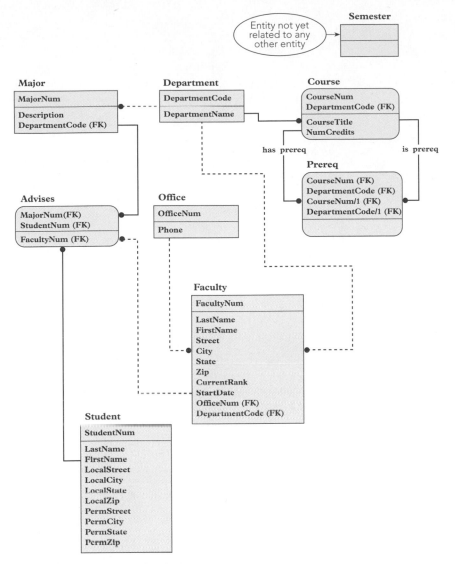

FIGURE A-10 Cumulative design after User View 2

User View 3—Report card: At the end of each semester, the system must produce a report card for each student. Report cards are fairly complicated documents in which the appropriate underlying relations are not immediately apparent. In such a case, it's a good idea to first list all the attributes in the report card and assign them appropriate names, as shown in Figure A-11. After identifying the attributes, you should list the functional dependencies that exist between these attributes. The information necessary to determine functional dependencies must ultimately come from the user, although you can often guess most of them accurately.

NOTE

Notice that there are duplicate names in the list. CreditsEarned, for example, appears three times: once for the course, once for the semester, and once for the cumulative number of credits earned by the student. You could assign these columns different names at this point. The names could be CreditsEarnedCourse, CreditsEarnedSemester, and CreditsEarnedCumulative. Alternatively, you could assign them the same name with an explanation of the purpose of each one in parentheses, as shown in Figure A-11. Of course, after you have determined all the tables and assigned columns to them, you must ensure that the column names within a single table are unique.

Department
CourseNum
CourseTitle
Grade
CreditsTaken   (Course)
CreditsEarned   (Course)
GradePoints   (Course)
CreditsTaken   (Semester)
CreditsEarned   (Semester)
GPA   (Semester)
TotalPoints   (Semester)
CreditsTaken   (Cumulative)
CreditsEarned   (Cumulative)
GPA   (Cumulative)
TotalPoints   (Cumulative)
SemesterCode
StudentNum
LastName
FirstName
Address
City
State
Zip
LocalAddress
LocalCity
LocalState
LocalZip

FIGURE A-11   Attributes on a report card from Marvel College

Assume the system's users have verified the attributes listed in Figure A-11 and your work is correct. Figure A-12 shows the functional dependencies among the attributes you identified on the report card. The student number alone determines many of the other attributes.

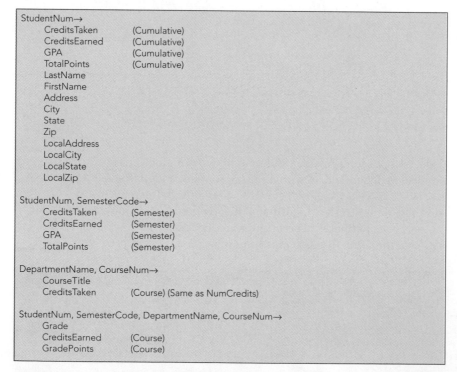

FIGURE A-12   Functional dependencies among the attributes on a report card

In addition to the student number, the semester must be identified to determine credits taken and earned, grade point average (GPA), and total points each semester. The combination of a department name (such as Computer Science) and a course number (such as 153) determines a course title and the number of credits.

Finally, the student number, the semester (semester and year), the department, and the course (department and course number) are required to determine an individual grade in a course, the credits earned from the course, and the grade points in a course. (The semester is required because the same course might be offered during more than one semester at Marvel College.)

> **NOTE**
>
> There is a parenthetical comment after CreditsTaken in the section determined by DepartmentName and CourseNum. It indicates that CreditsTaken is the same as NumCredits, which is a column already in the cumulative design. Documenting that the name you have chosen is a synonym for a name already in the cumulative design is a good practice.

The next step is to create a collection of tables that will support this user view. A variety of approaches will work. You could combine all the attributes into a single table, which you then would convert to third normal form. (In such a table, the combination of department, course number, course title, grade, and so on, would be a repeating group.) Alternatively, you could use the functional dependencies to determine the following collection of relations:

```
Student (StudentNum, LastName, FirstName, PermStreet, PermCity,
    PermState, PermZip, LocalStreet, LocalCity, LocalState,
    LocalZip, CreditsTaken, CreditsEarned, GPA, TotalPoints)
StudentSemester (StudentNum, SemesterCode, CreditsTaken,
    CreditsEarned, GPA, TotalPoints)
Course (DepartmentCode, CourseNum, CourseTitle, NumCredits)
StudentGrade (StudentNum, SemesterCode, DepartmentName,
    CourseNum, Grade, CreditsEarned, GradePoints)
```

All these relations are in third normal form. The only change you should make involves the DepartmentName attribute in the StudentGrade table. In general, if you encounter an attribute for which there exists a determinant that is not in the table, you should add the determinant. In this case, DepartmentCode is a determinant for DepartmentName, but it is not in the table; so you should add DepartmentCode. In the normalization process, DepartmentName will then be removed and placed in another table whose key is DepartmentCode. This other table will be merged with the Department table without the addition of any new attributes. The resulting StudentGrade table is as follows:

```
StudentGrade (StudentNum, SemesterCode, DepartmentCode,
    CourseNum, Grade, CreditsEarned, GradePoints)
```

Before representing this design in DBDL, examine the StudentSemester entity. Some of the attributes it contains (CreditsTaken, CreditsEarned, GPA, and TotalPoints) refer to the current semester, and all appear on a report card. Assume after further checking that you find that all these attributes are easily calculated from other fields on the report card. Rather than storing these attributes in the database, you can ensure that the program that produces the report cards performs the necessary calculations. For this reason, you will remove the StudentSemester table from the collection of tables to be documented and merged. (If these attributes are also required by some other user view in which the same computations are not as practical, they might find their way into the database when that user view is analyzed.)

> **Q & A**
>
> **Question:** Determine the tables and keys required for User View 3. Merge the result into the cumulative design and draw the E-R diagram for the new cumulative design.
>
> **Answer:** Figure A-13 shows the new cumulative design.

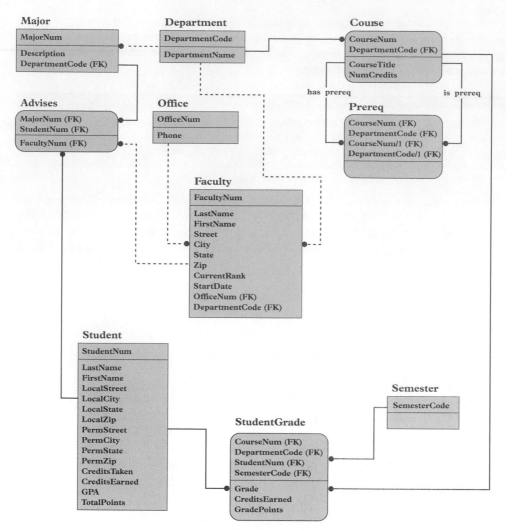

FIGURE A-13 Cumulative design after User View 3

User View 4—Class list: The system must produce a class list for each section of each course. Space is provided for the grades. At the end of the semester, the instructor enters each student's grade and sends a copy of the class list to the records office. Assume after examining the sample class list report (see Figure A-2) that you decide to create a single table (actually an unnormalized table) that contains all the attributes on the class list, with the student information (number, name, class standing, and grade) as a repeating group. (Applying the tips for determining the relations to support a given user view would lead more directly to the result; but for the sake of developing the example, assume you haven't done that yet.) The unnormalized table created by this method would be as follows:

```
ClassList (DepartmentCode, DepartmentName, SemesterCode,
    CourseNum, CourseTitle, NumCredits, SectionLetter,
    ScheduleCode, Time, Room, FacultyNum, FacultyLastName,
    FacultyFirstName, (StudentNum, StudentLastName,
    StudentFirstName, ClassStanding, Grade) )
```

NOTE

Because attribute names within a single table must be unique, it is not permissible to assign the attribute name LastName to both the faculty and student last names. Thus, the attributes that store the last and first names of a faculty member are named FacultyLastName and FacultyFirstName, respectively. Similarly, the attributes that store the last and first names of a student are named StudentLastName and StudentFirstName, respectively.

Note that you have not yet indicated the primary key. To identify a given class within a particular semester requires the combination of a department code, course number, and section letter or, more simply, the schedule code. Using the schedule code as the primary key, however, is not adequate. Because the information from more than one semester will be on file at the same time and because the same schedule code could be used in two different semesters to represent different courses, the primary key must also contain the semester code. When you remove the repeating group, this primary key expands to contain the key for the repeating group, which, in this case, is the student number. Thus, converting to first normal form yields the following design:

```
ClassList (DepartmentCode, DepartmentName, SemesterCode,
    CourseNum, CourseTitle, NumCredits, SectionLetter,
    ScheduleCode, Time, Room, FacultyNum, FacultyLastName,
    FacultyFirstName, StudentNum, StudentLastName,
    StudentFirstName, ClassStanding, Grade)
```

Converting to third normal form yields the following collection of tables:

```
Department (DepartmentCode, DepartmentName)
Section (SemesterCode, ScheduleCode, DepartmentCode, CourseNum,
    SectionLetter, Time, Room, FacultyNum)
Faculty (FacultyNum, LastName, FirstName)
StudentClass (SemesterCode, ScheduleCode, StudentNum, Grade)
Student (StudentNum, LastName, FirstName, ClassStanding)
Course (DepartmentCode, CourseNum, CourseTitle, NumCredits)
```

NOTE

Because the last name of a faculty member is now in a separate table from that of the last name of a student, it is no longer necessary to have different names. Thus, FacultyLastName and StudentLastName have been shortened to LastName. Similarly, FacultyFirstName and StudentFirstName have been shortened to FirstName.

Q & A

Question: Why was the grade included?
Answer: Although the grade is not actually printed on the class list, it will be entered on the form by the instructor and sent to the records office for posting. The grade verification report differs from the class list only in that the grade is printed. Thus, the grade will ultimately be required and it is appropriate to deal with it here.

Q & A

Question: Determine the tables and keys required for User View 4. Merge the result into the cumulative design and draw the E-R diagram for the new cumulative design.
Answer: Figure A-14 shows the new cumulative design.

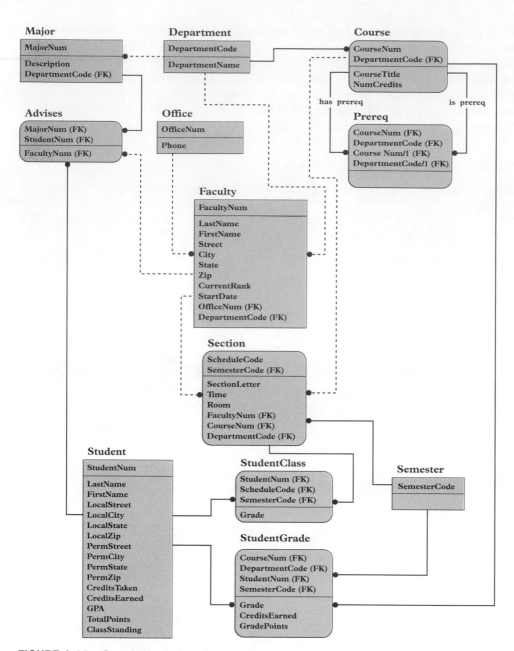

FIGURE A-14 Cumulative design after User View 4

User View 5—Grade verification report: After the records office processes the class list, it returns the class list to the instructor with the grades entered in the report. The instructor uses the report to verify that the records office entered the students' grades correctly. Because the only difference between the class list and the grade verification report is that the grades are printed on the grade verification report, the user views will be quite similar. In fact, because you made a provision for the grade when treating the class list, the views are identical and no further treatment of this user view is required.

User View 6—Time schedule: List all sections of all courses offered during a given semester. Each section has a unique four-digit schedule code. The time schedule lists the schedule code; the department offering the course; the course's number, section letter, and title; the instructor teaching the course; the time the course meets; the room in which the course meets; the number of credits generated by the course; and the prerequisites for the course. In addition to the information shown in the figure, the time schedule includes the date the semester begins and ends, the date final exams begin and end, and the last withdrawal date. The attributes on the time schedule are as follows: term (which is a synonym for semester), department code, department name, location, course number, course title, number of credits, schedule code, section letter, meeting time, meeting place, and instructor name.

You could create a single relation containing all these attributes and then normalize that relation, or you could apply the tips presented in Chapter 8 for determining the collection of relations. In either case, you ultimately create the following collection of relations:

```
Department (DepartmentCode, DepartmentName, Location)
Course (DepartmentCode, CourseNum, CourseTitle, NumCredits)
Section (SemesterCode, ScheduleCode, DepartmentCode, CourseNum,
    SectionLetter, Time, Room, FacultyNum)
Faculty (FacultyNum, LastName, FirstName)
Semester (SemesterCode, StartDate, EndDate, ExamStartDate,
    ExamEndDate, WithdrawalDate)
```

NOTE

Actually, given the attributes in this user view, the Section relation would contain the instructor's name (LastName and FirstName). There was no mention of instructor number. In general, as you saw earlier, it's a good idea to include determinants for attributes whenever possible. In this example, because FacultyNum determines LastName and FirstName, you add FacultyNum to the Section relation, at which point the Section relation is not in third normal form. Converting to third normal form produces the collection of relations previously shown.

Q & A

Question: Determine the tables and keys required for User View 6. Merge the result into the cumulative design and draw the E-R diagram for the new cumulative design.

Answer: Figure A-15 shows the new cumulative design.

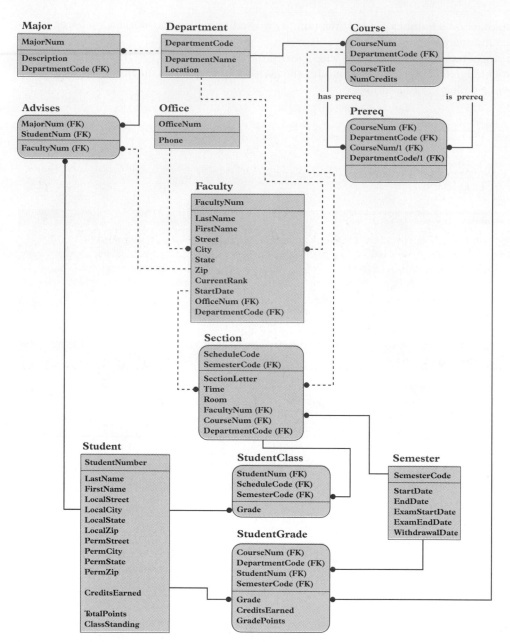

FIGURE A-15 Cumulative design after User View 6

User View 7—Registration request form: A student uses this form to request classes for the upcoming semester. Students indicate the sections for which they want to register by entering the sections' schedule codes; for each section, students may also enter a code for an alternate section in case the requested primary section is full. The collection of tables to support this user view includes a Student table that consists of the primary key, StudentNum, and all the attributes that depend only on StudentNum. These attributes include LastName, FirstName, and LocalStreet. Because all attributes in this table are already in the Student table in the cumulative design, this user view will not add anything new and there is no need for further discussion of it here.

The portion of this user view that is not already present in the cumulative design concerns the primary and alternate schedule codes that students request. A table to support this portion of the user view must contain both a primary and an alternate schedule code. The table must also contain the number of the student making the request. Finally, to allow the flexibility of retaining this information for more than a single semester to allow registration for more than a semester at a time, the table must also include the semester in which the request is made. This leads to the following relation:

RegistrationRequest (<u>StudentNum</u>, <u>PrimaryCode</u>, AlternateCode, SemesterCode)

For example, if student 381124188 were to request the section with schedule code 2345 and then request the section with schedule code 2396 as an alternate for the FA10 semester, the row (381124188, 2345, 2396, "FA10") would be stored. The student number, the primary schedule code, the alternate schedule code, and the semester code are required to uniquely identify a particular row.

Q & A

Question: Determine the tables and keys required for User View 7. Merge the result into the cumulative design and draw the E-R diagram for the new cumulative design.

Answer: Figure A-16 shows the new cumulative design. Notice that two relationships join the Section table to the RegistrationRequest table, so you must name each of them. In this case, you use "primary" and "alternate," indicating that one relationship relates a request to the primary section chosen and that the other relationship relates the request to the alternative section when there is one.

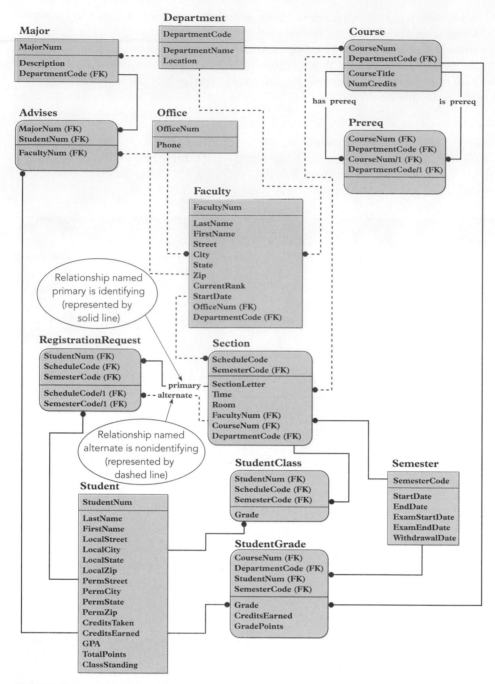

FIGURE A-16 Cumulative design after User View 7

NOTE

The foreign keys are the combination of PrimaryCode and SemesterCode as well as the combination of AlternateCode and SemesterCode. Because PrimaryCode and AlternateCode are portions of the foreign keys that must match the ScheduleCode in the Section table, they have been renamed ScheduleCode and ScheduleCode/1, respectively. Likewise, the second SemesterCode has been renamed SemesterCode/1.

User View 8—Student schedule: After all students are assigned to sections, the system produces a student schedule form, which is mailed to students so that they know the classes in which they have been enrolled. Suppose you had created a single unnormalized relation to support the student schedule. This unnormalized relation would contain a repeating group representing the lines in the body of the schedule as follows:

```
StudentSchedule (StudentNum, SemesterCode, LastName, FirstName,
    LocalStreet, LocalCity, LocalState, LocalZip, PermStreet,
    PermCity, PermState, PermZip, (ScheduleCode,
    DepartmentName, CourseNum, CourseTitle, SectionLetter,
    NumCredits, Time, Room) )
```

At this point, you remove the repeating group to convert to first normal form, yielding the following:

```
StudentSchedule (StudentNum, SemesterCode, LastName, FirstName,
    LocalStreet, LocalCity, LocalState, LocalZip, PermStreet,
    PermCity, PermState, PermZip, ScheduleCode,
    DepartmentCode, CourseNum, CourseTitle, SectionLetter,
    NumCredits, Time, Room)
```

Note that the primary key expands to include ScheduleCode, which is the key to the repeating group. Converting this table to second normal form produces the following:

```
Student (StudentNum, LastName, FirstName, LocalStreet,
    LocalCity, LocalState, LocalZip, PermStreet, PermCity,
    PermState, PermZip)
StudentSchedule (StudentNum, SemesterCode, ScheduleCode)
Section (SemesterCode, ScheduleCode, DepartmentCode, CourseNum,
    CourseTitle, SectionLetter, NumCredits, Time, Room)
Course (DepartmentCode, CourseNum, CourseTitle, NumCredits)
```

Removing the attributes that depend on the determinant of DepartmentCode and CourseNum from the Section table and converting this collection of tables to third normal form produces the following tables:

```
Student (StudentNum, LastName, FirstName, LocalStreet,
    LocalCity, LocalState, LocalZip, PermStreet, PermCity,
    PermState, PermZip)
StudentSchedule (StudentNum, SemesterCode, ScheduleCode)
Section (SemesterCode, ScheduleCode, DepartmentCode, CourseNum,
    SectionLetter, Time, Room)
Course (DepartmentCode, CourseNum, CourseTitle, NumCredits)
```

Merging this collection into the cumulative design does not add anything new. In the process, you can merge the StudentSchedule table with the StudentClass table.

User View 9—Full student information report: List complete information about a student, including his or her majors and all grades received to date. Suppose you attempted to place all the attributes on the full student information report into a single unnormalized relation. The table has two separate repeating groups: one for the different majors a student might have and the other for all the courses the student has taken.

NOTE

Several attributes, such as name and address, would not be in the repeating groups. All these attributes are already in the cumulative design, however, and are not addressed here.

The table with repeating groups is as follows:

```
Student (StudentNum, (MajorNum, DepartmentCode, LastName,
    FirstName), (SemesterCode, DepartmentCode, CourseNum,
    CourseTitle, NumCredits, Grade, GradePoints) )
```

Recall from Chapter 5 that you should separate repeating groups when a relation has more than one. If you don't, you will typically have problems with fourth normal form. Separating the repeating groups in this example produces the following:

```
StudentMajor (StudentNum, (MajorNum, DepartmentCode, LastName, FirstName) )
StudentCourse (StudentNum, (SemesterCode, DepartmentCode,
    CourseNum, CourseTitle, NumCredits, Grade, GradePoints) )
```

Converting these tables to first normal form and including FacultyNum, which is a determinant for LastName and FirstName, produces the following:

```
StudentMajor (StudentNum, MajorNum, DepartmentCode, FacultyNum,
    LastName, FirstName)
StudentCourse (StudentNum, SemesterCode, DepartmentCode,
    CourseNum, CourseTitle, NumCredits, Grade, Grade Points)
```

The StudentCourse table is not in second normal form because CourseTitle and NumCredits depend only on the DepartmentCode, CourseNum combination. The StudentMajor table is not in second normal form either because DepartmentCode depends on MajorNum. Removing these dependencies produces the following tables:

```
StudentMajor (StudentNum, MajorNum, FacultyNum, LastName,FirstName)
Major (MajorNum, DepartmentCode)
StudentCourse (StudentNum, SemesterCode, DepartmentCode,
    CourseNum, Grade, GradePoints)
Course (DepartmentCode, CourseNum, CourseTitle, NumCredits)
```

Other than the StudentMajor table, all these relations are in third normal form. Converting the StudentMajor table to third normal form produces the following tables:

```
StudentMajor (StudentNum, MajorNum, FacultyNum)
Faculty (FacultyNum, LastName, FirstName)
```

Merging this collection into the cumulative design does not add anything new. (You can merge the StudentMajor table with the Advises table without adding any new attributes.)

User View 10—Work version of the time schedule: This report is similar to the original time schedule (see Figure A-3), but it is designed for the college's internal use. It shows the current enrollments in each section of each course, as well as each section's maximum enrollment. The only difference between the work version of the time schedule and the time schedule itself (see User View 6) is the addition of two attributes for each section: current enrollment and maximum enrollment. Because these two attributes depend only on the combination of the semester code and the schedule code, you would place them in the Section table of User View 6; and after the merge, they would be in the Section table in the cumulative design. The cumulative design thus far is shown in Figure A-17.

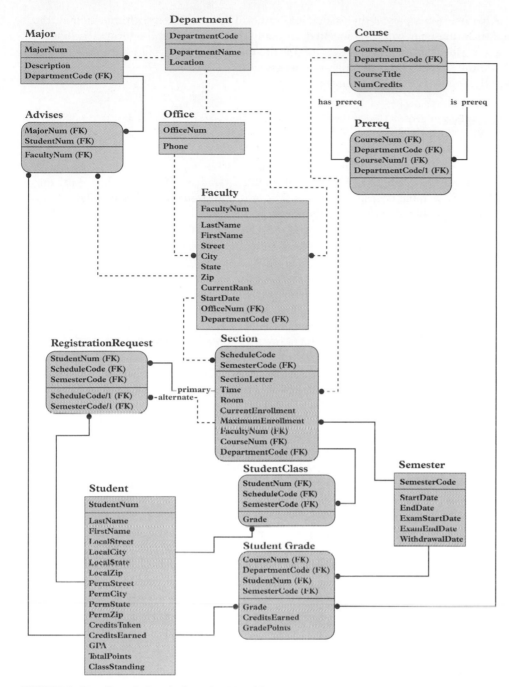

FIGURE A-17 Cumulative design after User View 10

Because the process of determining whether a student has had the prerequisites for a given course involves examining the grades (if any) received in these prior courses, it makes sense to analyze the user view that involves grades before treating the user view that involves enrollment.

User View 11—Post grades: For each section of each course, the system must post the grades that are indicated on the class list submitted by the instructor and produce a grade verification report. There is a slight problem with posting grades—grades must be posted by section to produce the grade report (in other words, you must record the fact that student 381124188 received an A in the section of CS 162 whose schedule code was 2366 during the fall 2010 semester). On the other hand, for the full student information report, there is no need to have any of the grades related to an actual section of a course. Further, because section information, including these grades, is kept for only two semesters, grades would be lost after two semesters if they were kept only by section because section information would be purged at that time.

A viable alternative is to post two copies of the grade: one copy will be associated with the student, the term, and the section; and the other copy will be associated with only the student and the term. The first copy would be used for the grade verification report; the second, for the full student information report. Report cards would probably utilize the second copy, although not necessarily.

Thus, you would have the following two grade tables:

```
GradeSection (StudentNum, DepartmentCode, CourseNum,
     ScheduleCode, SemesterCode, Grade)
GradeStudent (StudentNum, DepartmentCode, CourseNum,
     SemesterCode, Grade)
```

Because the DepartmentCode and CourseNum in the GradeSection table depend only on the concatenation of ScheduleCode and SemesterCode, they will be removed from the GradeSection table during the normalization process and placed in a table whose primary key is the concatenation of ScheduleCode and SemesterCode. This table will be combined with the Section table in the cumulative design without adding new fields. The GradeSection table that remains will be merged with the StudentClass table without adding new fields. Finally, the GradeStudent table will be combined with the StudentGrade table in the cumulative design without adding any new fields. Thus, treatment of this user view does not change the cumulative design.

User View 12—Enrollment: When a student attempts to register for a section of a course, you must determine whether the student has received credit for all prerequisites to the course. If the student is eligible to enroll in the course and the number of students currently enrolled in the section is less than the maximum enrollment, enroll the student. With the data already in place in the cumulative design, you can determine what courses a student has taken. You can also determine the prerequisites for a given course. The only remaining issue is the ability to enroll a student in a course. Because the system must retain information for more than one semester, you must include the semester code in the table. (You must have the information that student 381124188 enrolled in section 2345 in SP11 rather than in FA10, for example.) The additional table is as follows:

```
Enroll (StudentNum, SemesterCode, ScheduleCode)
```

The primary key of this table matches the primary key of the StudentClass table in the cumulative design. The fields occur in a different order here, but that makes no difference. Thus, this table will be merged with the StudentClass table. No new fields are to be added, so the cumulative design remains unchanged.

User View 13—Purge: Marvel College retains section information, including grades earned by the students in each section, for two semesters following the end of the semester, at which time this information is removed from the system. Periodically, certain information that is more than two terms old is removed from the database. This includes all information concerning sections of courses, such as the time, room, and instructor, as well as information about the students in the sections and their grades. The grade each student received will remain in the database by course but not by section. For example, you will always retain the fact that student 381124188 received an A in CS 162 during the fall semester of 2010; but once the data for that term is purged, you will no longer know the precise section of CS 162 that awarded this grade.

If you examine the current collection of tables, you will see that all the data to be purged is already included in the cumulative design and that you don't need to add anything new at this point.

FINAL INFORMATION-LEVEL DESIGN

Now that you are finished examining the user views, Marvel College can review the cumulative design to ensure that all user views have been met. You should conduct this review on your own to make certain that you understand how the requirements of each user can be satisfied. You will assume that this review has taken place and that no changes have been made. Therefore, Figure A-17 represents the final information-level design.

At this point, Marvel College is ready to move on to the physical-level design process. In this process, the appropriate team members will use the information-level design you produced to create the design for the specific DBMS that Marvel College selects. After it has done so, it will be able to create the database; load the data; and create the forms, reports, queries, and programs necessary to satisfy its requirements.

1. Discuss the effect of the following changes on the design for the Marvel College requirements:
   a. More than one instructor might teach a given section of a course, and each instructor must be listed on the time schedule.
   b. Each department offers only a single major.
   c. Each department offers only a single major, and each faculty member can advise students only in the major that is offered by the department to which the faculty member is assigned.
   d. Each department offers only a single major, and each faculty member can advise students only in the major that is offered by the department to which the faculty member is assigned. In addition, a student can have only one major.
   e. There is an additional transaction requirement: given a student's name, find the student's number.
   f. More than one faculty member can be assigned to one office.
   g. The number of credits earned in a particular course cannot vary from student to student or from semester to semester.
   h. Instead of a course number, course codes are used to uniquely identify courses. (In other words, department numbers are no longer required for this purpose.) However, it is still important to know which courses are offered by which departments.
   i. On the registration request form, a student may designate a number of alternates along with his or her primary choice. These alternates are listed in priority order, with the first one being the most desired and the last one being the least desired.

2. Complete an information-level design for Holt Distributors.

   General description. Holt Distributors buys products from its vendors and sells those products to its customers. The Holt Distributors operation is divided into territories. Each customer is represented by a single sales rep, who must be assigned to the territory in which the customer is located. Although each sales rep is assigned to a single territory, more than one sales rep can be assigned to the same territory.

   When a customer places an order, the computer assigns the order the next available order number. The data entry clerk enters the customer number, the customer purchase order (PO) number, and the date. (Customers can place orders by submitting a PO, in which case, a PO number is recorded.) For each part that is ordered, the clerk enters the part number, quantity, and quoted price. (When it is time for the clerk to enter the quoted price, the computer displays the price from the master price list. If the quoted price is the same as the actual price, the clerk takes no special action. If not, the clerk enters the quoted price.)

   When the clerk completes the order, the system prints the order acknowledgment/picking list form shown in Figure A-18 and sends it to the customer for confirmation and payment. When Holt Distributors is ready to ship the customer's order, this same form is used to "pick" the merchandise in the warehouse and prepare it for delivery.

ORDER ACKNOWLEDGMENT / PICKING LIST

10/12/2010 Order 12424

HOLT DISTRIBUTORS
146 NELSON PLACE
BRONSTON, MI 49802

SOLD
TO: Smith Rentals
      153 Main St.
      Suite 102
      Grandville, MI 49494

SHIP
TO: A & B Supplies
      2180 Halton Pl.
      Arendville, MI 49232

Customer	P.O. No.	Order Date	Sales Rep
1354	PO335	10/02/2010	10-Brown, Sam

Quantity						
Order	Ship	B/O	Item Number	Description	Price	Amount
6			AT414	Lounge Chair	$42.00	$252.00
4			BT222	Arm Chair	$51.00	$204.00

FIGURE A-18   Order acknowledgment/picking list for Holt Distributors

An order that hasn't been shipped (filled) is called an open order; an order that has been shipped is called a released order. When orders are released, the system prints an invoice, sends it to the customer, and then increases the customer's balance by the invoice amount. Some orders are completely filled; others are only partially filled, meaning that only part of the customer's order was shipped. In either case, when an entire order or a partial order has been shipped, the order is considered to have been filled and is no longer considered an open order. (Another possibility is to allow back orders when the order cannot be completely filled. In this case, the order remains open, but only for the back-ordered portion.) When the system generates an invoice, it removes the order from the open orders file. The system stores summary information about the invoice (number, date, customer, invoice total, and freight) until the end of the month. A sample invoice is shown in Figure A-19.

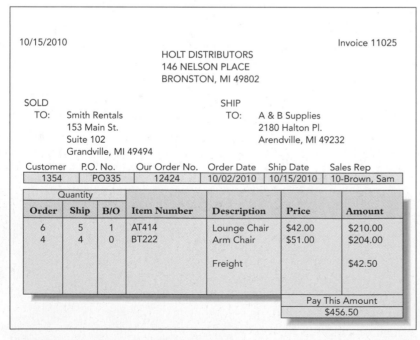

10/15/2010 Invoice 11025

HOLT DISTRIBUTORS
146 NELSON PLACE
BRONSTON, MI 49802

SOLD
TO: Smith Rentals
      153 Main St.
      Suite 102
      Grandville, MI 49494

SHIP
TO: A & B Supplies
      2180 Halton Pl.
      Arendville, MI 49232

Customer	P.O. No.	Our Order No.	Order Date	Ship Date	Sales Rep
1354	PO335	12424	10/02/2010	10/15/2010	10-Brown, Sam

Quantity						
Order	Ship	B/O	Item Number	Description	Price	Amount
6	5	1	AT414	Lounge Chair	$42.00	$210.00
4	4	0	BT222	Arm Chair	$51.00	$204.00
				Freight		$42.50

Pay This Amount
$456.50

FIGURE A-19   Invoice for Holt Distributors

Appendix A

Most companies use one of two methods to accept payments from customers: open items and balance forward. In the open-item method, customers make payments on specific invoices. An invoice remains on file until the customer pays it in full. In the balance-forward method, customers have balances. When the system generates an invoice, the customer's balance is increased by the amount of the invoice. When a customer makes a payment, the system decreases the customer's balance by the payment amount. Holt Distributors uses the balance-forward method.

At the end of each month, the system updates and ages customers' accounts. (You'll learn about month-end processing requirements and the update and aging processes in the following sections.) The system prints customer statements, an aged trial balance (described in the report requirements section), a monthly cash receipts journal, a monthly invoice register, and a sales rep commission report. The system then removes cash receipts and invoice summary records from the database and sets month-to-date (MTD) fields to zero. When the system processes the monthly data for December, it also sets the year-to-date (YTD) fields to zero.

Transaction requirements. The following transaction requirements are required by Holt Distributors:

a. Enter and edit territories (territory number and name).

b. Enter and edit sales reps (sales rep number, name, address, city, state, zip, MTD sales, YTD sales, MTD commission, YTD commission, and commission rate). Each sales rep represents a single territory.

c. Enter and edit customers (customer number, name, first line of address, second line of address, city, state, zip, MTD sales, YTD sales, current balance, and credit limit). A customer can have a different name and address to which goods are shipped (called the "ship to" address). Each customer has a single sales rep who is located in a single territory. The sales rep must represent the territory in which the customer is located.

d. Enter and edit parts (part number, description, price, MTD and YTD sales, units on hand, units allocated, and reorder point). Units allocated is the number of units that are currently present on some open orders. The reorder point is the lowest value acceptable for units on hand without the product being reordered. On the stock status report, which will be described later, an asterisk indicates any part for which the number of units on hand is less than the reorder point.

e. Enter and edit vendors (vendor number, name, address, city, state, and zip). In addition, for each part supplied by the vendor, enter and edit the part number, the price the vendor charges for the part, the minimum order quantity that the vendor will accept for this part, and the expected lead time for delivery of this part from this vendor.

f. Order entry (order number, date, customer, customer PO number, and order detail lines). An order detail line consists of a part number, a description, the number ordered, and the quoted price. Each order detail line includes a sequence number that is entered by the user. Detail lines on an order must print in the order of this sequence number. The system should calculate and display the order total. After all orders for the day have been entered, the system prints order acknowledgment/picking list reports (see Figure A-18). In addition, for each part ordered, the system must increase the units allocated for the part by the number of units that the customer ordered.

g. The invoicing system has the following requirements:

1. Enter the numbers of the orders to be released. For each order, enter the ship date for invoicing and the freight amount. Indicate whether the order is to be shipped in full or in part. If an order is to be partially shipped, enter the number shipped for each order detail line. The system will generate a unique invoice number for this invoice.

2. Print invoices for each of the released orders. (A sample invoice is shown in Figure A-19.)

3. Update files with information from the printed invoices. For each invoice, the system adds the invoice total to the current invoice total. It also adds the current balance and the MTD and YTD sales for the customer that placed the order. The system also adds the total to the MTD and YTD sales for the sales rep who represents the customer. In addition, the system multiplies the total by the sales rep's commission rate and adds this amount to the MTD commission earned and the YTD commission earned. For each part shipped, the system decreases units on hand and units allocated by the number

of units of the part or parts that were shipped. The system also increases the MTD and YTD sales of the part by the amount of the number of units shipped multiplied by the quoted price.

4. Create an invoice summary record for each invoice printed. These records contain the invoice number, date, customer, sales rep, invoice total, and freight.

5. Delete the released orders.

h. Receive payments on account (customer number, date, and amount). The system assigns each payment a number, adds the payment amount to the total of current payments for the customer, and subtracts the payment amount from the current balance of the customer.

Report requirements. The following is a list of the reports required by Holt Distributors:

a. **Territory List:** For each territory, list the number and name of the territory; the number, name, and address of each sales rep in the territory; and the number, name, and address of each customer represented by these sales reps.

b. **Customer Master List:** For each customer, list the customer number, the bill-to address, and the ship-to address. Also list the number, name, address, city, state, and zip of the sales rep who represents the customer and the number and name of the territory in which the customer is located.

c. **Customer Open Order Report:** This report lists open orders organized by customer. It is shown in Figure A-20.

10/16/2010   **HOLT DISTRIBUTORS**   PAGE 1
CUSTOMER OPEN ORDER REPORT

Order Number	Item Number	Item Description	Order Date	Order Qty	Quoted Price
Customer 1354 - Smith Rentals					
12424	AT414	Lounge Chair	10/02/2010	1	$42.00
Customer 1358 - • • • • • • • •					
•	•	•	•	•	•
•	•	•	•	•	•
•	•	•	•	•	•
•	•	•	•	•	•

FIGURE A-20   Open order report (by customer)

d. **Item Open Order Report:** This report lists open orders organized by item and is shown in Figure A-21.

10/16/2010   **HOLT DISTRIBUTORS**   PAGE 1
ITEM OPEN ORDER REPORT

Item Number	Item Description	Customer Number	Customer Name	Order Number	Order Date	Order Qty	Quoted Price
AT414	Lounge Chair	1354	Smith Rentals	12424	10/02/2010	1	$42.00
		1358	Kayland Enterprises	12489	10/03/2010	8	$42.00
			Total on order -			9	
BT222	Arm Chair	1358	Kayland Enterprises	12424	10/03/2010	3	$51.00
			•	•	•	•	•
			•	•	•	•	•
			•	•	•	•	•

FIGURE A-21   Open order report (by item)

e. **Daily Invoice Register:** For each invoice produced on a given day, list the invoice number, invoice date, customer number, customer name, sales amount, freight, and invoice total. A sample of this report is shown in Figure A-22.

10/16/2010		**HOLT DISTRIBUTORS**					PAGE 1
		DAILY INVOICE REGISTER FOR 10/15/2010					
Invoice Number	Invoice Date	Customer Number	Customer Name	Sales Amount	Freight	Invoice Amount	
11025	10/15/2010	1354	Smith Rentals	$414.00	$42.50	$456.50	
•	•	•	•	•	•	•	
•	•	•	•	•	•	•	
•	•	•	•	•	•	•	
•	•	•	•	$2,840.50	$238.20	$3,078.70	

FIGURE A-22   Daily invoice register

f. **Monthly Invoice Register:** The monthly invoice register has the same format as the daily invoice register, but it includes data for all invoices that occurred during the selected month.

g. **Stock Status Report:** For each part, list the part number, description, price, MTD and YTD sales, units on hand, units allocated, and reorder point. For each part for which the number of units on hand is less than the reorder point, an asterisk should appear at the far right of the report.

h. **Reorder Point List:** This report has the same format as the stock status report. Other than the title, the only difference is that parts for which the number of units on hand is greater than or equal to the reorder point will not appear on this report.

i. **Vendor Report:** For each vendor, list the vendor number, name, address, city, state, and zip. In addition, for each part supplied by the vendor, list the part number, the description, the price the vendor charges for the part, the minimum order quantity that the vendor will accept for this part, and the expected lead time for delivery of this part from the vendor.

j. **Daily Cash Receipts Journal:** For each payment received on a given day, list the number and name of the customer that made the payment and the payment amount. A sample report is shown in Figure A-23.

10/05/2010	**HOLT DISTRIBUTORS**		PAGE 1
	DAILY CASH RECEIPTS JOURNAL		
Payment Number	Customer Number	Customer Name	Payment Amount
•	•	•	•
5807	1354	Smith Rentals	$1,000.00
•	•	•	•
			$12,235.50

FIGURE A-23   Daily cash receipts journal

k. **Monthly Cash Receipts Journal:** The monthly cash receipts journal has the same format as the daily cash receipts journal, but it includes all cash receipts for the month.

l. **Customer Mailing Labels:** A sample of the three-across mailing labels printed by the system is shown in Figure A-24.

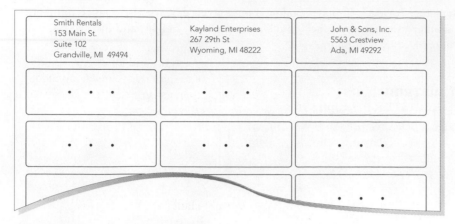

FIGURE A-24   Customer mailing labels

m. **Statements:** The system must produce a monthly statement for each active customer. A sample statement is shown in Figure A-25.

11/01/2010   **HOLT DISTRIBUTORS**
                 146 NELSON PLACE
                 BRONSTON, MI 49802

Smith Rentals   Customer Number: 1354
  153 Main St.   Sales Rep: 10 - Brown, Sam
  Suite 102
  Grandville, MI 49494   Limit: $5,000.00

Invoice Number	Date	Description	Total Amount
10945	10/02/2010	Invoice	$1,230.00
	10/05/2010	Payment	$1,000.00CR
11025	10/15/2010	Invoice	$456.50
	10/22/2010	Payment	$500.00CR

Over 90 $0.00	Over 60 $198.50		
Over 30 $490.20	Current $1,686.50	Total Due >>>>>>	$2,325.20
Previous Balance	Current Invoices	Current Payments	
$2,138.70	$1,686.50	$1,500.00	

FIGURE A-25   Statement for Holt Distributors

n. **Monthly Sales Rep Commission Report:** For each sales rep, list his or her number, name, address, MTD sales, YTD sales, MTD commission earned, YTD commission earned, and commission rate.

o. **Aged Trial Balance:** The aged trial balance report contains the same information that is printed on each customer's statement.

Month-end processing. Month-end processing consists of the following actions that occur at the end of each month:

a. Update customer account information. In addition to the customer's actual balance, the system must maintain the following records: current debt, debt incurred within the last 30 days, debt that is more than 30 days past due but less than 60 days past due, debt that is 60 or more days past due but less than 90 days past due, and debt that is 90 or more days past due.

The system updates the actual balance, the current invoice total, and the current payment total when it produces a new invoice or receives a payment; however, the system updates these aging figures only at the end of the month. The actual update process is as follows:

1. The system processes payments received within the last month and credits these payments to the past due amount for 90 or more days. The system then credits any additional payment to the 60 or more days past due amount, then to the more than 30 days past due amount, and then to the current debt amount (less than 30 days).

2. The system "rolls" the amounts by adding the 60 or more days past due amount to the 90 or more days past due amount and by adding the more than 30 days past due amount to the 60 or more days past due amount. The current amount becomes the new more than 30 days past due amount. Finally, the current month's invoice total becomes the new current amount.

3. The system prints the statements and the aged trial balances.

4. The system sets the current invoice total to zero, sets the current payment total to zero, and sets the previous balance to the current balance in preparation for the next month. To illustrate, assume before the update begins that the amounts for customer 1354 are as follows:

```
Current Balance:  $2,375.20    Previous Balance: $2,138.70
Current Invoices: $1,686.50             Current:    $490.20
Current Payments: $1,500.00             Over 30:    $298.50
                                        Over 60:    $710.00
                                        Over 90:    $690.00
```

The system subtracts the current payments ($1,500.00) from the over 90 amount ($690.00), reduces the over 90 amount to zero, and calculates an excess payment of $810.00. The system subtracts this excess payment from the over 60 amount ($710.00), reduces the over 60 amount to zero, and calculates an excess payment of $100.00. The system then subtracts the excess payment from the over 30 amount ($298.50) and reduces this amount to $198.50. At this point, the system rolls the amounts and sets the current amount to the current invoice total, producing the following:

```
Current Balance:  $2,375.20    Previous Balance:  $2,138.70
Current Invoices: $1,686.50             Current:  $1,686.50
Current Payments: $1,500.00             Over 30:    $490.20
                                        Over 60:    $198.50
                                        Over 90:      $0.00
```

The system then produces statements and the aged trial balance and updates the Previous Balance, Current Invoices, and Current Payments amounts, yielding the following:

```
Current Balance: $2,375.20     Previous Balance:  $2,375.20
Current Invoices:    $0.00              Current:  $1,686.50
Current Payments:    $0.00              Over 30:    $490.20
                                        Over 60:    $198.50
                                        Over 90:      $0.00
```

b. Print the monthly invoice register and the monthly cash receipts journal.

c. Print a monthly sales rep commission report.

d. Set all MTD fields to zero. If necessary, set all YTD fields to zero.

e. Remove all cash receipts and invoice summary records. (In practice, such records would be moved to a historical type of database for future reference. For the purposes of this assignment, you will omit this step.)

SQL REFERENCE

You can use this appendix to obtain details concerning important components and syntax for SQL. Items are arranged alphabetically. Each item contains a description and, where appropriate, both an example and a description of the query results. Some SQL commands also include a description of the clauses associated with them. For each clause, there is a brief description and an indication of whether the clause is required or optional.

ALTER TABLE

Use the ALTER TABLE command to change a table's structure. As shown in Figure B-1, you type the ALTER TABLE command, followed by the table name, and then the alteration to perform. (*Note:* In Access, you usually make these changes to a table in Design view rather than using ALTER TABLE.)

Clause	Description	Required?
ALTER TABLE *table name*	Indicates the name of table to be altered.	Yes
alteration	Indicates the type of alteration to be performed.	Yes

FIGURE B-1   ALTER TABLE command

The following command alters the Customer table by adding a new column named CustType:

```
ALTER TABLE Customer
ADD CustType CHAR(1)
;
```

The following command alters the Customer table by changing the length of the CustomerName column:

```
ALTER TABLE Customer
CHANGE COLUMN CustomerName TO CHAR(50)
;
```

The following command alters the Part table by deleting the Warehouse column:

```
ALTER TABLE Part
DELETE Warehouse
;
```

COLUMN OR EXPRESSION LIST (SELECT CLAUSE)

To select columns, use a SELECT clause with the list of columns separated by commas. The following SELECT clause selects the CustomerNum, CustomerName, and Balance columns:

```
SELECT CustomerNum, CustomerName, Balance
```

Use an asterisk in a SELECT clause to select all columns in the table. The following SELECT command selects all columns in the Part table:

```
SELECT *
FROM Part
;
```

Computed Fields

You can use a computation in place of a field by typing the computation. For readability, you can type the computation in parentheses, although it is not necessary to do so.

The following SELECT clause selects the CustomerNum and CustomerName columns as well as the results of subtracting the Balance column from the CreditLimit column:

```
SELECT CustomerNum, CustomerName, CreditLimit-Balance
```

Functions

You can use aggregate functions in a SELECT clause. The most commonly used functions are AVG (to calculate an average), COUNT (to count the number of rows), MAX (to determine the maximum value), MIN (to determine the minimum value), and SUM (to calculate a total).

The following SELECT clause calculates the average balance:

```
SELECT AVG(Balance)
```

CONDITIONS

A condition is an expression that can be evaluated as either true or false. When you use a condition in a WHERE clause, the results of the query contain those rows for which the condition is true. You can create simple conditions and compound conditions using the BETWEEN, LIKE, and IN operators, as described in the following sections.

Simple Conditions

A simple condition includes the field name, a comparison operator, and another field name or a value. The available comparison operators are = (equal to), < (less than), > (greater than), <= (less than or equal to), >= (greater than or equal to), and < > (not equal to).

The following WHERE clause uses a condition to select rows in which the balance is greater than the credit limit:

```
WHERE Balance>CreditLimit
```

Compound Conditions

Compound conditions are formed by connecting two or more simple conditions using one or both of the following operators: AND and OR. You can also precede a single condition with the NOT operator to negate a condition. When you connect simple conditions using the AND operator, all the simple conditions must be true for the compound condition to be true. When you connect simple conditions using the OR operator, the compound condition will be true whenever any of the simple conditions are true. Preceding a condition with the NOT operator reverses the truth or falsity of the original condition. That is, if the original condition is true, the new condition will be false; if the original condition is false, the new one will be true.

The following WHERE clause is true if those parts for which the warehouse number is equal to 3 *or* the number of units on hand is greater than 20:

```
WHERE Warehouse='3'
OR OnHand>20
```

The following WHERE clause is true if those parts for which *both* the warehouse number is equal to 3 *and* the number of units on hand is greater than 20:

```
WHERE Warehouse='3'
AND OnHand>20
```

The following WHERE clause is true if the warehouse number is not equal to 3:

```
WHERE NOT (Warehouse='3')
```

BETWEEN Conditions

You can use the BETWEEN operator to determine whether a value is within a range of values. The following WHERE clause is true if the balance is between 1,000 and 5,000:

```
WHERE Balance BETWEEN 1000 AND 5000
```

LIKE Conditions

LIKE conditions use wildcards to select rows. Use the percent sign (%) to represent any collection of characters. The condition LIKE '%Oxford%' will be true for data consisting of any character or characters followed by the letters "Oxford" followed by any other character or characters. Another wildcard is the underscore character (_), which represents any individual character. For example, 'T_m' represents the letter *T* followed by any single character followed by the letter *m* and would be true for a collection of characters such as Tim, Tom, or T3m.

Note: In Access SQL, the asterisk (*) is used as a wildcard to represent any collection of characters. (In MySQL, the percent sign (%) is used as a wildcard to represent any collection of characters.) Another wildcard in Access SQL is the question mark (?), which represents any individual character. Many versions of SQL, including MySQL, use the underscore (_) instead of the question mark to represent any individual character.

The following WHERE clause is true if the value in the Street column is Ox, ford, Oxford, or any other value that contains "Oxford":

```
WHERE Street LIKE '%Oxford%'
```

Access version:

```
WHERE Street LIKE '*Oxford*'
```

IN Conditions

You can use the IN operator to determine whether a value is in some specific collection of values. The following WHERE clause is true if the credit limit is 7,500, 10,000, or 15,000:

```
WHERE CreditLimit IN (7500, 10000, 15000)
```

The following WHERE clause is true if the part number is in the collection of part numbers located in warehouse 3:

```
WHERE PartNum IN
(SELECT PartNum
FROM Part
WHERE Warehouse='3')
```

CREATE INDEX

Use the CREATE INDEX command to create an index for a table. Figure B-2 describes the CREATE INDEX command.

Clause	Description	Required?
CREATE INDEX *index name*	Indicates the name of the index.	Yes
ON *table name*	Indicates the table for which the index is to be created.	Yes
column list	Indicates the column or columns on which the index is to be tested	Yes

FIGURE B-2   CREATE INDEX command

The following CREATE INDEX command creates an index named RepBal for the Customer table on the combination of the RepNum and Balance columns:

```
CREATE INDEX RepBal
ON Customer (RepNum, Balance)
;
```

CREATE TABLE

Use the CREATE TABLE command to create a table by describing its layout. Figure B-3 describes the CREATE TABLE command.

Clause	Description	Required?
CREATE TABLE *table name*	Indicates the name of table to be created.	Yes
(column and data type list)	Indicates the columns that make up the table along with their corresponding data types (see the "Data Types" section).	Yes

FIGURE B-3 CREATE TABLE command

The following CREATE TABLE command creates the Rep table and its associated columns and data types:

```
CREATE TABLE Rep
(RepNum  CHAR(2),
LastName  CHAR(15),
FirstName  CHAR(15),
Street  CHAR(15),
City  CHAR(15),
State  CHAR(2),
Zip  CHAR(5),
Commission  DECIMAL(7,2),
Rate  DECIMAL(3,2) )
;
```

Access version:

```
CREATE TABLE Rep
(RepNum  CHAR(2),
LastName  CHAR(15),
FirstName  CHAR(15),
Street  CHAR(15),
City  CHAR(15),
State  CHAR(2),
Zip  CHAR(5),
Commission  CURRENCY,
Rate  NUMBER )
;
```

Note: Unlike other SQL implementations, Access doesn't have a DECIMAL data type. To create numbers with decimals, you must use either the CURRENCY or NUMBER data type. Use the CURRENCY data type for fields that will contain currency values; use the NUMBER data type for all other numeric fields.

CREATE VIEW

Use the CREATE VIEW command to create a view. Figure B-4 describes the CREATE VIEW command.

Clause	Description	Required?
CREATE VIEW *view name* AS	Indicates the name of the view to be created.	Yes
query	Indicates the defining query for the view.	Yes

FIGURE B-4 CREATE VIEW command

The following CREATE VIEW command creates a view named Housewares, which consists of the part number, description, on hand, and price for all rows in the Part table on which the Class is HW:

```
CREATE VIEW Housewares AS
SELECT PartNum, Description, OnHand, Price
FROM Part
WHERE Class = 'HW'
;
```

DATA TYPES

Figure B-5 describes the data types that you can use in a CREATE TABLE command.

Data Type	Description
INTEGER	Stores integers, which are numbers without a decimal part. The valid data range is -2147483648 to 2147483647. You can use the contents of INTEGER fields for calculations.
SMALLINT	Stores integers but uses less space than the INTEGER data type. The valid data range is -32768 to 32767. SMALLINT is a better choice than INTEGER when you are certain that the field will store numbers within the indicated range. You can use the contents of SMALLINT fields for calculations.
DECIMAL(p,q)	Stores a decimal number p digits long with q of these digits being decimal places. For example, DECIMAL(5,2) represents a number with three places to the left and two places to the right of the decimal. You can use the contents of DECIMAL fields for calculations.
CHAR(n)	Stores a character string n characters long. You use the CHAR type for fields that contain letters and other special characters and for fields that contain numbers that will not be used in calculations. Because neither sales rep numbers nor customer numbers will be used in any calculations, for example, both of them are assigned CHAR as the data type. (Some DBMSs, such as Access, use TEXT rather than CHAR; but the two data types mean the same thing.)
DATE	Stores dates in the form DD-MON-YYYY or MM/DD/YYYY. For example, May 12, 2010, could be stored as 12-MAY-2010 or 5/12/2010.

FIGURE B-5   Data types

DELETE ROWS

Use the DELETE command to delete one or more rows from a table. Figure B-6 describes the DELETE command.

Clause	Description	Required?
DELETE FROM *table name*	Indicates the table from which the row or rows are to be deleted.	Yes
WHERE *condition*	Indicates a condition. Those rows for which the condition is true will be retrieved and deleted.	No (If you omit the WHERE clause, all rows will be deleted.)

FIGURE B-6   DELETE command

The following DELETE command deletes any row from the OrderLine table on which the part number is BV06:

```
DELETE
FROM OrderLine
WHERE PartNum='BV06'
;
```

DROP INDEX

Use the DROP INDEX command to delete an index, as shown in Figure B-7.

Clause	Description	Required?
DROP INDEX *index name*	Indicates the name of index to be dropped.	Yes

FIGURE B-7   DROP INDEX command

The following DROP INDEX command deletes the index named RepBal:

```
DROP INDEX RepBal
;
```

DROP TABLE

Use the DROP TABLE command to delete a table, as shown in Figure B-8.

Clause	Description	Required?
DROP INDEX *table name*	Indicates the name of table to be dropped.	Yes

FIGURE B-8 DROP TABLE command

The following DROP TABLE command deletes the table named SmallCust:

```
DROP TABLE SmallCust
;
```

GRANT

Use the GRANT statement to grant privileges to a user. Figure B-9 describes the GRANT statement.

Clause	Description	Required?
GRANT *privilege*	Indicates the type of privilege(s) to be granted.	Yes
ON *database object*	Indicates the database object(s) to which the privilege(s) pertain.	Yes
TO *user name*	Indicates the user(s) to whom the privilege(s) are to be granted.	Yes

FIGURE B-9 GRANT statement

The following GRANT statement grants the user named Johnson the privilege of selecting rows from the Rep table:

```
GRANT SELECT ON Rep TO Johnson
;
```

INSERT

Use the INSERT command and the VALUES clause to insert a row into a table by specifying the values for each of the columns. As shown in Figure B-10, you must indicate the table into which to insert the values and then list the values to insert in parentheses.

Clause	Description	Required?
INSERT INTO *table name*	Indicates the name of the table into which the row will be inserted.	Yes
VALUES *(values list)*	Indicates the values for each of the columns on the new row.	Yes

FIGURE B-10 INSERT command

The following INSERT command inserts the values shown in parentheses as a new row in the Rep table:

```
INSERT INTO Rep
VALUES
('16','Rands','Sharon','826 Raymond','Altonville','FL','32543',0.00,0.05)
;
```

INTEGRITY

You can use the ALTER TABLE command with an appropriate CHECK, PRIMARY KEY, or FOREIGN KEY clause to specify integrity. Figure B-11 describes the ALTER TABLE command for specifying integrity.

Clause	Description	Required?
ALTER TABLE *table name*	Indicates the table for which integrity is being specified.	Yes
integrity clause	CHECK, PRIMARY KEY, or FOREIGN KEY	Yes

FIGURE B-11   Integrity options

The following ALTER TABLE command changes the Part table so that the only legal values for the Class column are AP, HW, and SG:

```
ALTER TABLE Part
CHECK (Class IN ('AP','HW','SG') )
;
```

The following ALTER TABLE command changes the Rep table so that the RepNum column is the table's primary key:

```
ALTER TABLE Rep
ADD PRIMARY KEY(RepNum)
;
```

The following ALTER TABLE command changes the Customer table so that the RepNum column in the Customer table is a foreign key referencing the primary key of the Rep table:

```
ALTER TABLE Customer
ADD FOREIGN KEY (RepNum) REFERENCES Rep
;
```

JOIN

To join tables, use a SELECT command in which both tables appear in the FROM clause and the WHERE clause contains a condition to relate the rows in the two tables. The following SELECT statement lists the customer number, customer name, rep number, first name, and last name by joining the Rep and Customer tables using the RepNum fields in both tables:

```
SELECT CustomerNum, CustomerName, Customer.RepNum, FirstName, LastName
FROM Rep, Customer
WHERE Rep.RepNum = Customer.RepNum
;
```

Note: Many implementations of SQL also allow a special JOIN operator to join tables. The following command uses the JOIN operator to produce the same result as the previous query:

```
SELECT CustomerNum, CustomerName, Customer.RepNum, FirstName, LastName
FROM Rep
INNER JOIN Customer
ON Rep.RepNum = Customer.RepNum
;
```

REVOKE

Use the REVOKE statement to revoke privileges from a user. Figure B-12 describes the REVOKE statement.

Clause	Description	Required?
REVOKE *privilege*	Indicates the type of privilege(s) to be revoked.	Yes
ON *database object*	Indicates the database object(s) to which the privilege pertains.	Yes
FROM *user name*	Indicates the user name(s) from whom the privilege(s) are to be revoked.	Yes

FIGURE B-12   REVOKE statement

The following REVOKE statement revokes the SELECT privilege for the Rep table from the user named Johnson:

```
REVOKE SELECT ON Rep FROM Johnson
;
```

SELECT

Use the SELECT command to retrieve data from a table or from multiple tables. Figure B-13 describes the SELECT command.

Clause	Description	Required?
SELECT *column or expression list*	Indicates the column(s) and/or expression(s) to be retrieved.	Yes
FROM *table list*	Indicates the table(s) required for the query.	Yes
WHERE *condition*	Indicates one or more conditions. Only the rows for which the condition(s) are true will be retrieved.	No (If you omit the WHERE clause, all rows will be retrieved.)
GROUP BY *column list*	Indicates the column(s) on which rows are to be grouped.	No (If you omit the GROUP BY clause, no grouping will occur.)
HAVING *condition involving groups*	Indicates a condition for groups. Only groups for which the condition is true will be included in query results. Use the HAVING clause only if the query output is grouped.	No (If you omit the HAVING clause, all groups will be included.)
ORDER BY *column or expression list*	Indicates the column(s) on which the query output is to be sorted.	No (If you omit the ORDER BY clause, no sorting will occur.)

FIGURE B-13 SELECT command

The following SELECT command groups and orders rows by rep number. It displays the rep number, the count of the number of customers having this rep, and the average balance of these customers. It renames the count as NumCustomers and the average balance as AverageBalance. The HAVING clause restricts the reps to be displayed to only those having fewer than four customers.

```
SELECT RepNum, COUNT(*) AS NumCustomers, AVG(Balance) AS AverageBalance
FROM Customer
GROUP BY RepNum
HAVING COUNT(*)<4
ORDER BY RepNum
;
```

SELECT INTO

Use the SELECT command with an INTO clause to insert the rows retrieved by a query into a table. As shown in Figure B-14, you must indicate the name of the table into which the row(s) will be inserted and the query whose results will be inserted into the named table.

Clause	Description	Required?
SELECT field list	Indicates the list of fields to be selected.	Yes
INTO *table name*	Indicates the name of the table into which the row(s) will be inserted.	Yes
remainder of query	Indicates the remainder of the query (for example, FROM clause and WHERE clause) whose results will be inserted into the table.	Yes

FIGURE B-14 SELECT command with INTO clause

The following SELECT command with an INTO clause inserts rows selected by a query into the SmallCust table:

```
SELECT *
INTO SmallCust
FROM Customer
WHERE CreditLimit<=7500
;
```

SUBQUERIES

In some cases, it is useful to obtain the results you want in two stages. You can do so by placing one query inside another. The inner query is called a subquery and is evaluated first. After the subquery has been evaluated, the outer query can be evaluated.

The following command contains a subquery that produces a list of part numbers located in warehouse 3. The outer query then produces those order numbers in the OrderLine table that are on any rows containing a part number in the list.

```
SELECT OrderNum
FROM OrderLine
WHERE PartNum IN
(SELECT PartNum
FROM Part
WHERE Warehouse='3')
;
```

UNION

Connecting two SELECT commands with the UNION operator produces all the rows that would be in the results of the first command, the second command, or both.

The following query displays the customer number, last name, and first name of all customers that are represented by sales rep 65 or that have orders or both:

```
SELECT CustomerNum, CustomerName
FROM Customer
WHERE RepNum='35'
UNION
SELECT Customer.CustomerNum, CustomerName
FROM Customer, Orders
WHERE Customer.CustomerNum=Orders.CustomerNum
;
```

UPDATE

Use the UPDATE command to change the contents of one or more rows in a table. Figure B-15 describes the UPDATE command.

Clause	Description	Required?
UPDATE *table name*	Indicates the table whose contents will be changed.	Yes
SET *column = expression*	Indicates the column to be changed, along with an expression that provides the new value.	Yes
WHERE *condition*	Indicates a condition. The change will occur only on those rows for which the condition is true.	No (If you omit the WHERE clause, all rows will be updated.)

FIGURE B-15   UPDATE command

The following UPDATE command changes to 1445 Rivard the street address on the row in the Customer table on which the customer number is 524:

```
UPDATE Customer
SET Street='1445 Rivard'
WHERE CustomerNum='524'
;
```

APPENDIX **C**

"HOW DO I" REFERENCE

This appendix answers frequently asked questions about how to accomplish a variety of tasks using SQL. Use the second column to locate the correct section in Appendix B that answers your question.

How do I?	Review the Named Section(s) in Appendix B
Add columns to an existing table?	ALTER TABLE
Add rows?	INSERT
Calculate a statistic (sum, average, maximum, minimum, or count)?	1. SELECT 2. Column or Expression List (SELECT clause) (Use the appropriate function in the query.)
Change rows?	UPDATE
Create a data type for a column?	1. Data Types 2. CREATE TABLE
Create a table?	CREATE TABLE
Create a view?	CREATE VIEW
Create an index?	CREATE INDEX
Delete a table?	DROP TABLE
Delete an index?	DROP INDEX
Delete rows?	DELETE Rows
Drop a table?	DROP TABLE
Drop an index?	DROP INDEX
Grant a privilege?	GRANT
Group data in a query?	SELECT (Use a GROUP BY clause.)
Insert rows using a query?	SELECT INTO
Insert rows?	INSERT
Join tables?	Conditions (Include a WHERE clause to relate the tables.)
Order query results?	SELECT (Use an ORDER BY clause.)
Remove a privilege?	REVOKE
Remove rows?	DELETE Rows
Retrieve all columns?	1. SELECT 2. Column or Expression List (SELECT clause) (Type * in the SELECT clause.)
Retrieve all rows?	SELECT (Omit the WHERE clause.)

FIGURE C-1 How do I? reference

How do I?	Review the Named Section(s) in Appendix B
Retrieve only certain columns?	1. SELECT 2. Column or Expression List (SELECT clause) (Type the list of columns in the SELECT clause.)
Revoke a privilege?	REVOKE
Select all columns?	1. SELECT 2. Column or Expression List (SELECT clause) (Type * in the SELECT clause.)
Select all rows?	SELECT (Omit the WHERE clause.)
Select only certain columns?	1. SELECT 2. Column or Expression List (SELECT clause) (Type the list of columns in the SELECT clause.)
Select only certain rows?	1. SELECT 2. Conditions (Use a WHERE clause.)
Sort query results?	SELECT (Use an ORDER BY clause.)
Specify a foreign key?	Integrity (Use a FOREIGN KEY clause in an ALTER TABLE command.)
Specify a primary key?	Integrity (Use a PRIMARY KEY clause in an ALTER TABLE command.)
Specify a privilege?	GRANT
Specify integrity?	Integrity (Use a CHECK clause in an ALTER TABLE command.)
Specify legal values?	Integrity (Use a CHECK clause in an ALTER TABLE command.)
Update rows?	UPDATE
Use a calculated field?	1. SELECT 2. Column or Expression List (SELECT clause) (Enter a calculation in the query.)
Use a compound condition in a query?	Conditions
Use a compound condition?	1. SELECT 2. Conditions (Use simple conditions connected by AND, OR, or NOT in a WHERE clause.)
Use a condition in a query?	1. SELECT 2. Conditions (Use a WHERE clause.)
Use a subquery?	Subqueries
Use a wildcard?	1. SELECT 2. Conditions (Use LIKE and a wildcard in a WHERE clause.)
Use UNION operation?	UNION (Connect two SELECT commands with UNION.)

FIGURE C-1   How do I? reference (continued)

ANSWERS TO ODD-NUMBERED REVIEW QUESTIONS

CHAPTER 1—INTRODUCTION TO DATABASE MANAGEMENT

1. Redundancy is the duplication of data or the storing of the same data in more than one place. Redundancy wastes space, makes the updating of data more cumbersome and time-consuming, and can lead to inconsistencies.

3. An entity is a person, place, object, event, or idea for which you want to store and process data. An attribute, which is also called a field or column in many database systems, is a characteristic or property of an entity.

5. A database is a structure that can store information about multiple types of entities, the attributes of those entities, and the relationships among the entities.

7. An E-R diagram represents a database in a visual way by using a rectangle for each entity, using a line to connect two entities that have a relationship, and placing a dot at the end of a line to indicate the "many" part of a one-to-many relationship.

9. Database design is the process of determining the table structure of the desired database.

11. It is possible to get more information from the same amount of data by using a database approach as opposed to a nondatabase approach because all data is stored in a single database, instead of being stored in dozens of separate files, making the process of obtaining information quicker, easier, and even possible in certain situations.

13. The DBA (database administrator or database administration) is the central person or group in an organization in charge of the database and the DBMS that runs the database. The DBA attempts to balance the needs of individuals and the overall needs of the organization.

15. An integrity constraint is a rule that the data in a database must follow. A database has integrity when the data in it satisfies all established integrity constraints. A good DBMS should provide an opportunity for users to incorporate these integrity constraints when they design the database. The DBMS then should ensure that these constraints are not violated.

17. Data independence is the property that lets you change the structure of a database without requiring you to change the programs that access the database. With data independence, you easily can change the structure of the database when the need arises.

19. The more complex a product is in general (and a DBMS, in particular, is complex), the more difficult it is to understand and correctly apply its features. As a result of this complexity, serious problems may result from mistakes made by users and designers of the DBMS.

21. The great complexity of a database structure makes recovery more difficult. In addition, many users update the data at the same time, which means that recovering the database involves not only restoring it to the last state in which it was known to be correct, but also performing the complex task of redoing all the updates made since that time.

CHAPTER 2—THE RELATIONAL MODEL 1: INTRODUCTION, QBE, AND RELATIONAL ALGEBRA

1. A relation is a two-dimensional table in which (1) the entries in the table are single-valued; (2) each column has a distinct name; (3) all of the values in a column are values of the same attribute; (4) the order of the columns is immaterial; (5) each row is distinct; and (6) the order of the rows is immaterial.

3. An unnormalized relation is a structure that satisfies all the properties of a relation except the restriction that entries must be single-valued. It is not a relation.

5. In the shorthand representation, each table is listed, and after each table, all the columns of the table are listed in parentheses. Primary keys are underlined.

```
Branch (BranchNum, BranchName, BranchLocation, NumEmployees)
Publisher (PublisherCode, PublisherName, City)
Author (AuthorNum, AuthorLast, AuthorFirst)
Book (BookCode, Title, PublisherCode, Type, Price, Paperback)
Wrote (BookCode, AuthorNum, SequenceNum)
Inventory (BookCode, BranchNum, OnHand)
```

7. The primary key is the column or collection of columns that uniquely identifies a given row. The primary key of the Branch table is BranchNum. The primary key of the Publisher table is PublisherCode. The primary key of the Author table is AuthorNum. The primary key of the Book table is BookCode. The primary key of the Wrote table is the concatenation (combination) of BookCode and AuthorNum. The primary key of the Inventory table is the concatenation of BookCode and BranchNum.

9. Enter the criteria in the Criteria row for the appropriate field name.

11. Type the computation instead of a field name in the design grid. Alternatively, you can enter the computation in the Zoom dialog box.

13. Indicate the appropriate sort order (Ascending or Descending) in the Sort row of the design grid.

15. Include the field lists from both tables in the query design. Provided the tables have matching fields, a join line will connect the tables. Include the desired fields from either table in the design grid.

17. Use a delete query when you want to delete all rows satisfying some criteria.

19. Relational algebra is a theoretical way of manipulating a relational database. Relational algebra includes operations that act on existing tables to produce new tables, similar to the way the operations of addition and subtraction act on numbers to produce new numbers in the mathematical algebra with which you are familiar.

21. The PROJECT command selects only the specified columns.

23. The UNION command selects all rows that are in the first table, in the second table, or both.

25. The INTERSECT command selects all rows that are in both tables.

27. The PRODUCT command (mathematically called the Cartesian product) is the table obtained by concatenating every row in the first table with every row in the second table.

CHAPTER 3—THE RELATIONAL MODEL 2: SQL

1. To create a table in SQL, use a CREATE TABLE command. The word TABLE is followed by the name of the table to be created and then by the names and data types of the columns (fields) that make up the table. The data types you can use are INTEGER (large negative and positive whole numbers), SMALLINT (whole numbers from –32,768 to 32,767), DECIMAL (numbers that have a decimal part), CHAR (alphanumeric strings), and DATE (date values).

3. A compound condition is formed by connecting two or more simple conditions using one or both of the following operators: AND and OR. You can also precede a single condition with the NOT operator to negate a condition. When you connect simple conditions using the AND operator, all the simple conditions must be true for the compound condition to be true. When you connect simple conditions using the OR operator, the compound condition will be true whenever any of the simple conditions are true.

5. To use the LIKE or IN operators in an SQL query, include them in the WHERE clause. A character string containing one or more wildcards is included after the word LIKE. The word IN is followed by a list of values.

7. Use an SQL built-in function (COUNT, SUM, AVG, MAX, and MIN) by including it in the SELECT clause followed by the name of the field to which it applies.

9. To group data in SQL, include the words GROUP BY followed by the field or fields on which the data is to be grouped in the query results. If you group data, you only can include the fields on which you are grouping or statistics in the SELECT clause.

11. To qualify the name of a field in an SQL query, precede the field with the name of the table to which it belongs, followed by a period. It is necessary to qualify a field if the field name occurs in more than one of the tables listed in the FROM clause.

13. The update commands in SQL are INSERT, which inserts new rows in a table; UPDATE, which changes all the rows in a table that satisfy some condition; and DELETE, which deletes all the rows in a table that satisfy some condition.

CHAPTER 4—THE RELATIONAL MODEL 3: ADVANCED TOPICS

1. A view is an individual user's picture of the database. It is defined using a defining query. The data in the view never actually exists in the form described in the view. Rather, when a user accesses the view, his or her query is merged with the defining query of the view to form a query that pertains to the whole database.

3. a.
```
CREATE VIEW PartOrder AS
SELECT Part.PartNum, Description, Price, OrderNum, OrderDate,
NumOrdered, QuotedPrice
FROM Part, OrderLine, Orders
WHERE Part.PartNum = OrderLine.PartNum
AND Orders.OrderNum = OrderLine.OrderNum;
```
b.
```
SELECT PartNum, Description, OrderNum, QuotedPrice
FROM PartOrder
WHERE QuotedPrice > 100;
```
c.
```
SELECT Part.PartNum, Description, OrderNum, QuotedPrice
FROM Part, OrderLine
WHERE Part.PartNum = OrderLine.PartNum
AND QuotedPrice > 100;
```

5. The GRANT statement is used to assign privileges to users of a database. It relates to security because a user who does not have the privilege of accessing a certain portion of a database cannot access that portion of the database. The privileges that can be assigned include the privilege of selecting rows from a table, inserting new rows, and updating existing rows. The REVOKE command is used to revoke privileges.

7.
```
REVOKE SELECT ON Part FROM Stillwell;
```

9. a.
```
SELECT Name
FROM Systables
WHERE Creator = 'your name';
```
b.
```
SELECT Colname, Coltype
FROM Syscolumns
WHERE Tbname = 'Customer';
```
c.
```
SELECT Tbname
FROM Syscolumns
WHERE Colname = 'PartNum';
```

11. Null is a special value that represents missing information. Nulls are used when a value is either unknown or inapplicable. The primary key cannot accept null values. With a null value in the primary key, the primary key could not fulfill its main purpose of being the unique identifier for records in a table.

13. Adding an order to the Orders table on which the customer number does not match a customer number in the Customer table would violate referential integrity. In addition, changing the customer number on a record in the Orders table to a number that does not match a customer number in the Customer table would also violate referential integrity. If deletes do not cascade,

deleting a customer that has orders would violate referential integrity. If deletes cascade, such a customer can be deleted, in which case all orders for that customer will automatically be deleted.

15. Stored procedures are special files containing a collection of SQL statements that will be executed frequently. The statements in a stored procedure are compiled and optimized, enabling the stored procedure to execute as efficiently and as rapidly as possible. It also makes the execution of the commands in the stored procedure simpler than if the user had to type the command each time he or she wanted to use it.

CHAPTER 5—DATABASE DESIGN 1: NORMALIZATION

1. A column (attribute) B is functionally dependent on another column A (or possibly a collection of columns) if each value for A in the database is associated with exactly one value of B.
3. Column A (or a collection of columns) is the primary key for a relation (table) R, if: Property 1—all columns in R are functionally dependent on A; and Property 2—no subcollection of the columns in A (assuming A is a collection of columns and not just a single column) also has Property 1.
5. A table is in first normal form if it does not contain a repeating group.
7. A table is in third normal form if it is in second normal form and the only determinants it contains are candidate keys. If a table is not in third normal form, redundant data will cause wasted space and update problems. Inconsistent data might also be a problem.
9. An interrelation constraint is a condition that involves two or more relations. Requiring the value of a RepNum on a row in the Customer relation to match a value of RepNum in the Rep relation is an example of an interrelation constraint. The interrelation constraints are addressed through foreign keys.
11. Patient (<u>PatientNum</u>, HouseholdNum, PatientName)
    Household (<u>HouseholdNum</u>, HouseholdName, Street, City, State,
           Zip, Balance)
    Service (<u>ServiceCode</u>, Description, Fee)
    Appointment (<u>PatientNum</u>, <u>ServiceCode</u>, Date)
13. StudentNum → StudentName
    ActivityNum → ActivityName
    CourseNum → Description
    StudentNum → → ActivityNum
    StudentNum → → CourseNum

    Student (<u>StudentNum</u>, StudentName)
    Activity (<u>ActivityNum</u>, ActivityName)
    Course (<u>CourseNum</u>, Description)
    StudentActivity (<u>StudentNum</u>, <u>ActivityNum</u>)
    StudentCourse (<u>StudentNum</u>, <u>CourseNum</u>)

CHAPTER 6—DATABASE DESIGN 2: DESIGN METHOD

1. A user view is the view of data that is necessary to support the operations of a particular user. By considering individual user views instead of the complete design problem, the database design process is greatly simplified.
3. If the design problem were extremely simple, the overall design might not have to be broken down into a consideration of individual user views.
5. The primary key is the column or columns that uniquely identify a given row and that furnish the main mechanism for directly accessing a row in the table. An alternate key is a column or combination of columns that could have functioned as the primary key but was not chosen to do so. A secondary key is a column or combination of columns that is not any other type of key but is of interest for purposes of retrieval. A foreign key is a column or combination of columns in one table whose values are required to match the primary key in another table. Foreign keys furnish the mechanism through which relationships are made explicit.

7. Department (<u>DepartmentNum</u>, DepartmentName)
   Advisor (<u>AdvisorNum</u>, LastName, FirstName, DepartmentNum)
   FK   DepartmentNum → Department
   Course (<u>CourseCode</u>, Description)
   Student (<u>StudentNum</u>, LastName, FirstName, AdvisorNum)
   FK   AdvisorNum → Advisor
   StudentCourse (<u>StudentNum</u>, <u>CourseCode</u>, Grade)
   FK   StudentNum → Student
   FK   CourseCode → Course

9. a. No change is necessary.
   b. Store both the AdvisorNum and DepartmentNum columns in the Student table.

11. The method presented in this text is bottom-up; that is, an approach in which specific user requirements are synthesized into a design. By initially reviewing the requirements and determining a possible list of entities prior to following the steps in this method, you can gain the advantages to both top-down and bottom-up approaches.

13. Many answers are possible. Be sure the functional dependencies you represent are based on reasonable assumptions and that the tables you create are in third normal form based on these dependencies.

15. a. There is a many-to-many-to-many relationship between students, courses, and faculty members.
    b. There is a many-to-many relationship between students and courses. A faculty number functionally determines a course number; that is, each faculty member is associated with exactly one course.
    c. There are separate many-to-many relationships between students and courses, courses and faculty members, and students and faculty members.
    d. There is a many-to-many relationship between students and courses. A given student number-course number combination uniquely determines a faculty number. That is, when a student takes a course, a single faculty member teaches the course.
    e. There is a many-to-many relationship between students and courses. There is also a many-to-many relationship between courses and faculty members. A student number functionally determines a faculty number; that is, each student is associated with exactly one faculty member.

17. If one 3NF relation contains a column that is a determinant for a column in another 3NF relation with the same primary key, merging the relations will produce a relation that is not in third normal form. The following is an example of two such relations:

    Student1 (<u>StudentNum</u>, LastName, FirstName, AdvisorNum)
    Student2 (<u>StudentNum</u>, LastName, FirstName, AdvisorLast, AdvisorFirst)

    The following is the result of merging the Student1 and Student2 relations:

    Student (<u>StudentNum</u>, LastName, FirstName, AdvisorNum, AdvisorLast, AdvisorFirst)

    This table is not in third normal form because

    AdvisorNum → AdvisorLast, AdvisorFirst

CHAPTER 7—DBMS FUNCTIONS

1. In updating and retrieving data, users do not need to know how data is physically structured on disk nor what processes the DBMS uses to manipulate the data. These structures and manipulations are solely the responsibility of the DBMS.

3. Most PC-based DBMSs contain a catalog that maintains metadata about tables, fields, table relationships, views, indexes, users, privileges, and replicated data. Large, expensive DBMSs usually contain a data dictionary that serves as a super catalog containing the same metadata as a catalog and additional metadata such as that needed to track fragmented data.

5. A lost update could occur in a concurrent update situation when two users attempt to update the same data at the same time, and the DBMS does not does not have concurrency update features such as two-phase locking or timestamping.

7. A transaction is a set of steps completed by a DBMS that accomplishes a single user task; the DBMS must successfully complete all transaction steps or none at all for the database to remain in a correct state.

9. Deadlock occurs in a concurrent update situation when the first user is waiting for data that has been locked by a second user, and the second user is waiting for data that has been locked by the first user. Unless outside action occurs, each user could wait for the other's data forever. Deadlock occurs when each of the two users is attempting to access data that is locked by the other user.

11. Recovery is the process of returning the database to a state that is known to be correct from a state known to be incorrect.

13. A DBA uses forward recovery when a catastrophe destroys a database. Forward recovery consists of two steps. First, the DBA copies the most recent database backup over the live database. Second, the DBMS forward recovery feature uses the log to apply after images for committed transactions.

15. Security is the prevention of unauthorized access, either intentional or accidental, to a database.

17. Authentication refers to techniques for identifying the person who is attempting to access the DBMS. Three types of authentication are passwords, biometrics, and database passwords. A password is a string of characters assigned by the DBA to a user that the user must enter to access to the database. Biometrics identify users by physical characteristics such as fingerprints, voiceprints, handwritten signatures, and facial characteristics. A database password is a string of characters assigned by the DBA to a database that users must enter before they can access the database.

19. Permissions specify what kind of access a user has to objects in a database. A workgroup is a group of users, and a DBA usually assigns appropriate permissions to workgroups.

21. Privacy refers to the right of individuals to have certain information about them kept confidential. Privacy and security are related because it is only through appropriate security measures that privacy can be ensured.

23. Data independence is a property that lets you change a database structure without requiring you to change the programs that access the database.

25. Some utility services that a DBMS should provide include: services that let you change the database structure; services that let you add new indexes and delete indexes that are no longer needed; facilities that let you use the services available from your operating system; services that let you export data to and import data from other software products; services that provide support for easy-to-use edit and query capabilities, screen generators, report generators, and so on; support for both procedural and nonprocedural languages; and support for an easy-to-use menu-driven or switchboard-driven interface that allows users to tap into the power of the DBMS without having to resort to a complicated set of commands.

CHAPTER 8—DATABASE ADMINISTRATION

1. The DBA is the database administrator, or the person responsible for the database. The DBA is necessary because his responsibilities are critical to success in a database environment, especially when the database is shared among many users.

3. After the DBA determines the access privileges for each user, the DBA creates security policies and procedures, obtains management approval of the policies and procedures, and then distributes them to authorized users. The DBA uses the DBMS's security features, such as encryption, authentication, authorizations, and views, and uses special security programs, if necessary. Finally, the DBA monitors database usage to detect potential security violations and takes corrective action if a violation occurs.

5. Certain data, although no longer needed in the production database, must be kept for future reference. A data archive is a place for storing this type of data. The use of data archives lets an organization keep data indefinitely, without causing the database to become unnecessarily large. Data can be removed from the database and placed in the data archive, instead of just being deleted.

7. A shared lock permits other users to read the data. An exclusive lock prevents other users from accessing the data in any way.

9. Context-sensitive help means that if a user is having trouble and asks for help, the DBMS will provide assistance for the particular feature being used at the time the user asks for it.

11. The DBA installs the DBMS, makes any changes to its configuration when they are required, determines whether it is appropriate to install a new version of the DBMS when it becomes available, applies any vendor fixes to problems, coordinates problem resolution, and coordinates the users so that their needs are satisfied without unduly affecting other users.

13. The DBA does some of the training of computer users. The DBA coordinates other training, such as that which is provided by software vendors.

15. The production system is the hardware, software, and database for the users; the test system is a separate system that programmers use to develop new programs and modify existing programs.

17. If users access only certain fields in a table, splitting the table results in smaller tables than the original; the lower amount of data moves faster between disk and memory and across a network.

CHAPTER 9—DATABASE MANAGEMENT APPROACHES

1. A distributed database is a single logical database that is physically divided among computers at several sites on a network. A distributed database management system (DDBMS) is a DBMS capable of supporting and manipulating distributed databases.

3. A homogeneous DDDBS is one that has the same local DBMS at each site. A heterogeneous DDBMS is one that does not; there are at least two sites at which the local DBMSs are different. Heterogeneous systems are more complex.

5. Location transparency is the characteristic of a DDBMS that states that users do not need to be aware of the location of data in a distributed database. Data should be accessible at a remote site just as easily as it is at a local site; the only difference should be the response time.

7. Replication transparency is the characteristic that users do not need to be aware of any replication that has taken place in a distributed database. The DDBMS should handle updates to all copies of the data without users being aware of the steps taken by the DDBMS.

9. Fragmentation transparency is the characteristic that users do not need to be aware of any data fragmentation (splitting of data) in a distributed database. Users should feel as if they are using a single central database, even if data is stored at different sites.

11. In a well-designed distributed database, you can often increase its capacity by increasing the capacity at only one site. Also, you can increase capacity through the addition of new sites to the network.

13. Increased efficiency is an advantage in a distributed database because data available locally can be retrieved much more rapidly than data stored on a remote, centralized system.

15. Query processing is more complex in a distributed environment because of the difference between the time it takes to send messages between sites and the time it takes to access a disk.

17. With a two-phase commit, one site (often the site initiating the update) acts as coordinator. In the first phase, the coordinator sends messages to all other sites requesting they prepare to update the database (or acquire all necessary locks). Once the coordinator receives positive replies from all sites, the coordinator sends a message to each site to commit the update. At this point, each site must proceed with the commit process, and all sites must abort if any reply is negative. Two-phase commit guarantees consistency of the database.

19. The information-level design process is unaffected by the fact that the database is distributed.

21. The 12 rules against which you can measure DDBMSs are:

- Local autonomy: No site should depend on another site to perform its functions.

- No reliance on a central site: A DDBMS should not need to rely on one site more than any other site.

- Continuous operation: Performing any function should not shut down the entire distributed database.

- Location transparency: Users should feel as if the entire database is stored at their location.

- Fragmentation transparency: Users should feel as if they are using a single central database.

- Replication transparency: Users should not be aware of any data replication.

- Distributed query processing: A DDBMS must process queries as rapidly as possible even though the data is distributed.

- Distributed transaction management: A DDBMS must effectively manage transaction updates at multiple sites.

- Hardware independence: A DDBMS must be able to run on different types of hardware.

- Operating system independence: A DDBMS must be able to run on different operating systems.

- Network independence: A DDBMS must be able to run on different types of networks.

- DBMS independence: A DDBMS must be heterogeneous.

23. Placing the business functions on the client causes a client maintenance problem. Whenever programmers change the business functions, they must place the updated business functions on every client. Placing the business functions on the server causes a scalability problem and might cause the server to become a bottleneck and degrade the system's responsiveness to clients.

25. Scalability is the ability of a computer system to continue to function well even as utilization of the system increases.

27. Compared to a file server, the advantages of a client/server architecture are: lower network traffic; improved processing distribution; thinner clients; greater processing transparency; increased network, hardware, and software transparency; improved security; decreased costs; and increased scalability.

29. Dynamic Web pages are Web pages whose content changes in response to the different inputs and choices made by Web clients. You can use client-side extensions embedded in HTML documents or contained in separate files, or you can use server-side extensions usually contained in separately executed programs.

31. XML (Extensible Markup Language) is a metalanguage derived from a restricted subset of SGML and is designed for the exchange of data on the Web. Using XML, you can create text documents that follow simple, specific rules for their content, and you can define new tags that define the data in the document and the structure of the data, so that programs running on any platform can interpret and process the document.

33. XSL Transformations (XSLT) define the rules to process an XML document and change it into another document; this other document could be another XML document, an XSL document, an HTML or XHTML document, or most any other type of document.

35. A data warehouse is a subject-oriented, time-variant, nonvolatile collection of data in support of management's decision-making process.

37. A fact table in a data warehouse consists of rows that contain consolidated and summarized data.

39. Using OLAP software, users slice and dice data, drill down data, and roll up data.

41. You can measure OLAP systems against the following 12 rules: users should be able to view data in a multidimensional way; users should not have to know they are using a multidimensional database; users should perceive data as a single user view; the size and complexity of the warehouse should not affect reporting performance; the server portion of the OLAP software should allow the use of different types of clients; each data dimension should have the same structural and operational capabilities; missing data should be handled correctly and efficiently; OLAP should provide secure, concurrent access; users should be able to perform the same operations across any number of dimensions; users should not need to use special interfaces to make their requests; users should be able to report data results any way they want; and OLAP software should allow at least 15 data dimensions and an unlimited number of summary levels.

43. A domain is the set of values that are permitted for an attribute.

45. A method is an action defined for an object. A message is a request to execute a method. To execute the steps in a method, you must send a message to the object.

47. The Unified Modeling Language (UML) is an approach you can use to model all the various aspects of software development for object-oriented systems.

49. A visibility symbol in UML indicates whether other classes can view or update the value in an attribute in a class.

51. Generalization is the relationship between a superclass and a subclass.

Access delay A fixed amount of time required for every message sent over a network.

After image A record that the DBMS places in the journal or log that shows what the data in a row looked liked in the database after a transaction update.

Aggregate function A function to calculate the number of entries, the sum or average of all the entries in a given column, or the largest or smallest of the entries in a given column; also called *function*.

ALTER TABLE The SQL command that is used to change the structure of a table.

Alternate key A candidate key that was not chosen to be the primary key.

AND criterion Combination of criteria in which both criteria must be true.

Apache HTTP Server A free, open source Web server package that runs with most operating systems.

Application server In a three-tier client/server architecture, a computer that performs the business functions and serves as an interface between clients and the database server.

Archive See *data archive*.

Artificial key A column created for an entity to serve solely as the primary key and that is visible to users.

Association A relationship in UML.

Attribute A characteristic or property of an entity; also called a *field* or *column*.

Authentication A technique for identifying the person who is attempting to access a DBMS.

Authorization rule A rule that specifies which user has what type of access to which data in a database.

B2B See *business to business*.

Back-end machine See *server*.

Back-end processor See *server*.

Backup A copy of a database made periodically; the backup is used to recover the database when it has been damaged or destroyed; also called a *save*.

Backward recovery See *rollback*.

Batch processing The processing of a transaction file that contains a group, or "batch," of records to update a database or another file.

Before image A record that the DBMS places in the journal or log that shows what the data in a row looked like in the database before a transaction update.

Binary large object (BLOB) A generic term for a special data type used by relational DBMSs to store complex objects.

Binding The association of operations to actual program code.

Biometrics A technique to identify users of a database or other resource by physical characteristics such as fingerprints, voiceprints, handwritten signatures, and facial characteristics.

BLOB See *binary large object*.

Bottom-up design method A design method in which specific user requirements are synthesized into a design.

Boyce-Codd normal form (BCNF) A relation is in Boyce-Codd normal form if it is in second normal form and the only determinants it contains are candidate keys; also called *third normal form* in this text.

Business to business (B2B) E-commerce between businesses.

Calculated field A field whose value may be computed from other fields in the database; also called *computed field*.

Candidate key A minimal collection of columns (attributes) in a table on which all columns are functionally dependent but that has not necessarily been chosen as the primary key.

Cardinality The number of items that must be included in a relationship; also called *multiplicity*.

Cartesian product The table obtained by concatenating every row in the first table with every row in the second table.

Cascade delete A delete option in which related records are automatically deleted.

Cascade update An update option in which related records are automatically updated.

Catalog A source of data, usually stored in hidden database tables, about the types of entities, attributes, and relationships in a database.

Category The IDEF1X name for an entity subtype.

CHAR(n) The SQL data type for character data.

CHECK The SQL clause that is used to enforce legal-values integrity.

Class The general structure and actions of an object in an object-oriented system.

Class diagram A UML diagram that for each class, shows the name, attributes, and methods of the class, as well as the relationships between the classes in a database.

Client A computer that is connected to a network and that people use to access data stored on a server in a client/server system; also called a *front-end machine* or a *front-end processor*.

Client/server A networked system in which a special site on the network, called the *server*, provides services to the other sites, called the *clients*. Clients send requests for the specific services they need. Software, often including a DBMS, running on the server then processes the requests and sends only the appropriate data and other results back to the clients.

Client-side extension Instructions executed by a Web client to provide dynamic Web page capability. These extensions can be embedded in HTML documents or be contained in separate files that are referenced within an HTML document.

Client-side script See *client-side extension*.

Column A characteristic or property of an entity; also called an *attribute* or a *field*.

Command An instruction by a user that directs a database to perform a certain function.

Commit A special record in a database journal or log that indicates the successful completion of a transaction.

Communications network Several computers configured in such a way that data can be sent from any one computer on the network to any other. Also called a *network*.

Comparison operator See *relational operator*.

Complete category In IDEF1X, a collection of subtypes with the property that every element of the supertype is an element of at least one subtype.

Composite entity An entity in the entity-relationship model used to implement a many-to-many relationship.

Compound condition See *compound criteria*.

Compound criteria Two simple criteria (conditions) in a query that are combined with the AND or OR operators.

Computed field See *calculated field*.

Concatenation The combination of columns. To say a primary key is a concatenation of two columns means that a combination of values in both columns is required to uniquely identify a given row.

Concurrent update A situation when multiple users make updates to the same database at the same time.

Condition See *criterion*.

Context-sensitive help The assistance a DBMS provides for the particular feature being used at the time a user asks for help.

Cookies Small files written on a Web client's hard drive by a Web server.

Coordinator In a distributed network, the site that directs the update to the database for a transaction. Often, it is the site that initiates the transaction.

CREATE INDEX The SQL command that creates an index in a table.

CREATE TABLE The SQL command used to describe the layout of a table. The word *TABLE* is followed by the name of the table to be created and then by the names and data types of the columns (fields) that comprise the table.

Criteria The plural version of the word *criterion*.

Criterion A statement that can be either true or false. In queries, only records for which the statement is true will be included; also called a *condition*.

Cumulative design A design that supports all the user views encountered thus far in a design process.

Data archive A place where a record of certain corporate data is kept; also called an *archive*. Data that is no longer needed in a corporate database but must be retained for future reference is removed from the database and placed in the archive.

Data cube The perceived shape by a user of a multi-dimensional database in a data warehouse.

Data dictionary A catalog, usually found in large, expensive DBMSs, that stores data about the entities, attributes, relationships, programs, and other objects in a database.

Data file A file used to store data about a single entity. It's the computer counterpart to an ordinary paper file you might keep in a file cabinet, an accounting ledger, and so on. Such a file can be thought of as a table.

Data fragmentation   The process of dividing a logical object, such as the records in a table, among the various locations in a distributed database.

Data independence   The property that lets you change the structure of a database without requiring you to change the programs that access the database; examples of these programs are the forms you use to interact with the database and the reports that provide information from the database.

Data mining   The uncovering of new knowledge, patterns, trends, and rules from the data stored in a data warehouse.

Data warehouse   A subject-oriented, integrated, time-variant, nonvolatile collection of data in support of management's decision-making process.

Database   A structure that can store information about multiple types of entities, the attributes of these entities, and the relationships among the entities.

Database administration (DBA)   The individual or group that is responsible for a database.

Database administrator (DBA)   The individual who is responsible for a database, or the head of database administration.

Database design   The process of determining the content and structure of data in a database in order to support some activity on behalf of a user or group of users.

Database Design Language (DBDL)   A relational-like language that is used to represent the result of the database design process.

Database management system (DBMS)   A program, or a collection of programs, through which users interact with a database. DBMSs let you create forms and reports quickly and easily, as well as obtain answers to questions about the data stored in a database.

Database password   A string of characters assigned by the DBA to a database that users must enter before they can access a database.

Database server   In a three-tier client/server architecture and in other architectures, a computer that performs the database functions such as storing and retrieving data in a database.

DATE   The SQL data type for date data.

DBA   See *database administration.* (Sometimes the acronym stands for *database administrator.*)

DBDL   See *Database Design Language.*

DBMS   See *database management system.*

DDBMS   See *distributed database management system.*

Deadlock   A situation in which two or more database users are each waiting to use resources that are held by the other(s); also called *deadly embrace.*

Deadly embrace   See *deadlock.*

DECIMAL(*p,q*)   The SQL data type for decimal data.

Decrypting   A process that reverses the encryption of a database. Also called *decryption.*

Defining query   The query that is used to define the structure of a view.

DELETE   The SQL command used to delete a table. The word *DELETE* is followed by a FROM clause identifying the table. Use a WHERE clause to specify a condition. Any records satisfying the condition will be deleted.

Delete query   A query that deletes all records that satisfy some criterion.

Delimiter   A punctuation mark, such as a semicolon, that separates pieces of data.

Denormalizing   The conversion of a table that is in third normal form to a table that is no longer in third normal form. Denormalizing introduces anomaly problems but can decrease the number of disk accesses required by certain types of transactions, thus improving performance.

Department of Defense (DOD) 5015.2 Standard   A standard that provides data management requirements for the DOD and for companies supplying or dealing with the DOD.

Dependency diagram   A diagram that indicates the dependencies among the columns in a table.

Dependent entity   An entity that requires a relationship to another entity for identification.

Design grid   The portion of the Query Design window in Microsoft Access where you enter fields, criteria, sort orders, and so on.

Determinant   A column in a table that determines at least one other column.

Difference   When comparing tables, the set of all rows that are in the first table but that are not in the second table.

Dimension table   A table in a data warehouse that contains a single-part primary key, serving as an index into the central fact table, and other fields associated with the primary key value.

Disaster recovery plan   A plan that specifies the ongoing and emergency actions and procedures required to ensure data availability, even if a disaster occurs.

Distributed database   A single logical database that is physically divided among computers at several sites on a computer network.

Distributed database management system (DDBMS) A DBMS capable of supporting and manipulating distributed databases.

Division The relational algebra command that combines tables and searches for rows in the first table that match all rows in the second table.

Document Type Definition (DTD) A set of statements that specifies the elements (tags), the attributes (characteristics associated with each tag), and the element relationships for an XML document. The DTD can be a separate file with a .dtd extension, or you can include it at the beginning of an XML document.

Documenter An Access tool that prints documentation about the objects in a database.

Domain The set of values that are permitted for an attribute.

Drill down The process of viewing and analyzing lower levels of aggregation, or a more detailed view of the data.

DROP INDEX The SQL command that drops (deletes) an index from a table.

DROP TABLE The SQL command that drops (deletes) a table from a database.

DTD See *Document Type Definition*.

Dynamic Web page A Web page whose content changes in response to the different inputs and choices made by Web clients.

E-commerce See *electronic commerce*.

Electronic commerce (e-commerce) Business conducted on the Internet and Web.

Encapsulated In an object-oriented system, defining an object to contain both data and its associated actions.

Encryption A security measure that converts the data in a database to a format that's indecipherable to normal programs. The DBMS decrypts, or decodes, the data to its original form for any legitimate user who accesses the database.

Entity A person, place, object, event, or idea for which you want to store and process data.

Entity integrity The rule that no column (attribute) that is part of the primary key may accept null values.

Entity subtype Entity A is a subtype of entity B if every occurrence of entity A is also an occurrence of entity B.

Entity-relationship (E-R) diagram A graphic model for database design in which entities are represented as rectangles and relationships are represented as either arrows or diamonds connected to the entities they relate.

Entity-relationship (E-R) model An approach to representing data in a database that uses E-R diagrams exclusively as the tool for representing entities, attributes, and relationships.

Exclusive lock A lock that prevents other users from accessing the locked data in any way.

Existence dependency A relationship in which the existence of one entity depends on the existence of another related entity.

Extensible The capability of defining new data types in an OODBMS.

Extensible Hypertext Markup Language See *XHTML*.

Extensible Markup Language See *XML*.

Extensible Stylesheet Language See *XSL*.

Fact table The central table in a data warehouse that consists of rows that contain consolidated and summarized data.

Fat client In a two-tier client/server architecture, a client that performs presentation functions and business functions.

Field A characteristic or property of an entity; also called an *attribute* or a *column*.

File server A networked system in which a special site on the network stores files for users at other sites. When a user needs a file, the file server sends the entire file to the user.

First normal form (1NF) A table is in first normal form if it does not contain a repeating group.

Foreign key An column (attribute) or collection of columns in a table whose value is required either to match the value of a primary key in a table or to be null.

FOREIGN KEY The clause in a SQL CREATE TABLE or ALTER TABLE command that specifies referential integrity.

Form A screen object you use to maintain, view, and print data from a database.

Forward recovery A process used to recover a database by reading the log and applying the after images of committed transactions to bring the database up to date.

Fourth normal form (4NF) A table is in fourth normal form if it is in third normal form and there are no multivalued dependencies.

Fragmentation transparency The characteristic that users do not need to be aware of any data fragmentation (splitting of data) that has taken place in a distributed database.

FROM clause The part of an SQL SELECT command that indicates the tables in the query.

Front-end machine See *client*.

Front-end processor See *client*.

Function See *aggregate function*.

Functional dependence See *functionally dependent*.

Functionally dependent Column B is functionally dependent on column A (or on a collection of columns) if a value for A determines a single value for B at any one time.

Functionally determines Column A functionally determines column B if B is functionally dependent on A.

Generalization In UML, the relationship between a superclass and a subclass.

Global deadlock In a distributed database, deadlock that cannot be detected solely at any individual site.

GRANT The SQL statement that is used to grant different types of privileges to users of a database.

GROUP BY clause The part of an SQL SELECT command that indicates grouping.

Grouping The process of creating collections of records that share some common characteristic.

Growing phase A phase during a database update in which the DBMS locks all the data needed for a transaction and releases none of the locks.

HAVING clause The part of an SQL SELECT command that restricts the groups to be displayed.

Heterogeneous DDBMS A distributed DBMS in which at least two of the local DBMSs are different from each other.

HIPAA (Health Insurance Portability and Accountability Act) A federal law enacted in 1996 that specifies the rules for storing, handling, and protecting health care transactions.

Homogeneous DDBMS A distributed DBMS in which all the local DBMSs are the same.

Hot site A backup site that an organization can switch to in minutes or hours because the site is completely equipped with duplicate hardware, software, and data that the organization uses.

HTML (Hypertext Markup Language) A language used to create Web pages and derived from SGML.

HTTP (Hypertext Transfer Protocol) The data communication method used by Web clients and Web servers to exchange data on the Internet.

Hyperlink A tag in a Web page that links one Web page to another or links to another location in the same Web page.

Hypertext Markup Language See *HTML*.

Hypertext Transfer Protocol See *HTTP*.

IDEF1X A type of E-R diagram; or, technically, a language in the IDEF (Integrated Definition) family of languages that is used for data modeling.

Identifying relationship A relationship that is necessary for identification of an entity.

IIS (Internet Information Services) A Microsoft Web server package that comes with many versions of its operating systems.

Incomplete category In IDEF1X, a collection of subtypes with the property that there are elements of the supertype that are not elements of any subtype.

Independent entity An entity that does not require a relationship to another entity for identification.

Index A file that relates key values to records that contain those key values.

Index key The field or fields on which an index is built.

Information-level design The step during database design in which the goal is to create a clean, DBMS-independent design that will support all user requirements.

Inheritance The property that a subclass inherits the structure of the class as well as its methods.

INSERT The SQL command to add new data to a table. After the words *INSERT INTO*, you list the name of the table, followed by the word *VALUES*. Then you list the values for each of the columns in parentheses.

INTEGER The SQL data type for integer data.

Integrity A database has integrity if the data in it satisfies all established integrity constraints.

Integrity constraint A rule that must be followed by data in a database.

Integrity rules See *entity integrity, legal-values integrity*, and *referential integrity*.

Intelligent key A primary key that consists of a column or collection of columns that is an inherent characteristic of the entity.

Internet A worldwide collection of millions of interconnected computers and computer networks that share resources.

Internet Information Services (IIS) See *IIS*.

Interrelation constraint A constraint that involves more than one relation.

INTERSECT The relational algebra command for performing the intersection of two tables.

Intersection When comparing tables, an intersection is a new table containing all rows that are in both original tables.

INTO clause The SQL clause that inserts values into a table. An INTO clause consists of the word *INTO* followed by the name of the table to be created.

Intranet An internal company network that uses software tools typically used on the Internet and the World Wide Web.

Join In relational algebra, the operation in which two tables are connected on the basis of common data.

Join column The column on which two tables are joined. Also see *join*.

Join line In an Access query, the line drawn between tables to indicate how they are related.

Journal A file that contains a record of all the updates made to a database; also called a *log*. The DBMS uses the journal to recover a database that has been damaged or destroyed.

Journaling Maintaining a journal or log of all updates to a database.

Key The field on which data will be sorted; also called a *sort key*.

LAN See *local area network*.

Legal-values integrity The property that no record can exist in the database with a value in a field other than a legal value.

Live system See *production system*.

Local area network (LAN) A configuration of several computers connected together that allows users to share a variety of hardware and software resources.

Local deadlock In a distributed database, deadlock that occurs at a single site.

Local site From a user's perspective, the site in a distributed system at which the user is working.

Location transparency The property that users do not need to be aware of the location of data in a distributed database.

Locking A DBMS's denial of access by other users to data while the DBMS processes one user's updates to the database.

Log A file that contains a record of all the updates made to a database; also called a *journal*. The DBMS uses the log to recover a database that has been damaged or destroyed.

Logical key A primary key that consists of a column or collection of columns that is an inherent characteristic of the entity.

Major sort key See *primary sort key*.

Make-table query An Access query that creates a table containing the results of a query.

Mandatory role The role in a relationship played by an entity with a minimum cardinality of 1 (that is, there must be at least one occurrence of the entity).

Many-to-many relationship A relationship between two entities in which each occurrence of each entity can be related to many occurrences of the other entity.

Many-to-many-to-many relationship A relationship between three entities in which each occurrence of each entity can be related to many occurrences of each of the other entities.

Markup language A document language that contains tags that describe a document's content and appearance.

Message A request to execute a method. Also, data, requests, or responses sent from one computer to another computer on a network.

Metadata Data about the data in a database.

Metalanguage A language used to define another language.

Method An action defined for a object class.

Minor sort key See *secondary sort key*.

Multidependent In a table with columns A, B, and C, B is multidependent on A if each value for A is associated with a specific collection of values for B and, further, this collection is independent of any values for C.

Multidetermine In a table with columns A, B, and C, A multidetermines B if each value for A is associated with a specific collection of values for B and, further, this collection is independent of any values for C.

Multidimensional database The perceived structure by users of the data in a data warehouse.

Multiple-column index See *multiple-field index*.

Multiple-field index An index built on more than one field (column).

Multiplicity In UML, the number of objects that can be related to an individual object on the other side of a relationship; also called *cardinality*.

Multivalued dependence In a table with columns A, B, and C, there is a multivalued dependence of column B on column A (also read as "B is multidependent on A" or "A multidetermines B"), if each value for A is associated with a specific collection of values for B and, furthermore, this collection is independent of any values for C.

Natural join The most common form of a join.

Natural key   A primary key that consists of a column or collection of columns that is an inherent characteristic of the entity.

Network   See *communications network*.

Nonidentifying relationship   A relationship that is not necessary for identification.

Nonkey attribute   See *nonkey column*.

Nonkey column   An attribute (column) that is not part of the primary key.

Nonprocedural language   A language in which a user describes the task that is to be accomplished by the computer rather than the steps that are required to accomplish it.

Normal form   See *first normal form, second normal form, third normal form,* and *fourth normal form*.

Normalization process   The process of removing repeating groups to produce a first normal form table. Sometimes refers to the process of creating a third normal form table.

n-tier architecture   See *three-tier architecture*.

Null   A data value meaning "unknown" or "not applicable."

Object   A unit of data (set of related attributes) along with the actions that are associated with that data.

Object-oriented database management system (OODBMS)   A DBMS in which data and the methods that operate on that data are encapsulated into objects.

Office Open XML   A Microsoft file format that is a compressed version of XML and first used in the Office 2007 suite.

OLAP   See *online analytical processing*.

OLTP   See *online transaction processing*.

One-to-many relationship   A relationship between two entities in which each occurrence of the first entity is related to many occurrences of the second entity and each occurrence of the second entity is related to at most one occurrence of the first entity.

One-to-one relationship   A relationship between two entities in which each occurrence of the first entity is related to one occurrence of the second entity and each occurrence of the second entity is related to at most one occurrence of the first entity.

Online analytical processing (OLAP)   Software that is optimized to work efficiently with multidimensional databases in a data warehouse environment.

Online transaction processing (OLTP)   A system that processes a transaction by dealing with a small number of rows in a relational database in a highly structured, repetitive, and predetermined way.

OODBMS   See *object-oriented database management system*.

Optional role   The role in a relationship played by an entity with a minimum cardinality of zero (that is, there need not be any occurrences of the entity).

OR criterion   A combination of criteria in which at least one of the criteria must be true.

ORDER BY clause   The part of an SQL SELECT command that indicates a sort order.

Outer join   The form of a join in which all records appear, even if they don't match.

Partial dependency   A dependency of a column on only a portion of the primary key.

Password   A string of characters assigned by a DBA to a user that the user must enter to access a database.

Patriot Act   A federal law enacted in 2001 that specifies data retention requirements for the identification of customers opening accounts at financial institutions, allows law enforcement agencies to search companies' and individuals' records and communications, and expands the government's authority to regulate financial transactions.

Permission   The specification of the kind of access a user has to the objects in a database.

Persistence   The ability to have a program "remember" its data from one execution to the next.

Physical-level design   The step during database design in which a design for a given DBMS is produced from the final information-level design.

Polymorphism   The use of the same name for different operations in an object-oriented system.

Presidential Records Act   A federal law enacted in 1978 that regulates the data retention requirements for all communications, including electronic communications, of U.S. presidents and vice presidents.

Primary copy   In a distributed database with replicated data, the copy of the database that must be updated in order for the update to be deemed complete.

Primary key   A minimal collection of columns (attributes) in a table on which all columns are functionally dependent and that is chosen as the main direct-access vehicle to individual rows. Also see *candidate key*.

PRIMARY KEY   The SQL clause that is used in a CREATE TABLE or ALTER TABLE command to set a table's primary key field(s).

Primary sort key When sorting on two fields, the more important field; also called a *major sort key*.

Privacy The right of individuals to have certain information about them kept confidential.

Private visibility In UML, an indication that only the class itself can view or update the attribute value.

Procedural language A language in which a user specifies the steps that are required for accomplishing a task instead of merely describing the task itself.

Product The table obtained by concatenating every row in the first table with every row in the second table.

Production system The hardware, software, and database for the users; also called *live system*.

PROJECT The relational algebra command used to select columns from a table.

Protected visibility In UML, an indication that only the class itself or public or protected subclasses of the class can view or update the attribute value.

Public visibility In UML, an indication that any class can view or update the attribute value.

QBE See *Query-By-Example*.

Qualify To indicate the table (relation) of which a given column (attribute) is a part by preceding the column name with the table name. For example, *Customer.Address* indicates the column named Address within the table named Customer.

Query A question, the answer to which is found in the database; also used to refer to a command in a nonprocedural language such as SQL that is used to obtain the answer to such a question.

Query-By-Example (QBE) A data manipulation language for relational databases in which users indicate the action to be taken by completing on-screen forms.

RAID (redundant array of inexpensive/independent drives) A device used to protect against hard drive failures in which database updates are replicated to multiple hard drives so that an organization can continue to process database updates after losing one of its hard drives.

RDBMS See *relational DBMS*.

Record A collection of related fields; can be thought of as a row in a table.

Recovery The process of returning a database to a state that is known to be correct from a state known to be incorrect.

Redundancy Duplication of data, or the storing of the same data in more than one place.

Referential integrity The rule that if a table A contains a foreign key that matches the primary key of table B, then the value of this foreign key must either match the value of the primary key for some row in table B or be null.

Relation A two-dimensional table-style collection of data in which all entries are single-valued, each column has a distinct name, all the values in a column are values of the attribute that is identified by the column name, the order of columns is immaterial, each row is distinct, and the order of rows is immaterial. Also called a *table*.

Relational algebra A relational data manipulation language in which new tables are created from existing tables through the use of a set of operations.

Relational database A collection of relations (tables).

Relational DBMS (RDBMS) A DBMS that supports and manipulates a relational database.

Relational operator An operator used to compare values. Valid operators are=, <, >, <=, >=, <>, and !=. Also called a *comparison operator*.

Relationship An association between entities.

Remote site From a user's perspective, any site other than the one at which the user is working.

Repeating group Several entries at a single location in a table.

Replica A copy of the data in a database that a user can access at a remote site.

Replicate A duplicate of the data in a database that a user can access at a remote site.

Replication transparency The property that users do not need to be aware of any replication that has taken place in a distributed database.

Reserved word A word that is part of the SQL language.

REVOKE The SQL statement that is used to revoke privileges from users of a database.

Roll up View and analyze higher levels of aggregation.

Rollback A process to recover a database to a valid state by reading the log for problem transactions and applying the before images to undo their updates; also called *backward recovery*.

Row-and-column subset view A view that consists of a subset of the rows and columns in a table.

Sandbox See *test system*.

Sarbanes-Oxley (SOX) Act A federal law enacted in 2002 that specifies data retention and verification requirements for public companies, requires CEOs and CFOs to certify financial statements, and makes it a crime to destroy or tamper with financial records.

Save   See *backup*.

Scalability   The ability of a computer system to continue to function well as utilization of the system increases.

SEC Rule 17a-4   The rule of the Security and Exchange Commission that specifies the retention requirements of all electronic communications and records for financial and investment entities.

Second normal form (2NF)   A relation is in second normal form if it is in first normal form and no non-key attribute is dependent on only a portion of the primary key.

Secondary key   A column (attribute) or collection of columns that is of interest for retrieval purposes (and that is not already designated as some other type of key).

Secondary sort key   When sorting on two fields, the less important field; also called *minor sort key*.

Security   The prevention of unauthorized access to the database.

SELECT   The relational algebra command to select rows from a table. Also, the retrieval command in SQL.

SELECT clause   The part of an SQL SELECT command that indicates the columns to be included in the query results.

Server   A computer that provides services to the clients in a client/server system; also called a *back-end processor* or a *back-end machine*.

Server-side extension   Instructions executed by a Web server to provide dynamic Web page capability. These extensions are usually contained in separate files that are referenced within the HTML documents.

Server-side script   See *server-side extension*.

Session   The duration of a Web client's connection to a Web server.

SGML   See *Standard Generalized Markup Language*.

Shared lock   A lock that lets other users read locked data.

Shrinking phase   A phase during a database update in which the DBMS releases all the locks previously acquired for a transaction and acquires no new locks.

Simple condition   A condition that involves only a single field and a single value.

Single-column index   See *single-field index*.

Single-field index   An index built on a single field (column).

Slice and dice   In a data warehouse, selecting portions of the available data, or reducing the data cube.

SMALLINT   The SQL data type for integer data for small integers.

Smart card   Small plastic cards about the size of a driver's license that have built-in circuits containing processing logic to identify the cardholder.

Sort   The process of arranging rows in a table or results of a query in a particular order.

Sort key   The field on which data are sorted; also called a *key*.

SQL   See *Structured Query Language*.

Standard Generalized Markup Language (SGML)   A metalanguage used to create document markup languages; SGML became a standard in 1986. Languages based on the full SGML are used to manage large, complex reports and technical specifications for a variety of computer platforms, printers, and other devices.

Star schema   A multidimensional database whose conceptual shape resembles a star.

Stateless   A condition for a communication protocol, such as HTTP, in which the connection between the sender and the receiver, such as a Web server and a Web client, is closed once the sender responds to the sender's request and the sender retains no information about the request or the sender.

Statement history   The area of memory in MySQL that stores the most recently entered command.

Static Web page   A Web page that displays the exact same content for all Web clients.

Stored procedure   A file containing a collection of compiled and optimized SQL statements that are available for future use.

Structured Query Language (SQL)   A very popular relational data definition and manipulation language that is used in many relational DBMSs.

Stylesheet   A document that specifies how to process the data contained in another document and present the data in a Web browser, in a printed report, on a mobile device, in a sound device, or in other presentation media.

Subclass   A class that inherits the structure and methods of another class and for which you can define additional attributes and methods.

Subquery   In SQL, a query that appears within another query.

SUBTRACT   The relational algebra command for performing the difference of two tables.

Superclass In UML, a class that has subclasses.

Surrogate key A system-generated primary key that is usually hidden from users.

Synchronization The periodic exchange by a DBMS of all updated data between two databases in a replica set.

Synthetic key A system-generated primary key that is usually hidden from users.

Syscolumns The portion of the system catalog that contains column information.

Sysindexes The portion of the system catalog that contains index information.

Systables The portion of the system catalog that contains table information.

System catalog A structure that contains information about the objects (tables, columns, indexes, views, and so on) in a database.

Sysviews The portion of the system catalog that contains view information.

Table See *relation*.

Tag A command in a Web page that a Web browser processes to position and format the text on the screen or to link to other files.

TCP/IP (Transmission Control Protocol and Internet Protocol) The standard protocol for all communication on the Internet.

Test system The hardware, software, and database that programmers use to develop new programs and modify existing programs. Also called a *sandbox*.

Thin client In a client/server architecture, a client that performs only presentation functions.

Third normal form (3NF) A table is in third normal form if it is in second normal form and the only determinants it contains are candidate keys.

Three-tier architecture A client/server architecture in which the clients perform the presentation functions, a database server performs the database functions, and the application servers perform the business functions and serve as an interface between clients and the database server. Also called an *n-tier architecture*.

Timestamp The unique time when the DBMS starts a transaction update to a database.

Timestamping The process of using timestamps to avoid the need to lock rows in a database and to eliminate the processing time needed to apply and release locks and to detect and resolve deadlocks.

Top-down design method A design method that begins with a general database design that models the overall enterprise and then repeatedly refines the model to achieve a design that supports all necessary applications.

Transaction A set of steps completed by a DBMS to accomplish a single user task.

Transmission Control Protocol and Internet Protocol See *TCP/IP*.

Trigger An action that automatically occurs in response to an associated database operation such as INSERT, UPDATE, or DELETE.

Tuning The process of changing the database design to improve performance.

Tuple The formal name for a row in a table.

Two-phase commit An approach to the commit process in distributed systems in which there are two phases. In the first phase, each site is instructed to prepare to commit and must indicate whether the commit will be possible. After each site has responded, the second phase begins. If every site has replied in the affirmative, all sites must commit. If any site has replied in the negative, all sites must abort the transaction.

Two-phase locking An approach to locking that is used to manage concurrent update in which there are two phases: a growing phase, in which the DBMS locks more rows and releases none of the locks, and a shrinking phase, in which the DBMS releases all the locks and acquires no new locks.

Two-tier architecture A client/server architecture in which the clients perform the presentation functions, and a database server performs the database functions. In a fat client configuration, the clients perform the business functions, whereas in a thin client configuration, the database server performs the business functions.

UML See *Unified Modeling Language*.

Unified Modeling Language (UML) An approach to model all the various aspects of software development for object-oriented systems.

Uniform Resource Locator See *URL*.

Union A combination of two tables consisting of all records that are in either table.

Union compatible Two tables are union compatible if they have the same number of fields and if their corresponding fields have identical data types.

Unnormalized relation A structure that satisfies the properties required to be a relation (table) with one exception: repeating groups are allowed; that is, the entries in the table do not have to be single-valued.

UPDATE The SQL command to make changes to existing table data. After the word *UPDATE*, you indicate the table to be updated. After the word *SET*, you indicate the field to be changed, followed by an equals sign and the new value. Finally, you can include a condition in the WHERE clause, in which case, only the records that satisfy the condition will be changed.

Update anomaly An update problem that can occur in a database as a result of a faulty design.

Update query In Access, a query that updates the contents of a table.

UPS (uninterruptable power supply) A power source such as a battery or fuel cell, for short interruptions and a power generator for longer outages.

URL (Uniform Resource Locator) An Internet address that identifies where the Web page is stored—both the location of the Web server and the name and location of the Web page on that server.

User view The view of data that is necessary to support the operations of a particular user.

Utility services DBMS-supplied services that assist in the general maintenance of a database.

Validation rule In Access, a rule that data entered in a field must satisfy.

Validation text In Access, a message that is displayed when a validation rule is violated.

Victim In a deadlock situation, the deadlocked user's transaction that the DBMS chooses to abort to break the deadlock.

View An application program's or an individual user's picture of a database.

Visibility symbol In UML, a symbol preceding an attribute in a class diagram to indicate whether other classes can view or change the value in the attribute. The possible visibility symbols are public visibility (+), protected visibility (#), and private visibility (–). With public visibility, any other class can view or change the value. With protected visibility, only the class itself or public or protected subclasses of the class can view or change the value. With private visibility, only the class itself can view or change the value.

W3C (World Wide Web Consortium) An international organization that develops Web standards, specifications, guidelines, and recommendations.

Warm site A backup site that is equipped with an organization's duplicate hardware and software but not data.

Weak entity An entity that depends on another entity for its own existence.

Web (World Wide Web) A vast collection of digital documents available on the Internet.

Web browser A computer program that retrieves a Web page from a Web server and displays it on a Web client.

Web client A computer requesting a Web page from a Web server.

Web page A digital document on the Web.

Web server A computer on which an individual or organization stores Web pages for access on the Internet.

WHERE clause The part of an SQL SELECT command that indicates a condition the rows to be displayed must satisfy.

Workgroup In Access, a group of users who are assigned the same permissions to various objects in a database.

World Wide Web See *Web*.

World Wide Web Consortium See *W3C*.

XHTML (Extensible Hypertext Markup Language) A markup language that is stricter version of HTML and that is based on XML.

XML (Extensible Markup Language) A metalanguage derived from a restricted subset of SGML and designed for the exchange of data on the Web. You can customize XML tags to describe the data an XML document contains and how that data should be structured.

XML declaration An XML statement clause that specifies to an XML processor which version of XML to use.

XML schema A set of statements that specifies the elements (tags), the attributes (characteristics associated with each tag), and the element relationships for an XML document. The XML schema can be a separate file with a .xsd extension, or you can include it at the beginning of an XML document. It's a newer form of DTD that more closely matches database features and terminology.

XQuery A language for querying XML, XSL, XHTML, other XML-based documents, and similarly structured data repositories.

XSL (Extensible Stylesheet Language) A standard W3C language for creating stylesheets for XML documents.

XSL Transformations See *XSLT*.

XSLT (XSL Transformations) A language that defines the rules to process an XML document and change it into another document; this other document could be another XML document, an XSL document, an HTML or XHTML document, or most any other type of document.

SYMBOLS

& operator, 85
* (asterisk) symbol, 75, 86
= (equal to) operator, 37, 77
> (greater than) operator, 37, 77
>= (greater than or equal to) operator, 37, 77
< (less than) operator, 37, 77
<= (less than or equal to) operator, 37, 77
<> (not equal to) operator, 77
!= (not equal to) operator, 77
% (percent sign), 86
? (question mark), 87
[] square brackets, 41, 72
_ (underscore), 87

A

Access
    changing field properties in, 135–136
    creating views in, 120
    deleting fields in, 136
    Documenter tool, 140
    entering criteria in, 35–37
    legal-values integrity in, 134
    QBE in, 32–33
    queries, 33–35
    referential integrity in, 131–132
    specifying primary key in, 129–130
    SQL, 70, 72
    wildcard characters, 86–87
Access 2003, 70
Access 2007, 70
access delay, 277
access privileges. See user privileges
ADO.NET, 288
after image, 238
aggregate functions, 42–44, 90–92, 346
Alexamara Marina Group (case study)
    E-R diagram for, 25
    introduction to, 20–25
alternate keys, 155, 180, 182
ALTER TABLE command, 135, 345, 350–351
AND criterion, 37–38, 40
AND operator, 80–81, 83, 98, 110
Apache HTTP Server, 287
Application Program Interface (API), 288
application servers, 285
archiving, 259–260
artificial keys, 181
AS keyword, 92

associations, 303
asterisk (*) symbol, 75, 86
attributes, 4, 32
    class, 304
    nonkey, 157
authentication, 242–243
authorization rules, 243
authorizations, 243
Avg (average) function, 42, 44
AVG function, 90, 92

B

back-end processor, 284
backups, 238, 241–242, 263, 282
backup sites, 259
backward recovery, 241
before image, 238
Berners-Lee, Tim, 290
BETWEEN operator, 83–84, 346
binary large objects (BLOBs), 299
binding, 305
biometrics, 242
bottom-up design method, 196–197
Boyce-Codd normal form (BCNF), 161
built-in functions, 42–44, 90–92, 346
business-to-business (B2B) commerce, 290

C

calculated fields. See computed fields
candidate keys, 155, 161
cardinality, 217–218
Cartesian products, 62
cascade delete, 132
cascade update, 132
catalog. See system catalog
catalog services, 227–228
categories, 209–212
central computers, 275
CGI. See Common Gateway Interface
CHAR(n) data type, 72, 349
CHECK clause, 134
Chen, Peter, 213
class diagrams, 303–305
classes, 299–305
clients, 140, 284
    fat, 285
    thin, 285, 286
    Web, 287, 288–289

client/server systems, 140, 283–287
client-side extensions, 288
client-side scripts, 288
Codd, E. F., 129, 298
column names
    duplicate, 32
    qualifying, 32
    SQL, 72
columns, 4, 5, 31, 32
    determinants, 161
    foreign keys. *see* foreign keys
    functional dependencies among, 150–153, 156–160
    join, 58–59
    multivalued dependencies among, 167–170
    nonkey, 157
    primary keys. *see* primary keys
    selecting, 345
    specifying data type for, 72
commits, 238–239, 282
Common Gateway Interface (CGI), 288
communications networks, 276–277
comparison operators, 37, 77
complete categories, 212
complexity, 14
composite entities, 215–216
compound conditions/criteria, 37–40, 80–84, 98–99, 346
computed fields, 40–42, 84–86, 345–346
concatenation, 58, 85
CONCAT function, 85
concurrent updates, 225, 228–238, 263
    deadlock and, 236–237
    in distributed databases, 281
    lost updates and, 230–233
    in OODBMSs, 306
    timestamping and, 238
    two-phase locking, 233–236
conditions/criteria
    BETWEEN, 83–84
    IN, 87–88
    BETWEEN, 346–347
    IN, 347
    compound, 37–40, 80–84, 98–99, 346
    AND criterion, 37–38, 40
    to delete records, 53–54
    LIKE, 86–87, 347
    OR criterion, 37, 39
    simple, 35–37, 77–80, 346
confidential data, 243
consistency, 13
constraints
    in class diagrams, 304
    integrity, 13, 244–246
    interrelation, 172
context-sensitive help, 264
cookies, 289
cost factors, 264
COUNT function, 42–43, 90–91
CREATE INDEX command, 126–127, 347
CREATE PROCEDURE command, 140–141

CREATE TABLE command, 71–73, 103–104, 105, 111, 137, 347–348
CREATE VIEW command, 121, 348
criteria. *See* conditions/criteria
cumulative design, 178, 183–184
    design example, 185–190

D

data
    archiving, 259–260
    confidential, 243
    consistency, 13
    exchange of on Web, using XML, 290–293
    inconsistent, 157, 160
    information from, 12
    integrity, 13
    redundancy of, 2, 13, 159, 160
    sharing, 12–13
    update and retrieval of, 225, 226–227
data archives, 260
database administration (DBA), 13, 253–271
    access privileges, 254–257
    archiving, 259–260
    data dictionary management, 265
    DBMS evaluation and selection function, 261–265
    disaster planning, 258–259
    introduction to, 253–254
    maintenance function, 265
    performance tuning function, 267–270
    policy formulation and enforcement, 254–260
    security function, 257–258
    technical functions, 266–270
    training function, 265
database administrators, 13, 253–254
database design, 10
    bottom-up design method, 196–197
    cumulative design, 178, 183–184
    DBA role in, 266
    in distributed databases, 282
    entity subtypes and, 209–212
    E-R diagrams, 182–183
    E-R model, 213–218
    examples, 185–195
    gathering information for, 197–203
    information-level design, 177, 178–181, 183–184
    introduction to, 177–178
    many-to-many relationship considerations, 206–208
    Marvel College (example), 313–336
    merging tables and, 213
    normalization process and, 171–172, 180
    null values and, 208–212
    one-to-one relationship considerations, 203–206
    physical-level design, 177, 195–196
    process of, 177–178
    representation of user views, 179–180
    survey forms for, 197–198
    top-down design method, 196–197
    UML and, 302–305
    user views, 178
    using existing documents for, 198–203

Database Design Language (DBDL), 181–182
database management systems (DBMSs)
   See also distributed database management systems (DDBMSs)
   direct interaction with, 10
   evaluation and selection of, 261–265
   form creation, 10–11
   functions of, 225–248
     catalog services, 225, 227–228
     data integrity features, 225, 244–246
     data recovery, 225, 238–242
     security services, 225, 242–243
     support concurrent update, 225, 228–238
     support data independence, 226, 246–247
     support data replication, 226, 247–248
     update and retrieve data, 225, 226–227
     utility services, 226, 248
   indirect interaction with, 10
   introduction to, 9–12
   maintenance of, 265
   object-oriented, 299–306
   performance tuning, 267–270
   report creation, 11–12
   testing, 266–267
database passwords, 242–243
database processing
   advantages of, 12–14
   disadvantages of, 14
databases
   concurrent update of, 225, 228–238, 263
   defined, 5
   distributed, 264, 275–277
     advantages of, 279
     disadvantages of, 280–282
     rules for, 283
   introduction to, 4–9
   locking, 233–236
   multidimensional, 295–296
   relational, 29–32
   structure of, 6–8
   terminology, 4
   Web access to, 287–290
database servers, 285
database structure
   changing, 135–137
   querying, 138–140
data cubes, 295–297
data definition, 263
data dictionary, 228, 263
data dictionary management, 265, 282
data files, 5
data fragmentation, 278–279
data independence, 13–14, 124, 226, 246–247
data integrity features, 225, 244–246
data mining, 298
data recovery, 225, 238–242, 263
   backward recovery, 241
   in distributed databases, 281–282
   forward recovery, 240
   journaling, 238–240
   in OODBMSs, 306
   on PC-based DBMSs, 241–242

data replication, 226, 247–248, 264, 280
data restructuring, 263
data retrieval
   security issues, 128–129
   using indexes, 124–128
data types, 72, 244, 348–349
data warehouses, 276, 293–299
   architecture, 294
   structure and access, 295–298
Date, C. J., 283
DATE data type, 72, 349
dates, in conditions, 79
deadlock, 236–237, 281
deadly embrace, 236–237
DECIMAL(p,q) data type, 72, 349
decomposition process, 163–167
decryption, 242
defining query, 118
DELETE command, 103, 111, 349
delete queries, 53–54, 103
delimiters, 140–141
denormalization, 269–270
Department of Defense (DOD) 5015.2 Standard, 259
dependencies
   existence, 216–217
   functional, 150–153, 156–160, 198, 200–202
   multivalued, 167–170
dependency diagrams, 158, 161
dependent entities, 190, 218
DESC clause, 89
design grid, 33–35
design process. *See* database design
determinants, 161
diagrams
   class, 303–305
   dependency, 158, 161
   E-R. *see* entity-relationship (E-R) diagrams
difference operation, 60, 61–62
dimension tables, 295
disaster planning, 258–259
disaster recovery plans, 259
distributed database management systems (DDBMSs), 276–279
   characteristics of, 277–279
   fragmentation transparency, 278–279
   heterogeneous, 277
   homogeneous, 277
   location transparency, 277
   replication transparency, 277–278
distributed databases, 264, 275, 276–277
   advantages of, 279
   disadvantages of, 280–282
   rules for, 283
division operation, 63
documentation, 264
Documenter tool, 140
documents, information from existing, 198–203
Document Type Definition (DTD), 292
domain, 300
drill down analysis, 297–298

DROP INDEX command, 128, 349
DROP TABLE command, 137, 350
dynamic Web pages, 288

E

electronic commerce (e-commerce), 288, 290
encapsulation, 299, 305
encryption, 242
entities
    composite, 215–216
    defined, 4
    dependent, 190, 218
    determining, 179, 203
    in E-R diagrams, 183
    existence dependencies among, 216–217
    independent, 190
    properties of, 179
    in relational databases, 31
    relationships among, 5, 179–180
    weak, 218
entity integrity, 129–130
entity-relationship (E-R) diagrams, 9, 182–183,
    213–218
    cardinality in, 217–218
    composite entities, 216
    with crow's foot, 217
    with existence dependencies, 217
    many-to-many relationships, 214, 215
    many-to-many-to-many relationships, 214
    one-to-many relationships, 214, 215
entity-relationship (E-R) model, 213–218
entity subtypes, 209–212, 304–305
equal to (=) operator, 37, 77
E-R (entity-relationship) diagrams. See entity-
    relationship (E-R) diagrams
existence dependencies, 216–217
extensibility, 305
Extensible Hypertext Markup Language (XHTML),
    292–293
Extensible Markup Language (XML), 290–293
Extensible Stylesheet Language (XSL), 292
extensions
    client-side, 288
    server-side, 288

F

fact tables, 295
failure, impact of database, 14
FAQs, 355–356
fat clients, 285
field names
    changing, in views, 121
    qualifying, 97–98
    spaces in, 41
    in square brackets, 41
field properties, changing, 135–136

fields, 4, 32
    adding new, 246
    adding to query, 33–35
    changing length of, 246
    computed (calculated), 40–42, 84–86, 345–346
    concatenation of, 85
    creating new, 135–136
    deleting, 136–137
    format requirements, 244
    including all, in SQL query, 75–76
    index, 125–127
    legal values, 133–134
    null, 129, 131, 208–212
    sort key, 45
files, 5
file servers, 283–284
file size, 14
First function, 42
first normal form (1NF), 149, 155–156, 171
FOREIGN KEY clause, 131
foreign keys
    defining, in DBDL, 182
    determining, 180, 181
    in E-R diagrams, 183
    in fact tables, 295
    interrelation constraints and, 172
    one-to-one relationships and, 203–206
    referential integrity and, 131–133
format constraints, 244
forms, creation of, 10–11
forward recovery, 240
fourth normal form (4NF), 149, 169–171
fragmentation transparency, 278–279
FROM clause, 73, 97, 105–110
front-end processor, 284
functional dependencies, 150–153, 156–160
    identifying, from existing documents, 200–202
    splitting, 164
    in survey forms, 198
functions, 42–44, 90–92, 346

G

generalizations, 305
GIVING clause, 57
global deadlock, 281
GRANT statement, 128–129, 350
greater than (>) operator, 37, 77
greater than or equal to (>=) operator, 37, 77
GROUP BY clause, 94–96, 109
grouping, 44–45, 94–96
growing phase, 236

H

HAVING clause, 95–96, 109
Henry Books (case study)
    design example, 191–195
    introduction to, 14–20
heterogeneous DDBMs, 277
homogeneous DDBMs, 277
hot sites, 259
hyperlinks, 288
Hypertext Markup Language (HTML), 288, 290
Hypertext Transfer Protocol (HTTP), 287, 289

I

IDEF0, 182
IDEF1, 182
IDEF1X, 182, 209
IDEF2, 182
IDEF3, 182
IDEF4, 182
identifying relationships, 190
incomplete categories, 209, 211–212
inconsistent data, 157, 160
independent entities, 190
indexes, 124–128, 196, 246
index key, 126
information
    from data, 12
    from existing documents, 198–203
    from survey forms, 197–198
information-level design, 177, 178–181, 183–184
inheritance, 302, 305
IN operator, 87–88, 347
INSERT command, 102, 111, 137, 350
INTEGER data type, 72, 349
integrity, 13, 264, 350–351
    constraints, 13, 244–246
    features, 225
    legal-values, 133–134
integrity rules, 129–134
    entity integrity, 129–130
    referential integrity, 130–133
intelligent keys, 181
Internet, 287–290
Internet Information Services (IIS), 287
interrelation constraints, 172
INTERSECT command, 61
intersection operation, 60, 61
INTO clause, 103–104, 352–353
intranets, 264

J

Java Database Connectivity (JDBC), 288
join column, 58–59
JOIN command, 58–60
join line, 49

joins, 49–52
    natural, 60
    outer, 60
    in SQL, 97–100, 351
journaling, 238–240
journals, 238–240

K

keys
    alternate, 155, 182
    artificial, 181
    candidate, 155, 161
    DBMS support for, 195
    defining, in DBDL, 182
    foreign. see foreign keys
    identifying, 180–181
    intelligent, 181
    logical, 181
    natural, 181
    primary. see primary keys
    secondary, 180–181, 182
    sort, 45–49, 88–90
    surrogate, 181
    synthetic, 181

L

Last function, 42
legal compliance, with data archiving, 259–260
legal values, 244
legal-values integrity, 133–134
less than (<) operator, 37, 77
less than or equal to (<=) operator, 37, 77
LIKE operator, 86–87, 347
local area networks (LANs), 264
local deadlock, 281
local sites, 277
location transparency, 277
locking, 233–236, 237
logical keys, 181
logs, 238–240
lost updates, 230–233

M

major sort key, 45, 46–49, 89
make-table queries, 54–56
mandatory roles, 218
many-to-many relationships, 180, 206–208, 214
many-to-many-to-many relationships, 208, 214
markup languages, 288, 290, 292–293
Marvel College (design example), 313–336
    database requirements, 313–316
    information-level design, 316–336
MAX function, 90, 92
Max (largest value) function, 42
messages, 277, 301–302

metadata, 227
metalanguages, 290
methods, 301–302
Microsoft Access. *See* Access
Microsoft Office suite, 10
MIN function, 90, 92
minor sort key, 45, 46–49, 89
Min (smallest value) function, 42
multidimensional database, 295–296
multiple-column index, 127–128
multiple-field index, 127–128
multiple tables, joining, 51–52, 99–100
multiplicity, 304
multivalued dependencies, 167–170
    avoiding, 170–171
MySQL, 70–71
    See also SQL (Structured Query Language)
    editing commands, 71
    getting started with, 70–71
    querying database structure in, 140
    wildcard characters, 86–87

N

natural joins, 60
natural keys, 181
networks, 276–277
    client/server systems, 283–287
    messages in, 277
nonidentifying relationships, 190
nonkey attributes, 157
nonkey columns, 157
nonprocedural languages, 248, 263
normal forms, 149
    Boyce-Codd normal form (BCNF), 161
    first normal form (1NF), 149, 155–156, 171
    second normal form (2NF), 149, 156–159
    third normal form (3NF), 149, 159–162, 168, 171, 213
    fourth normal form (4NF), 149, 169–171
normalization process
    database design and, 171–172, 180
    denormalization, 269–270
    functional dependence and, 150–153
    incorrect decompositions, 163–167
    introduction to, 149
    keys and, 153–155
    multivalued dependencies and, 167–170
not equal to (!=) operator, 77
not equal to (<>) operator, 77
NOT (not equal to) operator, 37, 80, 82–83
n-tier architecture, 285
null values, 129, 131, 208–212

O

object-oriented database management systems (OODBMs), 299–306
    defined, 299
    inheritance, 302
    methods and messages, 301–302
    objects and classes, 299–301
    rules for, 305–306
    UML and, 302–305
objects, 299–301, 305
Office Open XML, 293
ON clause, 128
one-to-many relationships, 5, 31, 180, 214, 215
one-to-one relationships, 180, 203–206
online analytical processing (OLAP) software, 295, 298–299
online transaction processing (OLTP) systems, 293–294
Open Database Connectivity (ODBC), 288
operators, comparison (relational), 37, 77
optional roles, 218
OR criterion, 37, 39, 100
Order, 7
ORDER BY clause, 88–90, 108, 109, 110
order of operations, in computed fields, 42
OR operator, 80, 81–82
outer joins, 60
OVER clause, 57–58

P

parentheses (), 42
partial dependencies, 158
passwords, 242
Patriot Act, 259
PC-based DBMSs
    locking on, 237
    recovery on, 241–242
percent sign (%), 86
performance issues, 264, 306
performance tuning, 267–270
persistence, 306
physical-level design, 177, 195–196
policy formulation and enforcement, 254–260
polymorphism, 305
portability, 264
Premiere Products
    background, 1–4
    database design example, 185–190
    database relationships, 31
    database requirements of, 3–4
    database structure, 32
    E-R diagram for, 9, 190
    orders spreadsheet, 1–3
    sample data, 6–7, 30

Presidential Records Act, 259
primary copy, 280
PRIMARY KEY clause, 129
primary keys, 153–155
  defined, 32, 153
  defining, in DBDL, 182
  as determinants, 161
  determining, 179, 180–181
  entity integrity and, 129
  in E-R diagrams, 183
  in fact tables, 295
  first normal form and, 171
  in many-to-many relationships, 180
  in one-to-one relationships, 203–206
  representation of, 32
  second normal form and, 158–159, 160
  specifying, 129–130
  types of, 181
primary sort key, 45, 46–49, 89
privacy, 243
private visibility, 304
procedural languages, 248, 263
productivity, 13
product operation, 62
PROJECT command, 57–58, 59–60
protected visibility, 304
public visibility, 304

Q

queries
  See also specific queries; SQL queries
  adding fields to, 33–35
  complexity of, in distributed databases, 280–281
  compound conditions with, 80–84
  compound criteria, 37–40
  computed fields in, 40–42, 84–86
  creating tables from, 103–104
  criteria, 35–37
  of data warehouses, 295–298
  defined, 32
  to define views, 118–120
  defining query, 118
  delete, 53–54, 103
  functions in, 42–44
  grouping, 44–45, 94–96
  joins, 49–52, 97–100
  make-table, 54–56
  in OODBMSs, 306
  QBE, 32–35
  simple conditions with, 77–80
  sorting, 45–49
  SQL, 69, 73–80
  stored procedures, 140–141
  subqueries, 93, 353
  update, 53, 101–103
Query-By-Example (QBE), 32–33
  compound criteria, 37–40
  computed fields, 40–42
  delete queries, 53–54
  functions, 42–44

  grouping, 44–45
  joins, 49–52
  make-table queries, 54–56
  simple criteria, 35–37
  simple queries, 33–35
  sorting, 45–49
  update queries, 53
Query Tools Design tab, 33
question mark (?), 87

R

RAID, 259
records, 32
  concurrent update of, 228–238
  deleting, 53, 103
  deleting, and referential integrity, 132–133
  grouping, 44–45, 94–96
  inserting, 102
  null, 60
  sorting, 45–49, 88–90
  updating, 53, 101–103
recovery process, 14, 238–242
redundancy, 2, 13, 159, 160
redundant array of inexpensive/independent drives
  (RAID), 259
referential integrity, 130–133
relational algebra, 56–63
  division operation, 63
  JOIN command, 58–60
  product operation, 62
  PROJECT command, 57–58, 59–60
  SELECT command, 57
  set operations, 60–62
  SUBTRACT command, 61–62
relational database management systems (RDBMSs).
  See database management systems (DBMSs)
relational databases
  See also database management systems (DBMSs);
    databases
  defined, 31
  introduction to, 29–32
  representing structure of, 32
relational model
  indexes in, 124–128
  integrity rules, 129–134
  introduction to, 29–32
  vs. object-oriented model, 300–301
  QBE, 32–33
  security and, 128–129
  stored procedures, 140–141
  structure changes and, 135–137
  system catalog, 137–140
  triggers, 141–142
  views in, 117–124
relational operators, 37
relations
  See also tables
  defined, 31
  splitting, 171–172
  unnormalized, 31, 155–156

relationships, 5
  adding, 246–247
  changing, 246–247
  between entities, 179–180
  identifying, 190
  mandatory roles, 218
  many-to-many, 180, 206–208, 214
  many-to-many-to-many, 208, 214
  nonidentifying, 190
  one-to-many, 5, 31, 180, 214, 215
  one-to-one, 180, 203–206
  optional roles, 218
  referential integrity and, 130–133
remote sites, 277
repeating groups, 31, 155–156, 168, 171
replication, 226, 247–248, 264
replication transparency, 277–278
reports, creation of, 11–12
requirements
  balancing conflicting, 13
  identifying user, 197–203
reserved words, 73
REVOKE statement, 128–129, 351–352
ROLLBACK command, 101
rollbacks, 241
roll up analysis, 298
row-and-column subset view, 122
rows, 7–8, 31, 32
  concatenation of, 58
  deleting, 349
  Total row, 42

S

Sarbanes-Oxley (SOX) Act, 259
saves, 238
scalability, 285
scripts
  client-side, 288
  server-side, 288
secondary keys, 180–181, 182
secondary sort key, 45, 46–49, 89
second normal form (2NF), 149, 156–160
security
  in client/server systems, 286
  databases and, 13
  as DBA function, 257–258
  in distributed databases, 282
  in relational model, 128–129
  spreadsheets and, 2
  views and, 124, 128
security services, 225, 242–243
  authentication, 242–243
  authorizations, 243
  encryption, 242
  evaluating, 264
  privacy, 243
  views, 243

SELECT command, 105–110, 118–120, 345–346, 351, 352
  data retrieval with, 73–76
  joining tables with, 97, 100
  in relational algebra, 57
SELECT INTO command, 137, 352–353
semicolon (;), 140, 141
SEQUEL, 69
servers, 140
  application, 285
  in client/server systems, 284–285
  database, 285
  file, 283–284
  Web, 287, 288–289
server-side extensions, 288
server-side scripts, 288
sessions, 289
set operations, 60–62
SGML. See Standardized Generalized Markup Language
shared lock, 263
SHOW COLUMNS command, 140
SHOW INDEX command, 140
SHOW TABLES command, 140
shrinking phase, 236
simple conditions, 77–80, 346
single-column index, 127
single-field index, 127
SMALLINT data type, 72, 349
smart cards, 242
sorting, 45–49, 88–90
sort keys, 45–49, 88–90
SOURCE command, 71
spaces, in field names, 41
spreadsheets, limitations of, 1–3
SQL queries
  compound conditions with, 80–84
  computed fields in, 84–86
  creating tables from, 103–104
  for database structure information, 139–140
  to define views, 118–120
  delete, 103
  grouping, 94–96
  joins, 97–100
  restricting rows included in, 95–96
  simple conditions, 77–80
  with special operators, 86–88
  subqueries, 93, 353
  union, 100–101, 353
  update, 101–103, 111, 353
SQL (Structured Query Language)
  ALTER TABLE command, 135, 345, 350–351
  BETWEEN operator, 83–84, 346
  built-in functions, 90–92
  CHECK clause, 134
  commands summary, 69, 105–111
  comparison operators, 77
  CONCAT function, 85

conditions, 346–347
CREATE INDEX command, 126–127, 347
CREATE TABLE command, 71–73, 103–104, 105, 111, 137, 347–348
CREATE VIEW command, 121, 348
data retrieval, 73–80
data types, 72, 348–349
DELETE command, 103, 111, 349
DROP INDEX command, 128, 349
DROP TABLE command, 137, 350
FROM clause, 73, 97
GRANT statement, 128–129, 350
GROUP BY clause, 94–96, 109
HAVING clause, 95–96, 109
IN operator, 87–88
INSERT command, 102, 111, 137, 350
INTO clause, 103–104
introduction to, 69, 70–71
LIKE operator, 86–87, 347
ORDER BY clause, 88–90, 108, 109, 110
reserved words, 41, 73
REVOKE statement, 351–352
SELECT command, 73–76, 97, 100, 105–110, 345–346, 351, 352
SELECT INTO command, 137, 352–353
sorting, 88–90
subqueries, 353
table creation, 71–73
UPDATE command, 102, 111, 353
WHERE clause, 73, 76–77, 83, 95–96, 97, 100, 105–110
square brackets [], 41, 72
Standardized Generalized Markup Language (SGML), 290
star schema, 295
stateless protocols, 289
statement history, 71
static Web pages, 288
StDev (standard deviation) function, 42
stored procedures, 140–141
Structured Query Language. See SQL
stylesheets, 292
subclasses, 302, 304–305
subqueries, 93, 353
SUBTRACT command, 61–62
Sum function, 42
SUM function, 90, 91–92
superclasses, 305
surrogate keys, 181
survey forms, 197–198
synchronization, 247–248
synthetic keys, 181
Syscolumns table, 138, 139
Sysindexes table, 138
Systables table, 138
system catalog, 137–140
System R, 69

T

table names, 72
tables, 5
    altering structure of, 135–137
    creation of, 71–73
    creation of, from query, 103–105, 111, 137, 347–348
    defining, in DBDL, 182
    deleting, 137, 350
    deleting fields from, 136–137
    denormalizing, 269–270
    difference of, 60, 61–62
    dimension, 295
    division of, 63
    fact, 295
    in first normal form, 155–156, 171
    in fourth normal form, 169–171
    incorrect decomposition of, 163–167
    intersection of, 60, 61
    joining, 49–52, 58–60, 97–100
    joining multiple, 51–52, 99–100
    make-table queries, 54–56
    merging, 213
    primary keys of. see primary keys
    product of, 62
    in relational databases, 29–32
    relationships in. see relationships
    representing user views as, 179–180
    in second normal form, 149, 156–160
    splitting, 268–269
    structure of, 6–8
    in third normal form, 159–162, 168, 171, 213
    union compatible, 61
    union of, 60–61, 100–101
    unnormalized relations, 31
    updating, 101–103, 111, 353
tags, 288
TCP/IP, 287
testing, DBA role in, 266
thin clients, 285, 286
third normal form (3NF), 149, 159–162, 168, 171
    denormalization from, 269–270
    merging tables and, 213
three-tier architecture, 285, 286, 288–289
timestamping, 238
top-down design method, 196–197
Total row, 42
training, 264, 265
transactions, 236, 238–239
Transmission Control Protocol/Internet Protocol (TCP/IP), 287
transparency
    in client/server systems, 286
    fragmentation, 278–279
    location, 277
    replication, 277–278

triggers, 141–142
tuples, 32
two-phase commit, 282
two-phase locking, 233–236
two-tier architecture, 285

U

UML diagrams, 303–305
unauthorized access, 257–258
underscore (_), 87
Unified Modeling Language (UML), 302–305
Uniform Resource Locator (URL), 287
uninterruptible power supply (UPS), 259
union compatible, 61
union operation, 60–61, 100–101
UNION operator, 353
unnormalized relations, 31, 155–156
update anomalies, 149, 157, 159, 160
UPDATE command, 102, 111, 353
update queries, 53, 101–103
updates, concurrent. *See* concurrent updates
UPS (uninterruptible power supply), 259
USE command, 71
user privileges
    GRANT statement, 128–129, 350
    policy and enforcement of, 254–257
    REVOKE statement, 128–129, 351–352
user views, 178
    design examples, 185–195
    merging into cumulative design, 183–190
    representation of, 179–180
utility services, 226, 248

V

validation rules, 134
validation text, 134
Var (variance) function, 42

vendor support, 264
views, 117–124
    See also user views
    advantages of, 124
    changing field names in, 121
    defining, 118–120
    row-and-column subset view, 122
    security services of, 243
visibility symbol, 304

W

warm sites, 259
weak entities, 218
Web access, 287–290
Web browsers, 287
Web clients, 287, 288–289
Web pages, 287, 288
Web servers, 287, 288–289
WHERE clause, 57, 60, 73, 76–77, 83, 95–97, 100, 105–110
wildcard characters, 86–87
World Wide Web Consortium (W3C), 290
World Wide Web (WWW), 287

X

XHTML, 292–293
XML, 290–293
XML declaration, 291
XML schemas, 292
XQuery, 293
XSL, 292
XSL Transformations (XSLT), 292

Z

zero value, 129